LAN Wiring

An Illustrated Network Cabling Guide

James Trulove

Second Edition

McGraw-Hill
New York · San Francisco · Washington, D.C. · Auckland · Bogotá
Caracas · Lisbon · London · Madrid · Mexico City · Milan
Montreal · New Delhi · San Juan · Singapore
Sydney · Tokyo · Toronto

McGraw-Hill

A Division of The **McGraw-Hill** Companies

1 2 3 4 5 6 7 8 9 0   PBT/PBT   0 9 8 7 6 5 4 3 2 1 0

ISBN 0-07-135776-9

The sponsoring editor for this book was Zoe G. Foundotos, the editing supervisor was Caroline Levine, and the production supervisor was Sherri Souffrance. It was set in Vendome ICG by Joanne Morbit of McGraw-Hill's Professional Book Group composition unit, Hightstown, N.J.

Printed and bound by Phoenix Book Tech.

McGraw-Hill books are available at special quantity discounts to use as premiums and sales promotions, or for use in corporate training programs. For more information, please write to the Director of Special Sales, McGraw-Hill, Professional Publishing, Two Penn Plaza, New York, NY 10121-2298. Or contact your local bookstore.

This book is printed on recycled, acid-free paper containing a minimum of 50% recycled, de-inked fiber.

CONTENTS

V

Contents

Contents

Contents

Contents

INTRODUCTION

The theory and practice of LAN wiring has evolved very rapidly over the past decade. Almost every enterprise in every field, from business to education to recreation, now finds computer networks at the core of their daily activities. Even in homes around the world, we are installing local area networks and using their connections for communication, education, and entertainment. Networks are present today in schools, libraries, stores, hotels, and restaurants. No doubt your favorite coffee shop is making the transition to an "Internet Café."

The Internet expansion actually is only one of a cornucopia of computer applications at the center of this work-and-play revolution. These applications range from everyday productivity applications, such as word processors and spreadsheets, to far-reaching communications applications, such as e.mail, web surfing, and even video conferencing. The "net" result is that absolutely every human enterprise can now benefit from networking.

In the local area network, tremendous pressure exists to support the very newest and fastest technologies. This need for speed requires very sophisticated new LAN wiring systems. Providing those systems is what this book is all about.

All of this wiring technology is new. Granted, it is based on tried-and-true telephone wiring concepts, but it is not at all the same technology that has been adequate for voice connections over the past 100 years. It is a brand new science, a leading edge technology for the millennium.

Knowledge Is Power

One of the more challenging problems that you may encounter when taking on a new project or a new area of responsibility is finding factual, authoritative information on the subject. If you are new to the field of LANs and LAN wiring, you may be presented with a profusion of confusing, incomplete, self-serving, and even misleading information on a variety of cable types and systems. You can find tons of information on every LAN topology imaginable and volumes of print on new technologies that are just around the corner. You need a reference that describes LAN wiring technology in detail to help you learn the basics, prepare for the higher LAN speeds, and decode the market hype.

On the other hand, you may be an old hand at networks or the owner (or inheritor) of a mature LAN wiring system and be faced with important decisions regarding future upgrades to your network. What you need is a road map to LAN wiring technology, and you need an index to all the resources you will need to successfully complete your project.

This book will supply you with clear, concise information on the science and art of LAN wiring. LAN wiring certainly is a science, as much as any endeavor can be. Once you have picked a network topology, such as Ethernet or Token-Ring, and have chosen a wiring method, such as 100BaseT or fiber, the basic implementation is pretty much cut-and-dried. By default, you will have already defined the general wire type, minimum performance category, connectors, and even the connector wiring. You will know what maximum distances your cable runs may be, what wire/fiber size and specifications to use, and what standards to apply to your installation. You will be able to refer to this volume for specific guidelines for standards, limits, and practices. Ah, but there is also much art added to the science.

The art of LAN wiring is in the realization that many factors must be taken into account to complete a successful installation. How well your particular installation approaches the exact-science ideal will be affected by such things as the details of workmanship, the influences of environmental factors, and even the quality of your wiring components. You must analyze your need for future enhancements and the potential of your particular choices to provide compatibility with new technologies in the future. In addition, the cost factors must be weighed against their benefits. Your job must come in on time, under budget, and be done right. It must work properly with whatever level of performance was specified, and it must last for years with little or no maintenance.

LAN Wiring, Second Edition, is designed to give you access to all the latest wiring technology. It will provide an introduction to LAN topologies, structured cabling philosophies, and management concepts. In addition, the book will serve as a complete reference for cable specifications, connecting hardware, wiring patterns, and internationally recognized standards.

Organization of LAN Wiring, Second Edition

This new edition is divided into four main parts, including the reference sections.

Part 1, LAN Wiring Systems, describes the philosophies and core technologies of LAN wiring systems, including the major LAN topologies. Structured cable plant design and planning are covered, along with such factors as mixing data and voice and general cost factors. The many types of cable used in LAN wiring are detailed, and the classic LAN wiring topologies and several popular universal cabling systems are described. A strong emphasis is placed on metallic wiring, as it is the dominant wiring method in current use. However, information on fiber optic wiring is also provided for those who are implementing fiber optic networks and for those applications, such as backbone and campus wiring, for which fiber is so well suited. The issues of future LAN wiring needs, migrating to higher speeds, and emerging technologies are also explored.

Part 2, LAN Wiring Technology, illustrates and explains the details of LAN wiring interconnection, from the telecommunications outlet to the telecommunications room. Various types of wiring devices are shown for all phases of the wiring system. Charts are provided for the standard wiring patterns and common variations. Proper methods of installation and cable termination are covered, with emphasis on standards and workmanship. The total LAN wiring system also includes user cables, patch cords, and cross-connects, which are described as well. New chapters on work group wiring, fiber optic cabling, gigabit cabling, and wireless LANs are added in the Second Edition.

Part 3, LAN Wiring Management, discusses the planning, testing, maintaining, and personnel training of LAN wiring systems. The location of telecommunications room, interconnection, grounding, and power conditioning are covered in this section of the book. Approaches for cable testing, troubleshooting, certification, and maintenance are shown. In addition, physical wire system monitoring and control are described. Finally, important information on education and credentials are covered in a new chapter on training and installer certification.

The fourth part is comprised of the *Appendix, Glossary,* and *Index* and provides reference information for the other three parts of the book. The Appendix, which is split into four sections, lists on-line resources, standards and standard organizations, and the various membership and training organizations. These references include the various recognized standards for LAN wiring, including standards for cabling, telecommunications spaces, marking and identification, electrical safety, fire safety, and the common LAN topological standards, such as Ethernet and Token-Ring. LAN wiring is a constantly changing field, and for that reason we have also included some on-line resources to keep your knowledge up to date.

Who Should Read This Book?

LAN Wiring, Second Edition is written to be accessible as both a tutorial and a reference. For example, LAN managers can learn the basics of connecting their LAN hardware, from design to implementation. The beginning LAN wiring technologist can learn the details of performing a proper LAN cabling installation. The expert finally will have a single reference to all the standards, wiring methods, and wiring details of LAN implementations. The manager will have a planning and estimating guide that clearly explains the terms, concepts, and details to specify in a contract with an outside vendor.

What information does this book provide? It covers planning, topologies, cabling systems, performance and safety standards, detailed connection and workmanship, and managing of the installed cable system. It primarily covers three modern technologies: twisted pair wiring systems, fiber optic cabling, and wireless networking. The reference sections also include information on legacy coaxial and proprietary cabling systems that may be encountered in the office environment. Targeted information is provided on structured cabling, wire technology, LAN topologies, and special application areas such as work groups and gigaspeed wiring.

When should you read this book? The information in *LAN Wiring, Second Edition,* is useful at any stage of LAN implementation, from early planning to management. If you are planning to install, modify, or upgrade a LAN wiring system, you should use this book to assist you in the myriad of decisions, specifications, estimations, and evaluations for your network. Nothing is more important than the provision of a stable, reliable, and capable wiring system for your LAN. If you already have a mature LAN wiring system, you can use this book as a ready reference for standards, troubleshooting, and expansion. You will be able easily to determine what additional performance and/or speed your existing wiring is capable of offering. In addition, you can use the numerous lists, standards, tables, and illustrations to help decode manufacturer's literature and proposals.

ACKNOWLEDGMENTS

Many people have contributed to this Second Edition and the prior volume. We would like to gratefully acknowledge the assistance of the following people: Donna Ballast, Jim Bordyn, Larry Campbell, Paul Hines, Bob Jensen, Paul Kish, Henreicus Koeman, Bryan Lane, Todd Metcalf, Malcolm Myler, Mark Odie, Todd Palmer, Bob Paradine, Stephen Paulov and staff, Steve Schmerber, Masood Sharrif, Marjorie Spencer, Jerry Summers, Scott Vessell, John Wyzalek, and many others including all the fine folks at McGraw-Hill: Zoe Foundotos, Carol Levine, Bob Healy and all the myriad of artists, copy editors, production people, and many other individuals who helped in the creation of this book.

JAMES TRULOVE
jtru@lan-wiring.com
www.lan-wiring.com

1

LAN Wiring
Systems

1

Designing LAN Wiring Systems

Chapter 1 Highlights

- A standard installation
- Installability
- Reusability
- Reliability and maintainability
- The structure of cabling
- Cost factors in cabling systems
- Mixing data and telephone
- Advances in LAN wiring technology

Local Area Network (LAN) wiring has developed into a highly sophisticated science with a tremendous impact on the performance, reliability, and maintainability of your network. While a LAN comprises many components, the underlying wiring system is the foundation upon which all else rests. Without the proper physical connections of copper wire, connectors, punchdown blocks, jumper cables, patch panels, user cables, and (perhaps) fiber optics, the network will not operate reliably.

It is therefore quite important that you properly plan, specify, and implement a LAN cabling system that will provide your organization with a reliable level of service. The use of the appropriate component parts and good workmanship in your wiring system will ensure that it can meet the performance standards that are expected for the type of LAN you are installing. In addition, you should always plan for as much future growth as is feasible, within obvious financial limits.

LAN wiring systems are in a constant state of change, as are all computer systems. As with computer technology, some of the changes in LAN wiring technology are evolutionary and some are revolutionary. For example, accepted LAN wiring techniques, components, and practices tend to *evolve* as manufacturers, installers, and users refine the existing wiring technology. The introduction of tighter cable link performance standards and the trend toward guarantees of component performance are perfect examples of evolutionary change. Likewise, the introduction of new outlet and patch panel designs is also evolutionary. You do not really see a major shift in an evolutionary new product; the refined product is just more convenient to use or has an enhanced level of performance.

On the other hand, some changes are revolutionary. A *revolutionary* change is one that breaks with the past. In technology, this implies a product or method that is no longer compatible with past items, but offers greatly increased capability. For example, an increase in LAN speed is revolutionary. For the most part, the technology required to increase from 10 Mbps to 100 Mbps is a clear break from the previous practice, and is highly desirable, but no longer compatible with the past. Revolutionary changes tend to be abrupt.

The main focus of a LAN wiring system design effort must be to keep up with evolutionary changes in the wiring technology that will enhance the value of your system, while attempting to plan for the next revolutionary jump in LAN technology.

When LAN wiring over twisted-pair wire first began, little was known about the importance of wire performance. The initial interest in "twisted pair Ethernet" was to be able to use one's existing telephone wiring for the installation of a network, thus saving the enormous cost of installing new cabling. However, we soon found out that the typical twist-per-foot cable specifications that were fine for telephone use were rather marginal for LAN frequencies. To make matters worse, deregulation of the telephone industry meant that cost-conscious installers sometimes used a cheap grade of telephone cable that had two twists per foot or less, making it inadequate for LAN data use. To use twisted pair with confidence, the LAN manager and the LAN installer needed to be certain that a particular cable would perform adequately. So, a performance specification began to evolve that categorized cable performance by intended use. An early division of cable manufacturing quality levels eventually led to a well-researched and detailed set of performance specifications in five categories. The LAN user now could know with certainty that a Category 3 cable will operate well on LANs up to 16 megahertz (MHz) bandwidth, and that a Category 5e cable will perform to 100 MHz (and even 1000 Mbps).

In this way, many connector and wire specifications have evolved, along with workmanship practices, to provide a platform of standards that virtually guarantee a properly performing LAN wiring system. This process is not difficult to understand, but the LAN manager, installer, and technician must have a thorough understanding of the standards, procedures, and specifications needed for a network cabling plan that complies with LAN operating standards. The careful planning of a standards-compliant cabling system is called *structured wiring*.

This chapter will give you an overview of the philosophies in designing, planning, and estimating a LAN wiring system. The items in these sections should be ready knowledge before you delve into the rest of the book. A central theme of this book is that any wiring design should be a standard installation that supports at least the minimum performance criteria, is simple to install and maintain, and provides adequate future expandability and growth.

A Standard Installation

Maintaining the proper standards in your cable installation is the key to getting the most from your LAN cabling system. The internationally recognized standards exist to ensure that the combination of wiring, connectors, hubs, and network adapters will all perform properly in a completed network. The newly revised wiring standards make special note of specific cable installation guidelines and workmanship practices that have been found necessary for the more demanding modern networks at speeds of 100 Mbps and beyond.[1]

There are standards that apply to LAN wiring covering everything from electrical performance to safety issues. Should you dream up an exotic wiring system that meets your creative needs? Should you try to use some existing telephone wiring and add whatever else you need to make the system work? Should you pick and choose from various manufacturers' catalogs? How do you sort out all the options?

One way to proceed is to choose a very conventional LAN wiring installation that uses the standards to your advantage. In such an installation, you will wire your cables in a manner that will support the widest variety of current and future applications. This means you must use the proper cable, install it properly, maintain lengths within the maximum proscribed distances, use only connectors and jacks that meet the category of operation you need, and use proper installation techniques. In addition, you will want to thoroughly test and document your cable system for the appropriate level of performance. Of course, any future changes and additions to your standard cable system must also be done the same way.

Designing a standards-compliant, conventional installation will have distinct advantages for you. For example, your properly installed cable will never contribute to a connectivity failure. You can be assured that any problems will be found in the network hardware or software that utilizes your cabling system, not in the cable itself. That is true, assuming no inadvertent change has occurred in a cable. Naturally, it is always possible that someone could damage a cable while doing unrelated work on an electrical, plumbing, HVAC, or even telephone problem. Remember also that user cables (from the workstation to the wall jack) are considered part of the cabling system, and their failure can indeed disable a network connection.

Your standard installation will also ensure that any trained cable installer will be able to easily expand, troubleshoot, or repair your cabling system. If you test your cable when it is installed, you will be able to retest any cable drop and compare the results to the original cable certification done at installation. You will also know that your cable system meets the electrical and fire protection standards required by local and national authorities.

---

[1]Throughout this book, we concentrate on the advanced structured wiring system embodied in the standards from the Telecommunications Industries Association (TIA/EIA-568-B and others) and the International Standards Organization (ISO/IEC 11801 and others).

The standards you will need to reference are described throughout this book and are listed in the appendix. The appendix also contains sources for copies of the pertinent standards.

Installability

Your LAN wiring system should be easy to install and maintain. A variety of wire types, connector types, termination devices, and patching devices exist to support network wiring. The best choices are those that: support your application, present and future, and are easy to properly install, using a minimum of special equipment.

However, you must make many additional choices concerning the particular wiring components to be used. For example, what types of jacks are the easiest and most reliable to use? Should you use a punch-down termination and cross-connect wiring in the wiring closet or should you terminate cables directly onto a patch panel? Where should the telecommunications rooms be located? How should the individual cables be run to avoid electrical interference and other performance-decreasing factors? What color should the cable and connector plates be?

Some guidance to these choices is provided by recognized installation standards, but many are a matter of personal preference. For example, some jacks use wire termination methods that require special tools, while others do not. The experience level of those who will install and maintain the cabling system is also a factor. Some items are more easily installed and use such things as color code marking to simplify the process. Outlet plates are available that have modular construction, allowing you to mix connections for data, phone, and even CATV.

Installability will also have an effect on system cost. Whether you are doing the wiring yourself or using a contractor, factors such as time-to-install and reliability will be important. The easier the wiring components are to install, the lower will be your cost of installation, and a contractor will charge less because the installation will take less time. Doing the work yourself will cost less as well. Installability, however, does not mean the taking of shortcuts. Your goal is to achieve a very installable cabling system that is easy to maintain and meets all appropriate standards and performance criteria for your present and foreseeable future needs.

The sections on wiring devices in Part II of this book will show you a variety of wire termination devices, such as jacks and patches, and will compare the different installation techniques that impact installability.

Reusability

Everybody involved with computers and networking knows by now that technology moves very quickly. Changes are constantly made to upgrade performance and throughput. (As we all know, nothing stays the same but change itself.)

You cannot always afford to install the true leading-edge technology, but you should at least design your cabling network to allow for a future upgrade without replacing the cable. To do this means that you must design a cabling system that totally meets your current needs. If the current need is 100BaseT, then the cable system should be specified so that it meets that requirement, as a minimum. If possible, it may be convenient to design your system so that it could support multiple applications, such as data and voice, or data and video conferencing.

Then you should assess what future standards for which you might reasonably be expected to use the cabling system. You should also take into account the useful life of your facility, which might be five years, or even 20 years. Are your offices in leased space? You might expect to move in less time than if you owned the space. It might be best to "over-wire" by putting in more cables or cable pairs than you currently need. You may wish to use a modular outlet jack that allows the connector to be changed without reterminating the wire.

It may be virtually impossible to make any accurate technological prediction more than five or 10 years out. Consider for a moment what technology changes have taken place over the last 10 years! Undoubtedly, someone, somewhere, made an accurate prediction. However, there were also several predicted technology directions that simply didn't come true. Of the predictions for the next decade, which should you choose?

The only practical approach to cabling futures is to determine a cabling method that meets at least the current, widely deployed technology requirements. You then should consider installing to meet cabling standards that are at least one level beyond those current requirements. If this is not cost-prohibitive, your cable system will be able to go through at least one generational upgrade without replacement. For example, if you currently need a 100BaseT or 100 megabit (Mbit) Token-Ring network cabling system, you should consider installing a cabling system that will support 1000 Mbit Gigabit technology. Since much of the current componentry is being manufactured so as to exceed that data rate, your system will probably support a second upgrade to 1000 Mbit rates, should that occur.

Changes in cabling technology are a challenge for any book author. To help you keep informed on the latest developments and standards, we have developed an Internet site to use as a resource on current LAN developments. Consult *www.lan-wiring.com,* or see Appendix A, "Online Resources" for more information.

Reliability and Maintainability

Does anyone ever want an unreliable network cabling system? No, of course not! You, or someone else if you use a third-party installer, will be making many decisions regarding the components and methods used to install your LAN cable. Many of these decisions will have a long-term effect on how well your network performs.

On the other hand, if you install cable for someone else, you will want to make sure that the installation will stand the test of time. Your reputation is on the line, and you may be responsible for repairs should failures occur at some later time.

A very good philosophy is to install LAN cabling in such a manner that, when network problems occur, it will never, *never* be the cable system that is at fault. If you adhere to this guideline, you will save countless hours of tedious and expensive troubleshooting. Even though wiring components are relatively simple physical devices, they have technically sophisticated operating characteristics that require expensive equipment and trained technicians to troubleshoot. The performance issues are greatly compounded if you use the wiring to its maximum capability.

Because many companies with LAN installations do not have access to sophisticated wire test equipment, many LAN managers start troubleshooting a problem by looking at the servers, routers, hubs, and workstations connected to the cabling system. They check the configurations of network software, drivers, and applications. They may even change out hardware, including computers, printers, hub cards, or whatever seems to be related to the problem. After all that thrashing about, you never want the cable to be at fault. Too much emotional and physical energy has been tied up in troubleshooting by the time they get to the cable. It had better not be a cable fault!

So, to prevent all of this energy from getting expended at you, just make certain the cable system is done right. Proper installation of LAN wiring has become considerably more difficult as LAN speeds push upward. You should install all your wire in accordance with the latest standards, particularly those that address cable routing, telecommunications room locations, handling of the cable itself, and workmanship. Proper planning, components, installation practices, and workmanship will make your cable system installation both reliable and maintainable.

The Structure of Cabling

Basic to the philosophy of modern LAN wiring is the concept of structured cabling. With the higher speeds of today's networks, it is recognized that the total length of cable that connects from the hub to a workstation or other device has a finite maximum length. The entire networking system must therefore be broken up into chunks that allow workstation (or station) wire to be concentrated, with each cable length short enough to support the desired data rate. Structured wiring standards have been developed to help the LAN user plan a wiring system that stays within the maximum wiring distance for various LAN topologies. For example, in the case of 100BaseT, this cable must be no more than 100 m (328 ft), including patch and equipment cords.

We achieve the needed wiring concentration by placing telecommunications rooms (wiring centers) at appropriate locations in a building and then interconnecting those wiring closets as needed to provide the total network connectivity for the building. Typically, a model of a multistory building is used to illustrate this structured concept, as shown in Fig. 1.1.

On each floor of our model, a telecommunications room (TR) concentrates all of the station cables for that floor. Each workstation location has a wall or surface-mounted jack. The network cable is terminated at that jack and runs directly to the wire center. This is called a *home run,* as there are no intermediate connections, splices, taps, or daisy-chains. The wire may run in wire trays or conduit, or be draped over supports (running over a drop ceiling, the type with push-up ceiling tiles, is no longer permitted). Larger floors may require more than one telecommunications room per floor.

At the telecommunications room (TR), each station wire is terminated on an appropriate punchdown termination, or directly onto a patch panel location. The punchdowns or patch panels may be mounted to a wall or in a freestanding rack or cabinet. In the telecommunications room, some type of network device, such as a hub or concentrator, is

Figure 1.1
The structured wiring
system.

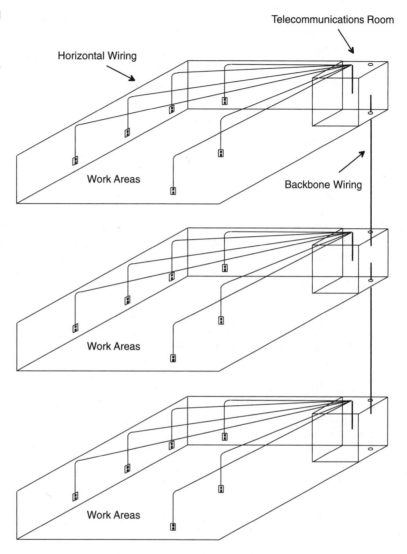

Figure 1.1
The structured wiring system.

connected to each station cable and electrically terminates the cable run. The hub passes the LAN signals on to other stations or wire centers for ultimate connectivity with the entire network. The process is essentially the same for different network topologies, although Token-Ring uses passive wiring concentrators called MultiStation Access Units (MSAUs), rather than the active hubs of 10/100/1000BaseT.

A telecommunications room is typically connected to the telecommunications rooms on other floors. This center-to-center wiring is usually done from floor to floor to floor, as a backbone, with LAN hubs on each floor. In some cases, it may be more effective to connect telecommunications rooms on several floors to a single backbone concentrator on one of the floors. Ideally, telecommunications rooms should be located directly above one another, to minimize the cable runs between them, but that varies from building to building. Fiber optic cable is sometimes a good choice for wiring between telecommunications rooms, as it totally eliminates grounding and bonding concerns that exist with metallic cable and can often be run longer distances than copper cable.

Obviously, there are differences between the wiring considerations for station wiring on a single floor and for TR-to-TR wiring between floors. For one thing, the fire protection requirements may be different for the vertical riser cabling used between floors. Physical securing of the riser cable may also be different, as may be the need for strength. For this reason, it is convenient to have terms for the two wiring paths that easily distinguish the concepts and physical characteristics of the cable.

Station wire from the wire center to the workstation termination is simply called *horizontal wiring*, a somewhat logical name as most of the wire run is horizontal across a floor of a building. The TR-to-TR cables are called *backbone wiring*, since they form a unifying structure between telecommunications rooms. Of course, a large floor of a building would have more than one telecommunications room, to adhere to the station wire maximums, but the wire between those telecommunications rooms would still be referred to as backbone. All the details of this structured wiring concept are covered in the next chapter.

Cost Factors in Cabling Systems

A variety of methods exist to implement LAN wiring schemes; most methods in common use will be described in succeeding chapters. These methods vary in three significant degrees, all of which have cost implications.

First, some LAN topologies require that certain types of wire and connections be used. For example, the Ethernet 10BaseT can use Category 3 twisted pair and jacks, while 100BaseTX and 1000BaseT require Category 5e twisted pair wire and jacks, and 100BaseFX and 1000BaseSX/LX require the proper mode fiber and connectors. Although your network requirements will frequently dictate which topology you must use, you may have several choices to make that will influence initial cabling cost as well as the cost of future expansion. In some cases, it would be possible to choose an older wiring technology such as 10BaseT on the basis of cost. However, when you consider the cost of a future upgrade to 100 megabits per second (Mbps), a Category 5e unshielded twisted pair installation would only support 10 Mbps 10BaseT today but 100 Mbps 100BaseTX (and Gigabit) for the future.

Second, LAN wiring methods vary in the extent to which they will support higher speed network data rates, including future standards that may not be fully implemented yet. For example, it is probably less expensive to install a Category 5e cabling system that supports your current need for 10/100 Mbps networking than it would be to install to an advanced Category 6 facility. Category 5e will support 1000 Mbps on four pairs, but that is its limit. You might never need the 1000/10,000 Mbps (1–10 Gbps) capability, particularly if your facility will be occupied for less than 5 years. What if you had to stay longer? Should you put in Category 5e jacks and cable to save money, or should you go ahead and install a Category 6 system now, in case the future comes along a little sooner than expected?

Third, you may choose combinations of cable and connection hardware that differ in features and cost. This factor may also influence the final, installed cost of your network cable system. Even among manufacturers that offer products certified to identical performance ratings, there is a lot of price variation. You may choose a big-name manufacturer whose cable and connectors cost more than the little guy's. Some manufacturers have products with extra features, such as modular snap-in construction or fancy color inserts. Some even offer special 10- or 15-year warranties if you use their cable, connectors, and certified installers. Others offer separate wire termination and jack modules that plug together. In many cases, the extra features will cost you extra money. You will have to judge the benefit of these extra cost features for your situation.

In the other sections of this book, we will try to point out the cost trade-offs of the devices that are described. You should be aware that any time a connector, cable, or other wiring device is referred to as "easier to use" or "flexible," there might be cost implications. Remember too that your goal is not necessarily to specify or install the cheapest wiring solution, but to find the balance among features, cost, and reliability that meets both your current and planned future needs.

Mixing Data and Telephone

Combining telephone and data on the same cable is sometimes a consideration for cost savings and efficiency. This so-called mixed media approach is an option that may work fine in a lower-speed network environment, such as 10BaseT or Token-Ring, but it may be fraught with problems at higher 100–1000 Mbps LAN speeds. (At the least, there are lots of "ifs.")

If your telephones are analog, your LAN speed is conventional, and the telephone cross-connect is in the same telecommunications room as the LAN hubs, then you probably could use the same cable for data and phone very successfully. This case assumes that you are using at least Category 3 Cable, the minimum data grade, and that the cable runs are not too close to the 90 m maximum.

However, if your telephones are digital, or you intend to use 100 or 1000 Mbps LAN data, or you are installing the LAN hubs in a separate telecommunications room from the telephone interconnects, then you should run two separate cables for data and voice. For that matter, if you know that the telephone installers are apt to be overly creative around the LAN wiring, you should run separate cables (color-coded for added protection). Table 1.1 shows some of the typical recommendations for telephone and data.

The reason for the concern for mixing voice and data is the phenomenon of crosstalk that occurs between pairs of the same cable. The higher frequency components of 100/1000 Mbps LAN signals, along with their greater attenuation per foot of cable run, make them much more susceptible to interference from other sources. Such sources clearly include digital telephones and other LAN signals. Even magnetic fields from fluorescent light fixtures can cause problems. At cable run distances that approach the maximum limit, these LAN signals are significantly more susceptible to problems from crosstalk, signal balance, interfering signals, magnetic coupling, and noise sources.

The danger of interference from telephone pairs is probably less for the analog signals of conventional telephones, since they limit transmitted frequencies to 4 kHz or less. However, both analog and digital telephones may use extra wire pairs for power. This power connection is unbalanced and may serve to increase the undesired coupling effects between pairs in the cable.

Some cabling designers feel that data and telephone wiring should never be mixed. There is some support for this view from the standards,

Table 1.1

Cabling Recom-
mendations for Var-
ious Applications

Application	Minimum cable performance categories		
	Cat. 2	**Cat. 3'**	**Cat. 5e**
Analog phone only	Yes	Yes	Yes
LAN/4/10/16 Mbps only	Yes (for runs under 100 ft)	Yes (for runs under 290 ft)	Yes (for runs under 350 ft)
Shared analog phone and LAN 4/10/16 Mbps	Yes† (if 4 pair cable)	Yes† (if 4 pair cable)	Yes† (if 4 pair cable)
Digital phone only	Not recommended	Yes	Yes
Shared digital phone and LAN/4/10/16 Mbps	No	Not recommended	Not recommended
LAN 100/1000 Mbps only	No	Yes/no‡	Yes
Shared digital phone and LAN 100/1000 Mbps	No	No	No

'Category 5e is required for 100BaseTX and 1000BaseT.
†May not operate at extended distances, particularly with Token-Ring 16 Mbps networks. Higher per-
formance category cables will provide better isolation between the phone and data signals.
‡Some 100 Mbps technologies are designed to operate on a minimum of Category 3 cable.

when interpreted strictly. None of the standards provide guidelines for
the application of two or more signals to the same cable, although some
users simply ignore the issue. The most widely used standard, TIA/EIA-
568-B, simply suggests two cables to each telecommunications outlet and
recommends that each be used for only one application. If you are going
to provide separate cables for phone and data, you can easily run them to
totally separate areas of a wiring closet, or even to separate wiring closets.

You will never cause a technical problem with either LAN or telephone systems by separating data and voice cables. On the other hand,
you can expect the installation cost to be higher in almost all cases.

Advances in LAN Wiring Technology

Many exciting advances have occurred in the art and practice of LAN
wiring. As we have learned more about the operation of our cabling system, both copper and fiber, at advanced frequencies and data rates, we

have been able to further refine the cabling components. These refinements have added orders of magnitude to our LAN capabilities. The norm in network copper cabling is now Category 5e, which has all but replaced Category 5 in the marketplace. However, the Category 6 specifications are out, with their operation to over 200 MHz and superior performance at the lower frequencies, and many are already specifying Category 6 as a minimum. Category 7, with operation to 600 MHz, is close behind, with some manufacturers already offering components that meet the presumed parameters of this technology. Copper cabling is capable of 1 Gbps operation at lowly Category 5e, with its 100 MHz bandwidth. Imagine what you can do with 600 MHz cable!

In the fiber arena, multimode 62.5/125 μm fiber is now in wide use. However, the high-bandwidth needs of gigabit operation have revealed concerns with run-of-the-mill fiber, and new formulations now abound to better support gigabit speeds at reasonable distances. In addition, new 50/125 μm multimode fiber constructions are now standardized, and many applications lend themselves to single-mode fiber. An entire catalog of new fiber connector types is available to better support our modular world. Although the fiber and connector types are not given the convenience of the copper world's categories, there is nevertheless a myriad of options, with commensurate capabilities.

Certified training in appropriate technology is a mandatory requirement for the network cable technician of today. The practical knowledge has become so detailed and the installation practices have become so crucial that every person involved in the LAN wiring process must possess specialized education and experience, as well as the corresponding credentials. In addition, every link in every LAN wiring system must be rigorously tested to the proper performance specifications, whether copper or fiber (or even wireless). Wiring has progressed from "anything goes as long as the pattern is right" to "it works only if the components and installation techniques are perfect."

Those of us who are involved in LAN wiring provide the superhighways over which the high-speed traffic of the network now ride, and a wild ride it is indeed. Over the past decade, we have seen network speeds move from a mere 10 Mbps all the way up to 1 Gbps—100 times the speed. What will occur over the next decade? More of the same, no doubt! Those who could not see the advantage of going from 10 Mbps up to 100 Mbps will now consider nothing less. Likewise, those who quickly adopted the bandwidth of 100 Mbps are now just as quickly moving their servers and backbones to 1 Gbps. For some, the future is moving too fast; for the rest of us, the future is now!

Structured Cabling System Design

Chapter 2 Highlights

- Basic terminology
- Twisted pair wiring
- Structured wiring
- Horizontal wiring
- Backbone wiring
- Location and routing
- Simple design example

The three keys to a successful LAN wiring installation are proper design, quality materials, and good workmanship. This chapter will cover the basics of proper wiring system design. Proper design involves the careful orchestration of several complex factors, applied in a standard fashion, to produce a successful installation plan that will meet your needs for today and for many years to come. These factors include the length of run, wire type, wire terminations, and routing. Many of the technical details are covered in other areas of this book. Here, we will give you an overview of how these factors come into play in creating a successful design.

In creating your cabling design, you will have to make many decisions. Most of these decisions can be reduced to a simple matrix, so that your overall performance requirements actually dictate the proper components to choose. In this way, you will ensure that the final wiring system design will provide the connectivity, performance, and growth that you need.

This chapter gives a brief explanation of structured cabling concepts with a summary of some of the design considerations involved and culminates in a simple design example.

In later chapters, we will explain many of the common wire types and wiring systems that are, or have been, used in LAN wiring. Although some of the older cabling methods are no longer being used in new installations, it is important that they be here for completeness and for those who must add to or modify older cabling systems. For the most part, we will talk about unshielded twisted pair wiring when citing design factors and installation techniques. This has been the area of heated activity in the specification of wire, installation and performance standards, and introduction of new technologies. This is clearly where the action is.

Fiber optic cabling is also a current technology, but its use in the workplace is limited at present. It is more expensive to buy, more difficult and expensive to install, requires more expensive workstation and hub interfaces, and generally exceeds the bandwidth required for the current and next generation of LAN data throughput. Fiber optic cabling does, however, have several unique and very beneficial characteristics that can be of great assistance in larger cabling designs. For that reason, a description of fiber optic cabling is included wherever that medium is particularly useful.

Basic Terminology

In order to talk about intricacies of a subject such as LAN wiring, we need a common lexicon. Unfortunately, the cabling industry has developed as a combination of many technologies and disciplines, each with its own vocabulary and terminology. Many of the twisted pair wiring techniques we use in today's LAN wiring were initially the domain of the telephone industry. In fact, the styles of wire, connectors, terminations, and even color codes that we use today were developed decades before anyone considered using telephone-type wire for data. Many of these telephone terms have been made directly a part of LAN wiring vocabularies. On the other hand, all of the terms that are used for LAN wiring signals, cables, standards, and termination equipment come to us from the computer networking industry. Now that we are "pushing the envelope" into the stratosphere of LAN speed performance, we are beginning to encounter even newer terms that may be unfamiliar to many of us, but would make any radio-frequency (RF) engineer smile.

In this book, we will attempt to be as consistent as possible in terminology so that you will always know what we mean by a particular word or phrase. We will particularly avoid jargon that is used only by a narrow segment of the industry, other than to explain special meanings of terms that you may encounter when dealing with contractors and suppliers. It will be necessary to use some terms almost interchangeably, such as the terms wiring and cabling. To not do so would be a disservice to you the reader, since you will certainly encounter all of these interchangeable terms in your everyday work. We will always try to be clear in our meaning.

A complete glossary appears at the end of this book. Each time we introduce a new term we will define it. Some LAN wiring terms are common words or phrases that have a particular meaning in the cabling industry. For the most part, after these are introduced, we will simply use them with their special meanings, without emphasis. In a few instances, special words or terms will be capitalized. Examples of this are Basic Link, Channel, and Permanent Link, which are very specific terms used in one of the service bulletins that revise, amplify, or clarify EIA/TIA standards. Such terms are not in general use and their meanings are specific to the definitions in a standards document. Acronyms and abbreviations are also part of the world of LAN wiring. The common terms will be formally introduced only once, although the less common ones may have their definitions repeated at the beginning of a section.

Specific, commonly accepted terms can be very useful in providing shortcuts to understanding and explanation of a complex system such as LAN wiring. However, jargon that is peculiar to a narrow part of an industry can be tedious and even confusing. We will therefore stick to more commonly used computer networking terms that are widely understood. If you have a technical background in a specific area, such as the telephone industry, you should easily understand the terms we use. Please do not be offended if we refer to LAN wire to the workstation as UTP cable, or horizontal cable, rather than IW (inside wire) or station wire. The former terms are much more specific to the LAN wiring that is our subject.

Twisted Pair Wiring

The basic wire type for most current LAN wiring installations is called *twisted pair.* Some refer to this wire as "unshielded" twisted pair (UTP) to differentiate it from "shielded" twisted pair. This style of wire is relatively inexpensive, is easy to connect, and provides self-shielding properties to minimize harmful interference to or from the cable. Twisted pair wire was originally used by the telephone industry to enable the widely separated individual wires that once populated telephone poles to be replaced by wire pairs in very close proximity. This twisted pairing allows more than one telephone conversation to be carried within the same cable jacket. The twisting causes the electromagnetic coupling from one wire to another to be canceled out and eliminates interfering crosstalk between pairs. Any significant amount of crosstalk coupling limits the length of cable that can be used. While the amount of coupling is insignificant at audio (voice) frequencies, it becomes much greater at the higher LAN frequencies. That is one of the reasons we do have to limit the length of LAN cabling. The construction of twisted pair cable is described in more detail in Chap 3.

The use of twisted pair cable is a key component in LAN wiring system design. The properties of this type of cable define the maximum usable length of a workstation cable at any given LAN data rate. The characteristics of LAN cable are carefully specified in the applicable standards. Particularly at the higher LAN speeds, these performance characteristics are the limiting factor. At 100 Mbps, we are pushing the capabilities of twisted pair wire. At 155 Mbps, we may be at the practical limit for a useful cable length to the workstation for two pair connections, but several new methods use all four pairs to jump to 1000 Mbps (1 Gbps) and are still not at the limit.

The Electronic Industries Association (EIA) and Telecommunications Industries Association (TIA) standards specify the maximum length from the wiring closet to the workstation as 100 m (90 m for the horizontal cable, or station cable to the wall jack, and 10 m for the patch cord plus the user cords). This is approximately 328 ft, and should be sufficient for most building installations. However, it does put one at the outer limit of performance where everything becomes critical, from routing and handling, to measurement and certification.

For example, several standards of the ATM (Asynchronous Transmission Mode) Forum specify twisted pair wiring for the medium. These new connectivity options range in speed from 25 Mbps to 155 Mbps and use the TIA/EIA-568-B 90 m (295 ft) horizontal cable link as the model. Some high-speed networking proposals use two pairs, while others use four pairs. The higher speeds may require the use of premium Category 5e cable and connecting hardware, but they still operate on conventional metallic twisted pair cable. Clearly, twisted pair wiring has a lot to offer for the future of networking.

Structured Wiring

Structured wiring based on the standards (such as TIA/EIA-568-B) assumes that the total wiring system will be divided into simple wiring units. These wiring units can be repeated as needed, combined into larger structures, and those structures interconnected to produce the overall wiring system. Although some standards exist for implementing particular LAN topologies, such as 10BaseT, over a wiring system, most of the standards simply require a generic, multipurpose, reusable wiring and cabling system that could be used for anything from voice to video. Of course, these standards are designed to allow LAN traffic to properly operate at specific data rates, and specify test parameters to ensure performance. However, the standards for the cabling itself are not at all specific to any type of LAN.

This is an enormous advantage to you, as the cable system designer, installer, or LAN manager. If you install a standards-based structured wiring system, you do not need to know anything at all about the exact type of LAN that will be installed in your wiring system. All you do need to know is what data rate will be expected of the LAN, because that will define the category of cable that must be used.

The most significant standard in LAN wiring is the "ANSI/TIA/EIA-568-B Commercial Building Telecommunications Wiring Standard." Along with related standards and bulletins, this standard defines a universal cabling system that meets a variety of needs. The TIA is really the moving force behind this standard, although it is approved by EIA and

ANSI and coordinated with many national and international standards. (By the way, one normally says the letters T, I, A individually, though you could certainly say it as a word.)

Any type of voice, data, or LAN signals can be connected to your structured cabling system. For example, the twisted pair wiring system that is specified by TIA/EIA 568-B will allow you to run analog or digital voice, 10BaseT Ethernet, 100BaseTX Ethernet, 100BaseT4 Ethernet, 100VG-AnyLAN, 1000BaseT, LocalTalk/AppleTalk (tm), Arcnet, 4 Mbps Token-Ring, 16 Mbps Token-Ring, 100 Mbps Token-Ring, CDDI (Copper FDDI at 100 Mbps), and ATM at 25, 51, 155, and 622 Mbps. Of course, ISDN (Integrated Services Digital Network), Switched 56 kbps, T1 or E1, and many other telephone services are also compatible with structured wiring systems. Also, several products to distribute video over twisted pair in NTSC format have been introduced, and you can even distribute IBM-5250 and -3270 terminals over the same wiring system.

Such a wide range of uses makes the term "generic" seem totally inadequate. Perhaps multipurpose or general use would better describe this structured cabling system.

The true versatility of the TIA 568-B wiring system is a powerful advantage to any organization that uses it. For this reason, you should be wary of compromises that limit your wiring system's utility. Although you may have a very specific purpose in mind, who can say to what use the wiring system will be put over its useful life of 10 or 20 years?

International standards also cover much of the same ground. For the most part, these standards are all coordinated. It would be tiresome indeed to constantly list all of the coordinated or referenced standards for each citing. Therefore, we will simply use the U.S. standard number in most cases. A fairly complete list of U.S., Canadian, European Community, and International standards is given in the appendix of this book. Standards that are coordinated are so noted.

The structural wiring system comprises several component structures. Within a building, the structures are called horizontal and backbone wiring.

Structured Wiring's Component Parts

Structured wiring defines a series of wiring elements that are assembled and connected to form telecommunications links (Fig. 2.1). The basic elements are the *horizontal wiring structure,* the *backbone wiring structure,* and a series of *cross-connect structures* that include the *main cross-connect,*

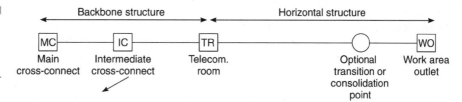

Figure 2.1
Structured wiring covers the cable runs from start to finish.

the *intermediate cross-connect,* and the *horizontal cross-connect.* These last three elements are usually contained in dedicated rooms within a building. For example, the horizontal cross-connect is contained in the telecommunications room, and is identified as such. The other rooms are often identified by their structure names, main cross-connect (MC) or intermediate cross-connect (IC). Backbone cable runs between the cross-connect structures, while horizontal wiring runs from the telecommunications room to the individual work areas.

This hierarchy is illustrated in Fig. 2.2. The structure holds within a single commercial building and between buildings, as on a campus. Although the structured wiring concept actually comes from the telephone industry, and thereby predates data uses of twisted pair cabling, its use is particularly critical to modern data networks. Structured wiring defines the precise maximum distances, types of connections, grade of cable and connectors, and installation practices necessary to provide a reliable and flexible data and voice cabling infrastructure. The embodiment of modern structured data/voice wiring is the TIA/EIA-568-B standard for North America and the ISO/IEC 11801 standard for the European community. For the most part, the balance of the countries around the world use one of these standards for their local cabling guidelines. Fortunately, these primary standards are becoming more and more coordinated through a process of international technical cooperation called *harmonization.*

The next few sections describe the major wiring strategies of horizontal and backbone cabling. Horizontal wiring takes the bulk of our attention in *LAN Wiring,* as it is the most complex structure, and offers the most challenges to designer and installer alike.

Horizontal Wiring Structure

The basic component of the structured wiring system is the *horizontal wiring* (cabling) structure. Horizontal cabling consists of the cable and connections from the wall or other jack at the user workstation outlet to the

Figure 2.2
Hierarchy of struc-
tured wiring.

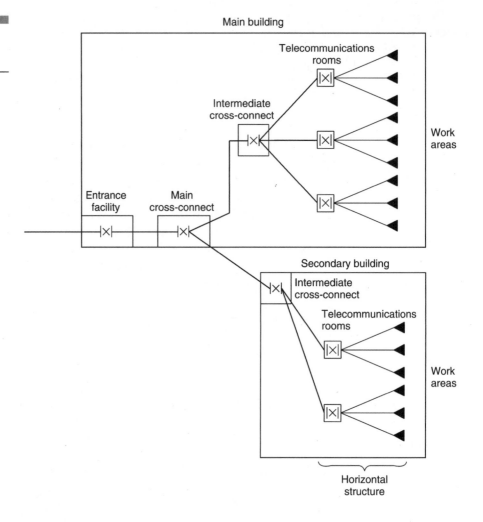

wire termination in the telecommunications room. As shown in Fig. 2.3, the horizontal wire may include a punchdown block, cross-connect wire, and a patch panel. However, it does not include hubs, routers, or network adapters. Horizontal cable is defined in TIA/EIA-568-B "Commercial Building Telecommunications Wiring Standard," and is the Basic Link (minus cross-connect and test cables) originally defined in TSB-67, "Transmission Performance Specifications for Field Testing of Unshielded Twisted-Pair Cabling Systems." In practice, horizontal wire is routed directly from the telecommunications room to the workstation, without any intermediate splice points, cable junctures, or taps, as might be common practice in traditional telephone wiring. This direct-wiring technique is referred to as a *home run*.

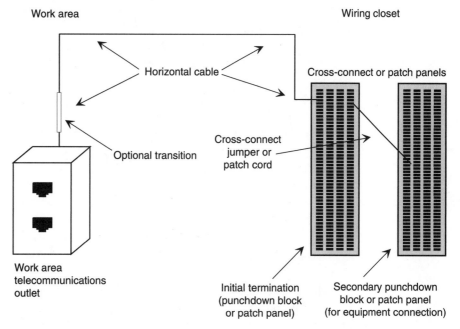

Figure 2.3
The horizontal wiring includes a telecommunications outlet, an optional transition, the horizontal cable, and alternatively a punchdown block and cross-connect wire, or a patch panel and patch cord.

Various wiring standards put special qualifications on horizontal wiring. Some of these may limit the types of components, such as cross-connects, that may be a part of horizontal wiring. More specifically, you should make sure that you use cable and components that are rated for the data rates you intend to use. For example, if you are installing a Category 5e (TIA/EIA-568-B) facility, you must use Category 5e rated jacks, cable, patches, and user cords. If you use a cross-connect field to connect between the station cables and a patch panel, the cross-connect wire must also be rated Category 5e.

In the horizontal wiring structure, the station cable is terminated in the area of each user workstation with an appropriate jack. The jack is typically an 8-pin modular female connector in a flush-mount wall plate that mounts in a single-gang utility outlet box or metal attachment ring secured in a hollow sheetrock wall. The 8-pin modular jack is sometimes referred to as an "RJ-45," because the connector/jack components are the same. However, RJ-45 actually applies to a special purpose jack configuration that is not used in LAN or standard telephone wiring. Several types of 8-pin jacks are available, and may be single (simplex) jacks, dual (duplex) jacks, or modular snap-in jacks of up to 6 jacks per plate. Solid walls, plaster walls, or hard-to-reach locations may use a surface mounted jack enclosure (such as a "biscuit block," or a flush-mount jack in a surface mount box) and appropriate surface raceway to reach the user outlet location.

Modular furniture requires special consideration, because the workstation cable may need to be dropped from the ceiling or run from an adjacent wall into wire channels within the modular furniture units. Some modular furniture provides special channels for telephone and compute wiring that is separated from the AC power wiring channels. Caution should be used when sharing the power wiring channel with data cabling. Not only is there a potential safety concern from having the LAN wire so close to power wire voltages, but there may be performance problems, particularly for Category 5e performance. Some guidelines specify a 2-in or more separation from power wires, while others specify a greater distance. In addition, local and national installation codes generally prohibit low-voltage wiring from being installed in the same conduit as power wires. Some local authorities may consider the open wiring channels of modular furniture to be in fact a conduit and therefore subject to this rule. Modular channels with a divider avoid this problem.

Horizontal wiring is specified according to performance categories defined in TIA/EIA-568-B. Both cable and connecting hardware (jacks and plugs) must meet performance guidelines appropriate for their use. Not all cable or components meet the minimum standards required for LAN use. One of the factors affecting performance in outlet jacks is the method of wire termination. The station cable connects to the jack by means of either screw terminals or some sort of insulation-displacement termination. Screw terminations were acceptable in Category 3 installations, but are prohibited in Category 5e or 6 installations, because of performance constraints. Category 3 is adequate for 10 or 16 MHz bandwidth networks, Category 4 for 20 MHz, and Category 5e for 100 MHz. For the most part, there are no networks that require Category 4 specifications, although it was used if you desired a "high-grade" Category 3 installation. Some references suggest Category 4 for Token-Ring networks, but the legacy standards required only Category 3. Category 5e is now the minimum specification.

Insulation-displacement terminations use a split metal contact upon which the insulated wire is forced, thereby cutting through the insulation and making metal-to-metal contact with the copper wire. Several types of insulation-displacement connectors (IDCs) exist, including modular snap-together IDC jacks, modular 110-type IDC jacks, and older 66-type IDC jack plates. Except for the 66-type jacks, all are commonly available in models that are certified to Category 5e performance standards. Category 5e 66-blocks do exist, but the installer may have difficulty maintaining the required twist all the way to the termination point.

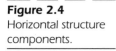

Figure 2.4
Horizontal structure
components.

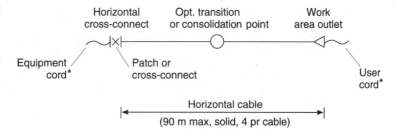

*Note: Σ (equipment cord, user cord, patch, or cross-connect jumper) ≤ 10m

From the workstation outlet jack, the horizontal station cable is run to the designated wiring closet. This cable run is designed to be within (possibly well within) the 90-m maximum specified by the standards. It may be routed with other individual station cables in wiring trays, through support rings over support beams or frames, through firewalls, or even in conduit. If you have special performance criteria, such as Category 5e or 6, you should take precautions to avoid fluorescent light baluns (the transformer in each light fixture), electrical wiring, heavy equipment (such as an elevator or other larger motor), and metal beams or metal walls. In general, the electrical wiring and metal items may be crossed (preferably at right angles), but not run alongside. Many authorities suggest that a 2-ft separation be maintained from these sources. EIA/TIA 569 provided specific guidelines, but these were dropped in TIA/EIA-569-A.

Station wire must be of a category that conforms to applicable performance standards. Obviously, a Category 5e installation requires Category 5e certified cable. However, a Category 3 installation could actually use Category 5e or 6 certified cable, since either would exceed the requirement. In some cases, existing telephone station wire may be able to perform at the lower data rates of Category 3, which can be verified through testing. If the wire was installed after about 1984, it probably has enough twists per foot to approximate current Category 3 standards. Prior to that year, telephone twisted pair cable was sometimes manufactured with less than two twists per foot, unless it was made to meet more stringent AT&T premises distribution standards. A cable scanner that measures impedance and near-end crosstalk (NEXT) should give you an indication whether such cable will work adequately. You should be aware that none of the current TIA/EIA standards encourage you to "steal" pairs from an existing telephone station cable, but you may be able to retrofit an existing installation if absolutely necessary.

Figure 2.5
Telecommunications
room components.

Station wire must also have an insulation type that meets safety and fire-protection standards. Fire protection and safety standards are specified by Underwriters Laboratories (UL), the National Fire Protection Association's (NFPA's) National Electrical Code (NEC), Canadian Standards Association (CSA), International Standards Organization (ISO), and various European Community (EC) and individual country standards in other parts of the world. These standards are meant to exclude flammable cables (or ones that exhaust toxic gasses when placed in a flame) from any plenum spaces. A *plenum* is an air duct or a return air duct that is used for forced air circulation, such as from a heating or air conditioning (HVAC) system. Any air space used for forced air circulation, other than an actual room intended for work areas, is a plenum (which means that the area above drop-in ceiling tiles is plenum space if separate ducted air returns are not used). A common arrangement is to use a plastic return-air grid in place of a ceiling tile, or to use fluorescent lighting fixtures with a row of vent openings along the edges, to provide air return into the above-ceiling area and eventually back to the HVAC air handler. All of this space is air plenum and requires special plenum rated cables. Cables and cable ratings are covered in more detail in Chap. 3, "Wire and Cable Technology for LANs," and App. C, "LAN Wiring Standards."

The telecommunications room (wiring closet) is the gathering point for all of the home-run cables that serve workstations of a particular floor or area in your building. The size and location of wiring closets are actually specified in standards, such as TIA/EIA-569-A. This standard refers to several types of wiring closets, such as the Telecommunications Room (TR), the Intermediate Cross-connect (IC), and the Main Cross-connect (MC). Such a variety of wiring closets is appropriate for telephone wiring, since it is much more extensible than LAN cabling. However, since the topic of this book is restricted to LAN wiring, we will simplify this to just "telecommunications room." For our purposes, then, all wiring outward from the patch or other station wire termination point to the workstation

Figure 2.6
Punchdowns.

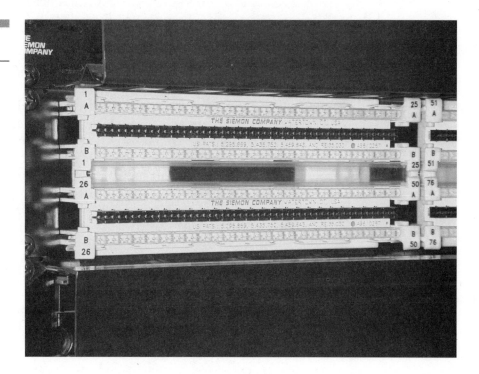

outlet jack will be considered simply horizontal cabling and will conform to those standard practices that are appropriate for LAN wiring.

As mentioned previously, horizontal cabling includes the wire terminations in the telecommunications room (TR). You will typically bring all the individual station cables into the wiring closet at the same entry point, then down a wall onto a wiring board. Wiring boards are often large 4-by-8 ft sheets of ³/₄-in plywood, mounted securely to a wall of the TR, and painted an appropriate color. Of course, TIA/EIA-569-A has a color code for every type of cross-connect, but you can probably use a neutral color of your choice on the wiring board. An alternative telecommunications room arrangement uses freestanding equipment racks or rails. Cables are routed in cable trays or in bundles across the top of the racks and down into each rack to the point of cable termination. This is particularly useful for very large installations where much equipment and many station terminations exist. Care should be taken in securing the cables and cable bundles to ceilings, walls, racks, and wire termination devices. In Category 5e/Category 6 installations, tightly binding the cables with tie-wraps should be avoided because it may deform the cable enough to distort its electrical characteristics. Likewise, you should maintain minimum bend radius with individual cables or with cable bundles. Avoid sharp corners and other abrasive edges.

Station wires may be terminated in the TR on a punchdown block, such as a 66M or 110-type block, or they may be terminated directly onto a patch panel. (Fig. 2.7) If a punchdown block is used, cross-connect wire must be run between the station wire termination and any other connection device, such as a patch panel. One alternative is to use a punchdown block with a built-in connector (typically a so-called 50-pin Telco connector). A fan-out adapter cable, often called a *hydra* or *octopus cable*, connects from the connectorized punchdown block and splits to six (for standard 4-pair installations), eight, or even twelve 8-pin modular connectorized cables that can individually plug directly into a hub. To get from the punchdown to a patch panel, a 50-pin jumper cable could be used instead of the octopus cable. Some manufacturers also offer punchdown blocks with built-in 8-pin modular jack connections. Although some of these devices may be certified to Category 4 or 5, they are much more appropriate for Category 3 installations.

Direct patch panel termination is often used for Category 5e/Category 6 installations. This method provides the advantage of simple one-location termination. There are no cross-connects, punchdown blocks, jumper cables, fan-out adapters or Telco connectors. All of these are potential trouble spots for Category 5e/Category 6, since each wire termi-

Figure 2.7
Patch panel
termination for
fiber-optic cables.

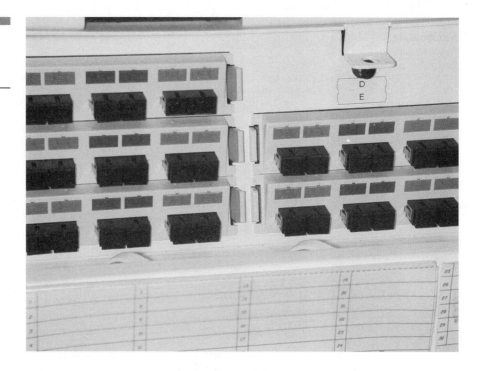

nation introduces "untwist" and all components must be certified. Installation testing is also simplified with patch panel termination, because all station cables compose a Basic Link[1] and can easily be scanned to certification standards. The only possible disadvantage to patch panel termination is that multiple-use applications of the station cables are difficult, because each cable terminates into one patch panel location. Some suppliers offer plug-in adapters with dual cable terminations, but this still limits the station cable to two four-wire jack terminations. This disadvantage is probably not too serious, because it allows one to totally comply with the wiring standards, and special applications can always be handled as an exception with adapter cables.

Backbone Wiring Structure

Backbone wiring may be within one building or between buildings on a campus. All wiring between telecommunications rooms (MCs, ICs, and

---

[1]The TIA Basic Link is being harmonized with a revision of the ISO/IEC Permanent Link. For more information, see Chapter 5.

Figure 2.8
Horizontal testing
structures and links.

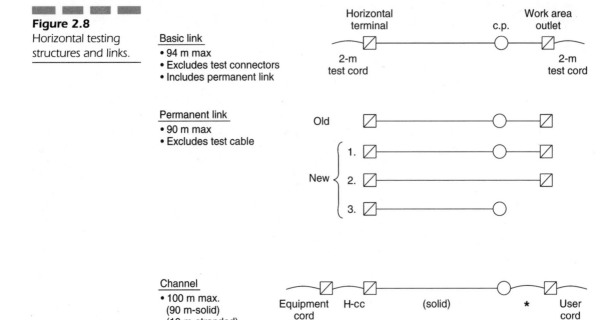

Basic link
• 94 m max
• Excludes test connectors
• Includes permanent link

Permanent link
• 90 m max
• Excludes test cable

Channel
• 100 m max.
  (90 m-solid)
  (10 m-stranded)

Key
☑ Connector-mating pair
○ Consolidation point (c.p.) or transition
── Solid-conductor cable
⌒ Stranded cable permitted
★ User cord runs continuously from consolidation point,
    but reduces horizontal length from H-cc to c.p.

TRs) is referred to as *backbone wiring*. In many cases, this backbone wiring will be between wiring closets on the same floor of a building, but the classic case is the multifloor model we saw in Fig. 1.1. Some standards refer to this as *vertical wiring,* in contrast to horizontal wiring and because of the confusion with logical LAN backbone segments (which may or may not run on backbone wiring). We will simply call it backbone wiring, and point out those few instances when we are referring specifically to a LAN standard's specification for logical backbone wiring.

Backbone wiring is usually simpler than horizontal wiring. It is customary to place LAN hubs or concentrators in each telecommunications room to connect to all the workstation wires on served by that telecommunications room. Backbone wiring then connects the hubs, one to another.

Backbone wiring may be implemented in a daisy-chain (hub-to-hub-to-hub) fashion, or in a star fashion, as shown in Fig. 2.9. In a network topology that must guard against too many repeater jumps, such as Ethernet, the star connection for backbone wiring provides advantages. Since telecommunications rooms in large buildings are frequently in a services shaft, one above the other, distances between telecommunications rooms are minimal. Thus the backbone wiring from each floor may be concentrated in a single master telecommunications room and the number of repeater jumps minimized. For other topologies, such as Token-Ring, it might be more convenient to daisy-chain the ring-in/ring-out from floor to floor.

In any event, the maximum cabling distances must still be observed. Backbone cabling may be terminated in the same manner as horizontal cabling. The backbone cable should be of a category and type that meets your performance requirements and the safety and fire protection requirements for your installation. If your installation requires Category 5e horizontal cable, for example, you should use the Category 5e cable for your backbone wiring. It does not hurt to use a higher category of cable than needed, so you might use Category 6 anyway. If you anticipate using your wiring system for multiple applications, double or triple each backbone wiring run. This will give you ample room for growth and the ability to rapidly respond to new requirements.

Some standard require that all riser cable (cable that goes vertically between floors) be riser rated, while some local authorities may require

Figure 2.9
Backbone wiring may be implemented in a daisy-chain (hub-to-hub-to-hub) fashion, or in a star fashion.

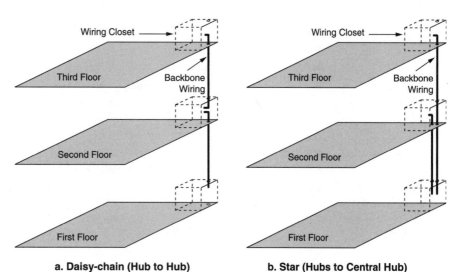

a. Daisy-chain (Hub to Hub)   b. Star (Hubs to Central Hub)

the tougher fire specifications of plenum cable. In any event, any openings you make between floors, whether in cable ports or direct opening, should have fire-shop material properly installed. This is particularly important in taller buildings. Consult the NEC, UL standards, and local standards for more guidance.

When wiring between far flung telecommunications rooms or between floors, electrical grounding and bonding requirements should be observed. EIA/TIA-607 describes the proper practices in detail.

As fewer cables are used for backbone wiring, you may be tempted to terminate the cables directly into an 8-pin modular male connector plug. Backbone cabling uses solid copper wire, and since most 8-pin modular plugs are designed for stranded wire, you could easily create a future problem. Just imagine an intermittent network failure that cures itself when you touch the backbone cable plug! Solid-wire 8-in modular connectors are certainly available, but they are impossible to distinguish without close examination and most installers recommend against their use. A better approach is to simply use the same patch panel or punchdown that terminates the station cables to terminate the backbone cables. Mark the backbone cable patch position clearly to differentiate it.

You may wish to more clearly identify backbone wiring by using cable colors, cross-connect/patch panel colors, and special markings. These colors and markings, identified in EIA/TIA-569-A, are covered in later chapters.

If your backbone wiring actually involves floor-to-floor cabling in vertical shafts or cable ports, you may need to use special riser cable that has a fire-retardant sheath, certified to meet NEC low-flame requirements. The NEC requires that riser cable meet UL flammability tests to be used in high-rise buildings. Local regulations may supplement or amend these requirements. Local building inspectors should be able to advise you about the requirements in your area.

As a last word on backbone wiring, fiber optic cable may provide some significant advantages over copper. Fiber optic cable is not subject to electrical or magnetic interference, and thus may be run in locations, such as elevator shafts or alongside power lines, that would cause problems for metallic cable. In addition, a cabling system may be electrically isolated between buildings or between floors of the same building by using fiber optic cable. Metallic cable, on the other hand, must be carefully installed with adequate consideration of grounding and bonding. Fiber optic cable may also be used for multiple applications with the use of fiber optic multiplexers (WDM, wave-division multiplexing). Thus, it would be possible for one fiber optic connection to carry Ethernet,

Token-Ring, T1, ATM, and FDDI all at the same time. Fiber optic cabling must meet the same plenum and riser specifications as metallic cable.

Location and Routing

The importance of location and routing of your cabling system components depends upon the level of cable facility that you intend to install. Category 3 installations have much more flexibility in routing than Category 5e/Category 6 installations. For networks that must provide 100 to 1000 Mbps speeds, and have horizontal runs that approach the 90-m limit, you must be very attentive to installation details. Location and routing of the station cables are critical to high-performance operation.

Telecommunications rooms should be located so that they are within 90 m "as the cable runs" from the proposed workstation outlets that are to be served. In some cases, the number of telecommunications rooms may have to be increased or relocated to stay below this distance. A scaled plan drawing should be prepared and cable runs planned out and measured (on the drawing) to be certain. Allow an extra $3^1/_2$ to 5 m for each vertical drop from the ceiling to the wall outlet and 2 to $3^1/_2$ m for the drop from the ceiling to the patch panel. Remember that the vertical drop must start well above the ceiling grid, because rules forbid placing the wire directly on the ceiling grid. If your patch panel is not wall-mounted, allow additional distance for the cable run to the rack or cabinet. Remember also that cable is normally run at right angles on the wiring board or in the rack. It is amazing how much extra cable all of these right angles require.

EIA/TIA-569-A specifies the minimum size and suggested locations for "telecommunication spaces" such as telecommunications rooms. It would be ideal if you could adhere to those standards, but some buildings may not allow such luxuries. Keep in mind that the network managers and technicians need to have quick and easy access to the patches and hubs in the telecommunications rooms. If you absolutely must share space, place the wire termination point in a corner of the room that will not be blocked by storage or office equipment. Place it at an easily reachable level where there is adequate light, ventilation, and power. Several manufacturers make wall-mounted cabinets that conceal and protect your wiring and hubs.

All LAN cables should be routed away from potential sources of interference. Avoid fluorescent lights, motors, pipes, structural steel, and power lines. The National Electrical Code, TIA/EIA-568-B, and ancillary documents specify the separation distances, both for safety and for interference. With the exception of fluorescent lights and electric motors, most interfering conductors can be crossed at right angles, but you should still maintain a minimum separation for safety reasons.

Some authorities advocate direct physical support Category 5e/Category 6 LAN cables. This requires the use of cable trays, ducts, cable hangers, or conduit for the entire route of each cable. In many situations, these extensive precautions are simply impossible or cost-prohibitive. However, in large buildings with suspended ceiling grids, it may be desirable to use a cable tray routing system to help organize and protect the LAN cable. The key element is to eliminate any deformation in the insulation or twist integrity of the cable. That generally means that you must avoid long unsupported spans of wire, sharp bends, kinks, binds, nicks, and tight tie-wraps.

A Simple Design Example

To illustrate a basic data-only wiring system design, we will show a cabling plan for a simple two-story building using the concepts we have covered in this chapter.

Our example will have 24 workstations per floor. We will use Category 5e specifications, direct patch panel termination, and will have one telecommunications room per floor. We will specify the components, installation, and testing and will develop a sample bill of materials.

Figure 2.10 shows the layout of each floor. We will specify flush-mounted jacks for the user outlets. Each user will have a 3-m Category 5e cable from the wall outlet to the workstation. We will use Category 5e plenum cable with 4 24-gauge twisted pairs for the station cable that runs from each user outlet to the wiring closet. At the wiring closet, we will use a wall-mounted 24-position patch panel that will directly terminate each station cable. A 1-m Category 5e patch cord will be provided to connect each patch position to the network hub or MAU. In some situations, longer patch cords should be specified. Plan the placement of your hubs before ordering patch cords. Between the two floors we will run one riser cable of the same type as the station cables, if local codes permit.

Of course, the LAN network will also need appropriate LAN hubs or MSAUs, network adapters, servers, routers and software, but our concern here is simply the wiring system that connects all these devices.

Figure 2.10
A typical floor layout showing the telecommunications outlets and direct cable runs to the telecommunications room. You should allow 25–50% more cable for cable trays and obstacle avoidance.

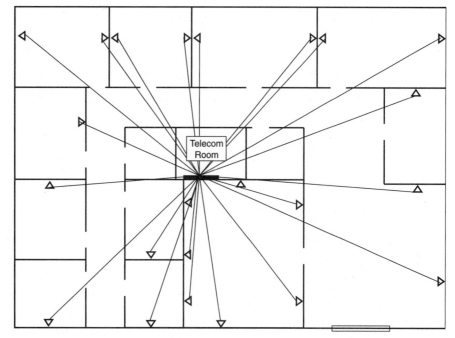

The bill of materials is shown in Table 2.1 If you have to obtain the materials, you will also have some brand-name decisions to make. Some brands have special features, offer long warranties, or are easier to install. An experienced installer will have preferences that can guide your choices. There will also be some differences in quality and pricing. All the items should be certified by their manufacturers to meet the Category 5e performance criteria you have set. To implement the two-cables-per-workstation suggestion of TIA/EIA-568-B, you will need to double many of the quantities.

The wall plates, user and patch cords, and cable are generally available in a variety of colors. The wall plates and user cords are usually color coordinated to the office. The color choices are somewhat limited, so you may have to go with a neutral color such as gray or ivory. Surface raceway is also available in color choices. Patch cords technically fall under the miscellaneous category of EIA/TIA-569-A and are ideally yellow in color. However, if your wiring closet is actually a vacant wall in a common area, and a sea of yellow wire would distract from the decor, you may wish to specify a more neutral color. The EIA/TIA police will never know (we hope).

The instructions to the installers will be as follows. Pull the station cables as home runs, with no splices, from each workstation outlet

	Quantity	Description
TABLE 2.1		
Typical Bill of Materials	A/R	Cable, 4 twisted pair, 24 gauge, Category 5e, plenum-rated
	A/R	Cable, 4 twisted pair, 24 gauge, Category 5e, riser-rated, gray or white
	48	Plate for modular jack, ivory
	48	8-pin jack module, Category 5e, wired per T568A, ivory
	2	Patch panel, 24 position, Category 5e
	2	Mount for patch panel
	48	User cord, 3 m, Category 5e, stranded twisted pair, PVC (non-plenum), gray
	48	Patch cord, 1 m, Category 5e, stranded twisted pair, PVC (non-plenum), yellow
	2	Duct, fingered, 3 in × 3 in, 6-ft length (cut to 2-ft lengths for patch cord management)
	A/R	Duct, fingered, 2 in × 2 in, 6-ft length (for vertical cable entrance into wiring closet/patch panel area)

NOTE: Quantities shown are for two floors, 24 work areas per floor, and one LAN cable per work area. Some standards, such as TIA/EIA-568-B, recommend two cables per work area. An existing telephone cable at each work area would yield the total of two cables.

location to the telecommunications rooms on each floor, avoiding electrical wires, pipes, motors, and fluorescent light fixtures. The station cables will be run from each outlet through hollow walls to the space above the ceiling grid. Outlets may be surface mounted on solid walls and the cable placed in surface conduit to reach the area above the ceiling grid. Cables should be run supported as much as possible with no tight tie-wraps, no kinks, and respecting the minimum bend radius for Category 5e (approximately four times the cable diameter). Cable runs must not exceed 90 m from outlet to patch. The patches should be mounted at a convenient height on the wall of each wiring closet with adequate support for the cable bundle and no tight tie-wraps. The riser cable should be run in an existing cable port in a vertical shaft or utility area, if possible. Fire-stopping material should be placed where the riser cable runs between floors.

The cable pairs should be wired using the T568A pattern of TIA/EIA-568-B. Proper workmanship standards should be observed, including the minimum twist standard for connecting Category 5e cable. After installation, each station cable connection should be scanned from outlet to patch with a Category 5e cable scanner that complies with TIA/EIA-568-B (TSB 67). If a cable run is near the 90-m maximum length, the entire Channel should be scanned, including the patch and user cords that are to be used for the workstation connection. A printed report will be provided, including a plan diagram showing the numbering of all outlets. All cables, outlets, and patch panels are to be marked in accordance with EIA/TIA-606.

Well, that pretty much covers an ideal, concise wiring system specification. You could mention a few more standards, if you wish. As you will see in later chapters, there are quite a number of standards that more closely define your installation. Keep in mind that most of these standards are voluntary and exist to make absolutely certain your wiring system will be up to stringent performance specifications. Many of the standards overlap or are "coordinated," so that they actually say the same thing. Too much reference to standards may actually raise the price of your installation. Some of the standards are overkill for small simple networks. For example, it is doubtful that you could make use of half the detailed marking specifications in a 10- or 20-user network. However, if you are planning a 1000- or 2000-user wiring system, adherence to the marking standard will be extremely useful. Likewise, the color codes and even the routing guidelines may be less significant for smaller networks with shorter runs. You will have to be the judge. The standards do not imply that your network will fail if you do not observe all the guidelines; they simply say that it will work at the specified performance category if you do.

CHAPTER **3**

Wire and Cable Technology for LANs

Chapter 3 Highlights

- Basic cable types
- Twisted pair UTP & STP cable
- Coaxial cable
- Nonpaired cable
- Fiber optic cable
- Special purpose wire

Many types of wire and cable are used in LAN wiring. This chapter will describe the major types of cable that are currently in use. You should define the cable that will be used long before you begin your cable design. The cable chosen will often dictate the connectors, terminations, useful distances, routing, and LAN types that your wiring will support.

We will cover types of cable that are used for all types of networks, including Ethernet, Token-Ring, and Arcnet. Although we mention these network topologies and their variations, we will concentrate here on the cable, rather than on its use.

Basic Cable Types

All LAN cable can be divided into three basic types: twisted pair, coax, and fiber optic. We will cover each of these types in detail. We will also cover round and flat nonpaired wire often used for telephone sets. However, because nonpaired wire is not used for LAN wiring, it will be presented only so that you may easily recognize it and avoid its use.

Each basic cable type has several variations and different types may even be found in combination. An example of this is twisted pair cable, which may be found in unshielded and shielded varieties. There are also several types of nonpaired wire, such as round 4-wire cable, ribbon cable, and flat cable (often called "silver satin"). Some manufacturers, such as IBM, specify combination cables that contain both shielded and unshielded pairs. Even combination fiber optic and metallic cables are offered. However, for our purposes here, we will avoid all of the myriad of combinations and concentrate on the pure cable types.

Twisted Pair UTP and STP Cable

Twisted pair wiring is the most common type of LAN wiring cable in use today. It is versatile, easy to install, inexpensive, and has favorable performance characteristics. Twisted pair cable is available with or without shielding. It comes in a variety of colors, wire gauges, insulation, twisting, and outer sheath materials.

An illustration of twisted pair wire is shown in Fig. 3.1. A large number of pairs can be in the same cable sheath (the outer jacket). Telephone cables are commonly available in 2 pair, 4 pair, 6 pair, 25 pair, 100 pair, and

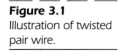

Figure 3.1
Illustration of twisted
pair wire.

even larger bundles. LAN twisted pair cable, however, usually comes as 4 pair cable.

The 4 pair cable is a topic of several standards, including TIA/EIA-568-B. It is the cable around which most of the important LAN cable specifications and performance tests are based. Most LAN topologies actually use only two of the four pairs, however, so some installations place two LAN connections on each 4 pair cable. Telephone connections often use two pairs and may sometimes be wired in the same cable with a 2 pair LAN connection. You should be cautious in robbing pairs from the LAN cable, because some of the 100 Mbps and higher LAN schemes may use all four pairs.

Another common cable that is found in twisted pair wire installations is the 25 pair jumper cable. This cable is preterminated in 50-pin male or female connectors (sometimes called a 50-pin Telco connector). In telephone wiring, the 25 pair jumper cable is convenient for connecting between the local exchange carrier's RJ-21X demarcation point and the user's punchdown blocks or Private Branch Exchange (PBX) switching equipment. In LAN wiring, the 25 pair jumper is commonly used between connectorized punchdown blocks and patch panels. Category 3 and Category 5 grade 25 pair jumpers are available, but some caution should be used in deploying them in a Category 5 installation. Some authorities are concerned with the combination of two or more 100 Mbps signals in the same cable sheath. A 25 pair jumper would theoretically allow you to combine as many as six such signals within the same sheath (6×4 pairs = 24 pairs with one unused pair). Also, each connector introduces more untwisted wire into a circuit and some of the standards limit the amount of untwist as well as the number of connectors in a link.

UTP/STP General Construction

Twisted pair cable consists of one or more pairs of insulated wires that are twisted together and joined in a common sheath. The main characteristics of twisted pair cable are wire gauge, stranding, twist pitch, insulation type, characteristic impedance, and sheath material. Each of these items may affect the suitability of a cable for a particular application.

Cables for LAN wiring may be made with either stranded or solid copper wire. Solid wire is normally used for cable runs that will be terminated on insulation-displacement connectors (IDCs), such as outlet jacks or punchdown blocks. Stranded wire is normally used for user and patch cords that are terminated with an insulation-piercing 6-pin modular (RJ-11 style) or 8-pin modular (RJ-45 style) modular connector (plug). You should never use solid copper wire with a modular-type plug, unless the plug is specially designed for solid wire (see Chap. 10). Some older installations for Category 3 and below use screw terminal jacks that technically can be used for stranded wire, although solid wire is the norm for station cable.

The twisting of the two wires causes interfering electromagnetic fields to couple equally to each wire in the pair, as shown in Fig. 3.2, thus effectively canceling out the resulting interfering signal. This mode of operation is referred to as "balanced" transmission. For proper cancellation it requires that a desired signal be applied to the wire pair by a balanced driving circuit and load. The equally coupled interfering signal is ignored by the balanced load, although common-mode component of the interfering signal may exist. A balanced circuit is shown in Fig. 3.3. As an additional benefit, electromagnetic emissions from a balanced twisted pair are reduced (note we said "balanced!"). This prevents the high frequencies of a LAN signal from interfering with other devices. However, some radiation of the signal does occur. The transmitted signal amplitude is kept low to maintain the spurious emissions within acceptable limits. Longitudinal balance of the entire Channel is important for low emissions as well as susceptibility to outside interference.

UTP/STP Wire Sizes

The conductor diameter in twisted pair wires is referred to by wire gauge. The common standard is the American Wire Gage (AWG). Table

Figure 3.2
Twisted wire causes interfering electromagnetic fields to couple equally to each wire in the pair and cancel out at the load.

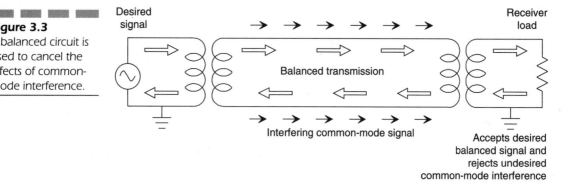

Figure 3.3
A balanced circuit is used to cancel the effects of common-mode interference.

3.1 shows a comparison of wire gauges to physical size. Smaller diameter conductors correspond to larger wire gauge numbers. Thus, AWG-26 is smaller than AWG-19. Telephone outside plant cabling is typically 24 or 26 gauge, although in rural areas 19 or 22 gauge may be used. The larger diameter wires have more physical strength, which is an advantage for wires that must be run between distant supports. Larger wires also have a lower resistance per unit length, which reduces resistance to direct current and lessens attenuation to alternating frequencies, such as voice. (At LAN frequencies, the capacitance is the primary contributor to attenuation.) The greater copper content of larger wire sizes also increases the cost of the wire.

At higher frequencies, such as those of a high-speed LAN, signal current concentrates at the outer diameter of the conductor, a phenomenon known as *skin effect*. In a stranded conductor, the outer skin is ill-defined, because of the many conducting strands that make up the wire, and attenuation is increased. For that reason, standards typically limit the length of stranded wire that is allowed in a Channel. Solid copper wire must be used for the horizontal cable portion of the Channel. Stranded wire may be used for user and patch cords, where a 20% increase in attenuation is allowed.

The common wire gauge that is specified by standards for LAN wiring use is 24 gauge solid copper wire. This wire size provides a good compromise between attenuation and cost. It is also an easy size to work with.

UTP/STP Electrical Characteristics

Twisted pair wire, like other types of transmission lines, has a characteristic impedance. This impedance is a result of the dielectric properties

TABLE 3.1

American Wire
Gage for Selected
Solid Bare Copper
Wire Sizes

Gage (AWG)	Nominal diameter		Area, circular mils	Nominal resistance, ohms/kft
	mm	inches		
10	2.60	0.1019	10380.0	0.9989
12	2.05	0.0808	6530.0	1.588
14	1.63	0.0641	4107.0	2.525
16	1.29	0.0508	2583.0	4.016
18	1.02	0.0403	1620.0	6.385
20	0.813	0.0320	1020.0	10.15
22	0.643	0.0253	640.4	16.14
24	0.511	0.0201	404.0	25.67
26	0.404	0.0159	253.0	40.81
28	0.320	0.0126	159.8	64.9
30	0.254	0.0100	100.5	103.2
32	0.203	0.0080	63.21	164.1
34	0.160	0.0063	39.75	260.9
36	0.127	0.0050	25.00	414.8

SOURCE National Institute of Standards and Technology.

of the insulation and the closeness of the conductors. The insulation's dielectric properties are a function of the type of material used. A variety of characteristic impedance values are available. The wire used for LAN twisted pair wiring has a characteristic impedance of 100 ohms, 120 ohms, or 150 ohms. Most of the UTP cable you will deal with will be rated at 100 ohms, as that is the standard for unshielded twisted pair wire in most parts of the world.

Shielded or screen twisted pair wire (STP) generally has a characteristic impedance of either 100 or 150 ohms. Newer STP constructions that are in wide use in Europe are of the 100 ohm variety. TIA/EIA-568-B specifies this 100 ohm construction for STP applications.

The specific type of 150 ohm STP that was defined in TIA/EIA-568-A is called *STP-A* to differentiate it from the other types of STP cable. This STP-A is often found in Token-Ring installations that use the IBM cabling system, as covered in Chap. 4. To interconnect between 100 ohm and 150 ohm wire, you must use a *balun transformer* to minimize unwanted reflection and power loss caused by the impedance mismatch. If a network with 100 ohm station cable connects to user equipment with 150 ohm interfaces, a special cable with a built-in balun, called a *media adapter,* must be used for proper operation. STP-A is typically rated at 300 MHz, in contrast to the 100 MHz rating of normal Category 5e and lower cable.

If you are familiar with telephone circuits, you may know that such circuits are generally characterized for operation at 600 or 900 ohms. This load impedance is designed to minimize current flow to subscriber telephone instruments, in addition to other factors. It is not representative of the characteristic impedance of the wire pair, which is not as significant at voice frequencies.

Why is the characteristic impedance important to LAN wiring? The reason is twofold. First, the maximum amount of power is transferred when the impedances of the source and load are matched. A greater power transfer extends the usable cable distance. Second, mismatches cause reflections of the signal, which can deteriorate the quality of the received LAN signal and cause unwanted emissions. At high LAN data rates and long cable distances, this might become critically important to performance.

As we will see later, a cable's characteristic impedance may vary over the frequency range of the LAN signal. The standards specify a maximum variation that must be met for the cable to meet minimum performance criteria.

The resistance and capacitance of a twisted pair cable cause an attenuation of the LAN signal that is proportional to length of cable. This signal attenuation is expressed in decibels (dB), indicates a power ratio, $10_{\log}$ (P_{in}/P_{out}), of the input versus output power expressed as a logarithm. The typical attenuation of a cable is given in dB per foot or meter so that you can calculate the total attenuation of a cable run by multiplying its per unit value by the total length. Fortunately, a standards-compliant installation will ensure that the total attenuation (including cable and connectors) is within the maximum limits necessary for proper LAN operation. The attenuation of a typical cable run is enough below the maximum allowed that an attenuation measurement usually reveals bad connections or partial cable breaks, rather than cable deficiencies.

The capacitance of twisted pair wiring is the factor that causes attenuation to increase steadily with frequency. The attenuation caused by capacitance is proportional to $1/(2\pi f)$. Thus, the greater the frequency, the less the parallel impedance between the wires, and the more the signal attenuation. Although this attenuation factor is linear, in practice, twisted pair cables exhibit peaks and valleys that can affect transmission performance at specific frequencies. For this reason, the new performance measurement standards for installed cable require numerous discrete measurements over the entire frequency range of a cable category.

Another characteristic of twisted pair cable is called *propagation delay.*[1] Propagation delay is defined as the time, usually in nanoseconds, for a signal pulse to travel from one end to the other end of a 100 m cable. The requirements for propagation delay are shown in Table 3.2. In multiple-pair cables, another parameter is measured, the *delay skew,* defined as the difference in propagation delay between pairs in the cable. Delay skew must be held to a minimum for LAN signaling methods that split up the transmitted data and send it simultaneously on more than one pair. Examples of this technique are 100BaseT4, 1000BaseT (T4 and TX), ATM-155, and ATM-622. If the delay skew is too great, the data pulses will arrive at improper times, and the receiver circuitry will be unable to reassemble the data packets.

Installation practices can adversely impact twisted wire cable performance. The impedance characteristics can be significantly disturbed by any distortion of the cable insulation or cable twist. These disturbances can be easily observed on a device called a time-domain reflectometer (TDR). When connected to a TDR, a cable with a normally flat response will show all sorts of lumps and bumps when the cable is sharply bent, stapled, tie-wrapped, and partially untwisted. These distortions in the cable may cause the installed cable to fail the final installed testing and should therefore be avoided.

UTP/STP Insulation

Two primary types of insulation materials are traditionally used in twisted pair LAN wiring, polyvinyl chloride (PVC) and fluorocarbon polymers. PVC is very flexible and is often used with stranded wire for LAN user cords and patch cords, where its flammability is not a disadvantage. PVC has dialectic properties that may make it unsuitable for

---

[1]Propagation delay is inversely related to the *velocity of propagation* of the cable. The velocity of propagation is normally given as a decimal fraction (or a percentage) of the speed of light in a vacuum. Although twisted pair cables have this parameter, it is a more common specification for coaxial cable. Propagation delay is more commonly found in twisted pair specs, although either parameter can easily be calculated from the other.

TABLE 3.2

Propagation Delay
and Delay Skew
Requirements*

	Category 5/TSB-95 Class D	Category 5e Class D (revised)	Category 6 Class E	Category 7 Class F
Propagation delay, ns	548	548	548	504
Delay skew, ns	50	50	50	20

*Category requirements are for TIA/EIA-568-B and anticipated revisions for Categories 6 and 7. Class requirements are for ISO/IEC 11801 and anticipated revisions.

wire insulation of Category 5e cable, although it can be used in cable jackets. Polyethylene (PE) is sometimes used for Category 5e cable and may be substituted on one pair of some plenum-rated cables. Fluorocarbon polymers include polytetrafluoroethylene (PTFE or TFE) and fluorinated ethylene-propylene (FEP), which are not as flexible as PC, but meet the stringent flammability testing for use in plenum spaces. These fluorocarbon polymers are sometimes referred to by the DuPont trademark Teflon®. Wire insulation and sheaths may be a copolymer of TFE and FEP. Another fluorocarbon polymer that is often used in cable jackets of plenum-rated cables is ethylene-chlorotrifluorethylene (ECTFE or HALAR®). This material does not have the proper dielectric qualities to be used as a wire insulation, but efforts are under way to formulate it for that purpose.

The increasing use of large amounts of cable for LAN wiring has resulted in the creation of new cable guidelines that emphasize the flammability of the cable sheath and insulation. Conventional wire insulations were found to be relatively flammable and a hazard when placed in the air plenums. Plenums, or air ducts, carry the heated or cooled air around buildings. Any cable that burned in a plenum could create toxic gas and smoke that would be a danger to people in other parts of a building. In addition, the fire could actually spread through plenums. Manufacturers soon developed cable insulations that were less flammable and could be used in plenums.

Early fire protection specifications referred to sections of the National Electrical Code (the NEC) or to flammability tests of Underwriters Laboratories (UL). Similar specifications were published by other countries. Several of the pertinent NEC articles are shown in Table 3.3. The NEC Article 800 is often referred to for LAN cabling and telecommunications cable. The NEC differentiates cable types by voltage class and by flammability. Class 2 (up to 150 volts) and Class 3 (up to 300 volts) are both covered by Article 725. Article 725 might appear to apply to LANs, but it actually refers the user to Article 800 for cable

TABLE 3.3

National Electrical
Code (NEC) Articles
for Low-Voltage
Wiring

Article	Types of cable
725	Remote signaling and power limited circuits
760	Fire protection signaling systems
770	Fiber optic cables
800	Communication cables
820	Coax cables

classified for communications use (which does include LAN cable). In addition, permitted use is graded by three levels of flammability: general use, riser use, and plenum use.

The confusing labeling of NEC and UL specifications led to a revision of the NEC in 1987 that instituted a system of alphanumeric designators for permissible cable-use ratings. The common cable-use codes are shown in Table 3.4. Cables may be specified that meet or exceed the requirements for a specified use. You may find cable that has dual markings, if the cable is certified for use in each classification.

UTP/STP Color Coding and Marking

Each insulated wire of a twisted pair LAN cable is colored differently. The colors form a standard code, so that each wire may be easily found and terminated properly. The color code for 4 pair LAN cables is shown in Table 3.5. Each pair of the cable has complementary colors. For example, pair 1 wires are coded white-blue and blue-white. The white-blue wire is a white wire with a blue stripe at intervals along the insulation of the wire. (The stripe is sometimes called a *tracer.*) Conversely, the blue-white wire is a blue wire with a white stripe. In a 4 pair cable, white will be the common color of all the wires, and the white wire will always be numbered (or punched down) first. Each pair of a 4 pair wire may be referred to by its color that is unique from the other three pairs. Thus, the "blue" pair contains the white-blue and the blue-white wires, while the "green" pair contains the white-green and green-white wires.

Color-coding and proper termination of each wire color is very important in LAN wiring because the signals are polarity sensitive; reversing a pair will cause a failure. The polarity of each pair is often referred to as *tip* and *ring,* which stems from the days of telephone plug-

TABLE 3.4

Common National
Electrical Code
(NEC) Cable-Use
Codes for Metallic
Communications
and Control Cable

NEC article	Code	Meaning	Allowable substitutions*
725	CL3P	Class 3 Plenum	MPP CMP FPLP
	CL3R	Class 3 Riser	CL3P MPR CMR FPLR
	CL3	Class 3	CL3P CL3R MP MPG CM CMG FPL PLTC
	CL3X	Class 3, limited use	CL3P CL3R CL3 MP MPG CM CMG FPL PLTC CMX
	CL2P	Class 2 Plenum	Cl3P
	CL2R	Class 2 Riser	CL3P CL2P CL3R
	CL2	Class 2	CL3P CL3R CL2P CL2R CL3 MP MPG CM CMG FPL PLTC
	CL2X	Class 2, limited use	CL3P CL3R CL2P CL2R CL2 CL3 CL3X MP MPG CM CMG FPL PLTC CMX
800	MPP	Multipurpose Plenum	None
	MPR	Multipurpose Riser	MPP
	MP, MPG	Multipurpose	MPP MPR
	CMP	Communications Plenum	MPP
	CMR	Communications Riser	MPP CMP MPR
	CM, CMG	Communications	MPP CMR MPG MP
	CMX	Communications limited use	CMG CM

*In general, a cable with a more strict usage code may be substituted in an application that allows a less strict usage code. For example, a cable rated to the stricter Class 2, Plenum (CL2P) code may be substituted in Class 2, Riser (CL2R) application. Likewise, CMP, MPP, and FPLP, which may be substituted for CL3P, will substitute for any code that allows CL3P to be substituted. The chart does not list all possible substitutions, only those to the next level.

SOURCE: 1996 NEC.

boards used to route calls. The switchboard plug consisted of three contact areas, referred to as the tip, ring, and sleeve, much like a modern stereo plug. The primary color was wired to the ring and the secondary to the tip. The sleeve was used for grounding.

The 25 pair jumper cable is another common cable that may be found in some LAN installations. The color code for 25 pair cables is

The FEP Controversy

The impact of computer networking on the plastic materials industry has been felt very directly by users of plenum-rated cable. Plenum cable uses insulation and sometimes jacketing made with fluorocarbon plastics for their smoke- and flame-retardant properties. Fluorinated ethylene-propylene (FEP) is often the material most used for insulating the individual conductors because of its superior dielectric properties at the very high frequencies of modern LAN cable.

The problem is that the increasing use of this popular plenum-rated material often outpaces the ability of the material manufacturers to produce sufficient quantities for the cable industry. Additionally, the rapid growth of LANs and the tendency to overwire new installations for future requirements has thus far made increases in manufacturing capacity only a short-term cure. Of course, some double-ordering from cable and material users has occurred, which has caused the appearance of shortages to be even more severe.

For the cable user, the FEP crisis may mean that supplies of cable are periodically lower than normal, or that the favored manufacturer, style, or even jacket color is not available on short notice. Deliveries may be extended, selection limited, and prices higher.

Several cable manufacturers have substituted other types of insulation on one or more of the four pairs in popular LAN cables to stretch the supply. Unfortunately, they are very restricted in the types of material that can be used for the very high-speed data rates of modern LAN systems. Common substitutions are polyethylene, polyolefin, and some special polyvinyl chloride compounds. Some of these materials, in pure form, would not be able to pass the UL flame-spread tests, but are acceptable when used in conjunction with three normal FEP pairs and a plenum-rated jacket.

Costs of plenum-rated cables of all types have increased because of these material shortages. Some question remains whether the substitution of alternate materials may affect the long-term stability of the cable. However, there is only limited evidence of problems at this point, as the alternative polyethylene and polyolefin materials are normally quite suitable for high frequency use in nonplenum applications. The current controversy centers around dielectric property differences that cause a delay skew failure in a mix such as 2-2 or 3-1, and the tendency of non-FEP plenum cables to flame up dramatically in simulations, even though they are able to pass the smoke-tunnel tests. Experimentation is ongoing to find other materials, such as from the chlorofluorocarbon family, that can meet plenum requirements and have adequate dielectric characteristics for Category 5e and 6 LAN use. You will see some of these materials on the market today.

At the current rate of growth of LAN networks, we will continue to experience periodic material shortages for years to come.

shown in Table 3.6. Although 25 pair and larger bulk pair cabling is often used in telephone wiring, ordinary telephone cabling is not appropriate for today's very high-speed data rates. It can function at Category 3 as a multipair jumper at short distances, such as between a patch panel and a connectorized punchdown block (both with 50-pin telco jacks). If

TABLE 3.5

Color Code for 4-Pair Cable Pairs

Pair #	Primary color *ring*	Secondary color (stripe) *tip*
1	Blue	White
2	Orange	"
3	Green	"
4	Brown	"

25 pair cable is used in a LAN application, it should have the proper twists-per-foot as cable for Category 3. Some manufacturers offer 25 pair jumpers that are said to meet Category 3 or Category 5e, but you should use caution if your ultimate application will need to support 100 Mbps or higher data rates. Placing multiple 100 Mbps circuits within the same 25 pair sheath may cause excessive crosstalk, resulting in excessive error rates or even link failure. The TIA is working on specifications for 25 pair cable which will allow its use for multioutlet centers, where modular furniture is used.

Cable sizes above 25 pairs are usually in multiples of the basic 25 pair bundle. Each group of 25 pairs is called a *binder group* and is marked within the larger bundle with a uniquely colored plastic binder that is spiral-wrapped along the bundle. Groups of four or more binder groups may be wrapped and included in the same outer jacket to form larger cables. The sheath of twisted pair cabling should be clearly marked by the manufacturer to show the level of performance that is guaranteed. The marking normally shows three performance indicators, the EIA/TIA Category, the UL/NEC permitted use, and the number and size of conductors. The manufacturer's name and part number are also normally shown. A typical marking might be:

4PR—24GA—CMP—Category 5e—XYZ Cable Company—PN 99999

This example of a cable marking would indicate a 4 pair, 24 AWG cable, tested for fire/smoke/voltage ratings to CMP standards of the NEC, certified to Category 5e TIA standards, manufactured by XYZ Cable Company, with a part number of 99999.

Because standards for cable have changed so much over the past few years, you may encounter older cables with a variety of markings for their transmission characteristics and fire and smoke ratings. Because the TIA standards have been continually refined, you might find cables that bear markings of the older standard revisions or the levels program (for example, TSB-40, Level 5, or EIA/TIA 568). The plenum rating might

TABLE 3.6

Color Code for 25-Pair Cable Pairs

Pair no.	Primary color *ring*	Secondary color (stripe) *tip*
1	Blue	White
2	Orange	"
3	Green	"
4	Brown	"
5	Slate	"
6	Blue	Red
7	Orange	"
8	Green	"
9	Brown	"
10	Slate	"
11	Blue	Black
12	Orange	"
13	Green	"
14	Brown	"
15	Slate	"
16	Blue	Yellow
17	Orange	"
18	Green	"
19	Brown	"
20	Slate	"
21	Blue	Violet
22	Orange	"
23	Green	"
24	Brown	"
25	Slate	"

be marked as "Article 800" or "UL 910," rather than the CMP wire use type. This cable is acceptable if already installed in appropriate applications (such as 10BaseT, for example), but should not be installed for new service. Most of this early cable will not meet the rigorous standards for today's Category 5e, which should really be the minimum quality cable for any new installations.

UTP/STP Shielding and Screening

Shielding is sometimes used with twisted pair cable to provide better noise immunity and lower emissions. Two types of shielding are commonly available for STP cable: the foil shield and the braided wire shield. The foil shield usually has a bare stranded wire, called a *drain wire*, that is in contact with the shield and provides the electrical connection for the foil. Braided shields may have a drain wire, or the braid may be formed into a wire or clamped to make electrical contact. Some cables may have combination foil/braid shields or double shields.

Screening is another term that is often applied to the type of electromagnetic shielding used with twisted pair cables. The term has gained new acceptance with the latest revisions to the international standards. Cables that are referred to as screened are generally constructed with an overall foil screen, which means that the foil screen wraps completely around all four pairs (or all 25 pairs in a multipair jumper cable). Screened cables are in much greater use in Europe than in North America, where their use is found mostly in high-interference locations. Screened cables require special plugs and jacks to allow for the screen interconnection. In addition, the connection of cable screens to a building ground (earth) can introduce problems with ground currents and ground loops. Particular attention must be paid to the information and practices contained in EIA/TIA-607, *Commercial Building Grounding and Bonding Requirements for Telecommunications*.

Screened cables are also available with individually screened pairs. In the current implementations of Category 7 cables, for example, individually screened pairs are further screened with an overall foil or braid shield. This type of construction provides the optimum in electromagnetic isolation between pairs, as well as providing EMI (electromagnetic interference) immunity and very low EMI emission characteristics. These cables may be characterized up to 1000 MHz (1 GHz), and are well suited for leading-edge developments in copper cable transmission.

Coaxial Cable

Coaxial cable, or coax, is the original LAN cable. It was first used for local area networking in Ethernet networks, IBM PC-NET broadband networks, and Arcnet networks. Coax is still in use in thousands of locations, even though many newer installations have converted to twisted pair.

Coax is primarily used for its self-shielding properties, low attenuation at LAN frequencies, and installation expense. The construction of the cable greatly reduces susceptibility to outside interfering signals and noise, besides minimizing the radiated emissions from the cable. Because coax was originally designed to carry radio frequencies, it has fairly low attenuation characteristics. The cable is readily available, relatively inexpensive, and allows a daisy-chain or tapped LAN connection that minimizes total cable length.

Coax cables may have solid or stranded center conductors, foil or braided shield, several types of insulating dielectric, and several types of outer jacket materials. The cable is available in a variety of standard sizes and impedances.

Several types of coax cable are available for LAN use. Table 3.7 shows the common types and their uses. As with twisted pair cable, coax is available in a variety of insulation types and may be used in plenum spaces and riser shafts. Coax is included in IEEE 802.3, Ethernet Version 2.0, the NEC, and other standard documents. It was formerly included in EIA/TIA-568 1991, but was dropped in TIA/EIA-568-A 1995 as a recommended cable type.

Coax General Construction

Coax cable gets its name from its construction. As you can see from Fig. 3.4, coax consists of a center conductor and a coaxially positioned outer shield conductor that are separated by an insulating plastic, called a dielectric. An outer jacket insulates the shield. The shield may be a foil-wrap with a drain wire or a wire braid. Some coax, such as that used for thick Ethernet, may have a double shield layer.

Theoretically, the positioning of the center conductor, surrounded completely by the concentric shield conductor, keeps all electromagnetic fields between the two conductors, as shown in Fig. 3.5. This mode of operation is referred to as "unbalanced," as opposed to the balanced arrangement of twisted pair cable. Other terms for the unbalanced

TABLE 3.7

Common Types of Coaxial Cable

Coaxial cable type	Rated use	Typical NEC type	Center cond. AWG	Nom. imped., ohms	Nom. vel. prop. (c), %	Typical O.D	
						mm	in
RG-6/U	Nonplenum	CL2	18 sol.	75	82	6.86	0.270
	Plenum	CMP	18 sol.	75	82	5.94	0.234
RG-8/U	Nonplenum	CL2	11 str.	50	78	10.24	0.403
	Plenum	CMP	10 str.	50	83	9.07	0.357
RG-11/U	Nonplenum	CM	14 sol.	75	78	10.29	0.405
			18 str.	75	66	10.29	0.405
	Plenum	CMP	14 sol.	75	83	8.84	0.348
FG-58/U	Nonplenum	CL2X	20 sol.	50	66	4.90	0.193
	Plenum	CMP	20 sol.	53.5	69.5	4.04	0.159
RG-58A/U	Nonplenum	CL2	20 str.	50	66	4.90	0.193
	Plenum	CL2P	20 str.	50	80	4.06	0.160
RG-59/U	Nonplenum	CM	20 sol.	75	78	6.15	0.242
	Plenum	CMP	20 sol.	75	82	5.38	0.212
RG-62/U	Nonplenum	CM, CL2	22 sol.	93	84	6.15	0.240
	Plenum	CMP	22 sol.	93	85	5.08	0.200
Ethernet 10Base2	Nonplenum	CL2	20 str.	50	80	4.70	0.185
	Plenum	CL2P	20 str.	50	80	4.06	0.160
Ethernet 10Base5	Nonplenum	CM, CL2	12 sol.	50	78	10.29	0.405
	Plenum	CMP	12 sol.	50	78	9.53	0.375

NOTE: In general, a cable with a more strict usage code may be substituted in an application that allows a less strict usage code. For example, a cable rated to the stricter Class 2, Plenum (CL2P) code may be substituted in a Class 2, Riser (CL2R) application. Likewise, CMP, MPP, and FPLP, which may be substituted for CL3P, will substitute for any code that allows CL3P to be substituted. The chart does not list all possible substitutions, only those to the next level.

mode include single-ended and bipolar. The shield is maintained at "ground" potential, while the center conductor is driven with the LAN signal.

Because the shield is grounded, interfering signals from outside the coax cable should be prevented from entering the cable and coupling to the center conductor. Grounding is quite important in coax cable installations. The shielding properties of the cable depend somewhat upon good ground connections (both signal and earth grounds). However, the interconnection of grounds from different parts of a building, or even between buildings, can cause problems. Grounding requirements for thick Ethernet coax are fairly severe, to ensure proper operation. Safety and signal integrity are also issues in grounding. The NEC and EIA/TIA-607 detail grounding and bonding requirements for coax LAN cable.

Figure 3.4
Coax consists of a center conductor and a coaxially positioned outer shield conductor separated by an insulating material, called a *dielectric*.

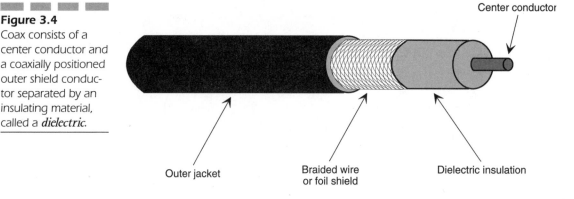

Center conductor

Outer jacket

Braided wire or foil shield

Dielectric insulation

Figure 3.5
The positioning of the center conductor, surrounded completely by the concentric shield conductor, keeps all electromagnetic fields between the two conductors.

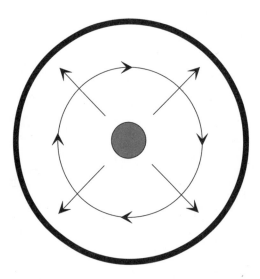

Ethernet LAN wiring uses two types of coax cable, often referred to as *thicknet* and *thinnet*. Thicknet is the original Ethernet coax trunk distribution cable, now called 10Base5. Its larger diameter gave rise to the nickname thicknet after much smaller coax began to be used. Thicknet cables are run in the walls or above the ceiling tiles near locations where workstations are planned. A transceiver mounts directly around the coax and makes connection to the center conductor through a hole drilled through the outer jacket, shield, and dielectric insulation. This arrangement is sometimes referred to as a *vampire tap*.

Thinnet is a newer Ethernet coax cable standard, 10Base2, that allows connection to the back of the workstation adapter. The adapter, in effect, contains a built-in transceiver. Thinnet uses BNC-type connectors and T-adapters. The cable is about $1/4$ inch (1 cm) in diameter and is much more flexible than thicknet cable. It is also less expensive, giving rise to another nickname, *cheapernet*. Distances for a thinnet Ethernet segment are limited, as compared to thicknet.

Arcnet is another LAN topology that originally used coax cable. Several types of coax have been used with Arcnet, including RG-11/U, RG-59/U, and RG-62/U. RG-11/U and RG-59/U are 75-ohm coax cables, while RG-62/U is a 93-ohm cable. The workstations are connected directly to the coax in a star arrangement. Each leg terminates in a passive or an active hub device that couples the LAN signal to all of the other connected workstations or servers.

Coax Wire Sizes

Wire sizes for coax cable are usually specified by "RG" number or manufacturer's part number. Thick Ethernet cable is a special case that is referred to by the names Ethernet trunk cable, backbone cable, or 10Base5 cable. A list of coax types used in LAN wiring is shown in Table 3.8. The RG numbering system is a standard coax cable rating system that identifies physical size, characteristic impedance, power handling, and other characteristics. Common LAN coax cables are Ethernet, RG-8/U, RG-11/U, RG-58A/U, RG-59/U, and RG-62/U. The outer jacket of RG-8/U and RG-11/U is roughly $1/2$ inch (12.7 mm), while RG-58A/U, 59/U, and 62/U are on the order of $1/4$ inch (6.4 mm). Plenum versions of these cables are usually a little smaller. The wire sizes of the center conductor and the diameter of the outer jacket (including dielectric and shield) are defined by the RG number and the insulation type. You must specify whether you want connectors for PVC or plenum insulation types, since the cable dimensions are not the same.

Coax Electrical Characteristics

Coax cables are available in a variety of standard characteristic impedances. Because the primary use for coax is the transmission of RF signals, the coax impedances often reflect the needs of RF equipment. These standard impedances are 50 ohms, 75 ohms, and 92 ohms. The dimensions and properties of the center conductor, dielectric, and shield combine to define the coax cable's characteristic impedance. The dielectric constant in flame-resistant cable is very different from that of conventional polyethylene or foam dielectric cable. It is for this reason that plenum cable typically has a smaller cable diameter.

The attenuation of coax cable less than 1.5 dB per 100 ft at 10 MHz. At 100 MHz, the typical attenuation is under 5 dB per 100 ft. Obviously, the useful coax cable length is reduced as the frequency goes up. Some types of networks base their maximum coax cable runs on the total allowable

TABLE 3.8

Types of Coax Used in LAN Wiring

Coax type	Impedance	LAN use	Special considerations
RG-6/U	75 ohms	Arcnet	Smaller and cheaper alternative to RG-11/U. Lower loss but more rigid than RG-59/U. Must use RG-6/U connectors.
RG-8/U	50 ohms	Ethernet Thicknet	May substitute for 802.3 cable in an emergency or special circumstances.
Rg-11/U	75 ohms	Arcnet	Heavy cable is difficult to work with. See RG-59/U.
RG-58A/U	50 ohms	Ethernet	Stranded center conductor required. Original "cheapernet" cable.
RG-59/U	75 ohms	Arcnet	Smaller and cheaper than RG-11/U, this cable is easier to install and readily available, as it is also used for CATV.
RG-62	93 ohms	Arcnet, IBM 3270	The famous IBM Terminal cable can be adapted to Arcnet.
Ethernet 10Base2	50 ohms	Ethernet Thinnet	Cable is marked for 802.3. Double-shielded cable similar to RG-8/U.
Ethernet 10Base5	50 ohms	Ethernet Thicknet	Cable is marked for 802.3. Similar to RG-58A/U with stranded center conductor.

loss from end to end (the loss budget), while other types of networks have timing constraints that limit the usable distance.

As with other wiring types, the capacitance of the cable is the primary contributor to attenuation at higher frequencies. Even so, coax has a greater maximum usable length than twisted pair. For example, a 10 Mbps Ethernet segment can be 500 m (1640.5 ft) using thicknet and 185 m (607 ft) using thinnet. (Note: segment lengths in Ethernet are limited by attenuation and frequency-based signal distortion, while network span is limited by timing constraints.) Arcnet, which runs at 2.5 Mbps, allows 2000 ft (610 m) for each coax leg with an active hub. For comparison, twisted pair wiring is limited to 100 m (328 ft).

The standard coax dielectrics are polyethylene and PTFE or FEP, although several other materials are also used. The loss characteristics of the dielectric material contribute to the loss at any given frequency. Low-loss dielectric materials and constructions, such as foam or helical dielectrics, are available. However, those low-loss cables are cost-prohibitive for normal LAN cable installations.

Coax Insulation

The insulation in coax cables includes the jacket as well as the insulating dielectric that separates the center conductor and the shield.

The primary types of insulation materials used with coax are polyethylene, polypropylene, various synthetic rubber compounds, polyvinyl chloride (PVC), and fluorocarbon polymers such as polytetrafluoroethylene (PTFE), fluorinated ethylene-propylene (FEP), and ethylene-chlorotrifluorethylene (ECTFE or HALAR®). Nonplenum coax cables generally use polyethylene as the dielectric that surrounds the center conductor and use noncontaminating synthetic resin for the outer jacket. Plenum-rated coax cables use a PTFE dielectric (such as Teflon®) or FEP and a similar outer jacket.

Coax Color Coding and Marking

The most common color for nonplenum coax cable is black. Most of the nonplenum cable that is used for LAN wiring will have a black synthetic rubber or PVC outer jacket with white markings. Plenum cable usually has a white or translucent jacket with black markings. A notable exception to this rule is classic "thick" Ethernet (10Base5) cable. Thick Ethernet cable is usually a very bright color such as yellow.

Coax cable should be marked in a similar manner as twisted pair. The cable will generally have a manufacturer's name and part number, the UL/NEC class or use marking, and the appropriate "RG" number. It may also have EIA/TIA markings or IEEE 802.3 markings. Coax use rating are covered in NEC Articles 800 and 820. If you use RG-58 type cable for thinnet, you should be aware that some of this type of cable does not meet 802.3 specifications and may not be usable at the maximum length for thinnet. You should be sure to specify RG-58A/U that is certified for 802.3 use. EIA/TIA-568 1991 contains specifications for coax, and those markings may be on the cable jacket.

Ethernet trunk cable (thicknet) will have marking bands every 2.5 m to indicate locations for transceiver tap installation.

Coax Shielding

A concentric shield conductor surrounds the center conductor and dielectric insulation in a coax cable. This outer shielding layer causes the electrical and magnetic fields to be contained within the cable and shields the center conductor from fields outside the cable.

Two types of shielding are generally used in coax cables, foil, and braid. The foil shield, sometimes called an *overall foil shield,* is a very thin metallic foil that totally surrounds the dielectric. The braid shield consists of a wire mesh of very fine wire conductors. As with the foil shield, the braid surrounds the dielectric and forms a shield. The braid is effective at LAN frequencies even though there are small gaps between the wires. The gaps are not significant compared to the wavelength of the LAN frequencies. Some coax, such as Ethernet trunk cable, uses a double-braid shield to further increase the shielding properties. Cables with braid shields are generally more expensive than cable with foil shields.

Nonpaired Cable

We will cover nonpaired cable in this section to help you identify it. Nonpaired cable is not proper for use as LAN wiring because it lacks the self-shielding properties of twisted pair wire. It both generates and is susceptible to interference, which means that it will probably violate the allowable emission standards of national and international organizations [with government entities such as the U.S. Federal Communications

Commission (FCC) this is also a legal violation]. Nonpaired cable also may not exhibit a consistent characteristic impedance and may create crosstalk between the pairs. This crosstalk happens even at voice frequencies, as you may know if you have two phone lines in your house and have older, nonpaired wiring. At LAN frequencies, the crosstalk between the transmit and receive pairs may mean that you will receive enough of your transmitted signal to cause an error.

You will undoubtedly encounter those that have used short lengths of nonpaired cable (such as the flat telephone cord that is used with RJ-11 connectors) with no apparent problems. Don't believe them. In all likelihood, they are actually experiencing undetected data errors that just slow down their network. In any event, such cable is a problem waiting to happen. What works at 2.5 MHz may be marginal at 10 MHz, impossible at 16 MHz, and a joke at 100 MHz. Don't let nonpaired cable be used anywhere in your network!

Nonpaired General Construction

Nonpaired wire may be divided into two general classifications, round cable and flat cable. Both consist of two or more insulated wire conductors and may have a covering jacket over the wires. As with paired cable, nonpaired cable may be shielded with either a foil or a braid conductor.

Round cable, as the name implies, consists of two or more wires that are contained within a round protective jacket. An example of round nonpaired wire is shown in Fig. 3.6. Three or more wires may sometimes be wound in a loose spiral within the jacket, but this is not the paired twisting that provides self-shielding. Round cable may have either stranded or solid conductors. The common variations are the 4-wire solid conductor cable used for inside residential wiring (called "4-wire," "inside wire," "IW," or "JK") and the multiconductor cable used for RS-232 data wiring. The 4-wire cable is famous for causing crosstalk when two phone lines are installed on the same cable. If you are lucky, and your house is newer, you will have telephone grade 4 pair wire. The older RS-232 terminal wiring could use nonpaired wire because the data rate was very low (9600 bps versus 10,000,000 bps for Ethernet!).

Flat cable is available in two common types: ribbon cable and flat jacketed cable. Ribbon cable is a jacketless design where the insulation of each wire is joined between the individual insulated, stranded conductors. Figure 3.7 shows an example of ribbon cable. The wires are laid side by side, forming a flat cable. The ribbon cable may have an overlaying shield and

Figure 3.6
An example of round nonpaired wire.

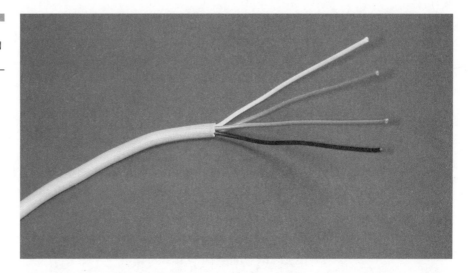

Figure 3.7
An example of ribbon cable.

outer jacket, although this is unusual. Ribbon cables are often used for connections to circuit boards within equipment. They generally are not used externally, because the lack of shielding exceeds FCC emission standards.

An example of flat, jacketed cable is the common telephone cord used with modular RJ-11 (style) and RJ-45 (style) connectorized equipment, as shown in Figure 3.8. It is sometimes called "silver satin" or "telephone zip

Figure 3.8
An example of flat-jacketed cable is the common telephone cord that is used with modular RJ-11 (style) and RJ-45 (style) connectorized equipment.

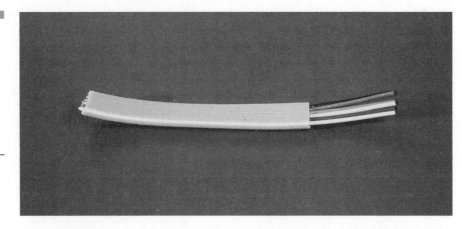

cord," although it is available in many colors and really doesn't zip apart to separate the conductors. Whatever you call it, it consists of two or more insulated, stranded conductors that are positioned side by side and covered with a jacket to form a flat cable. The conductors are run in parallel to each other and are not in any way twisted or paired. Silver satin can also cause crosstalk between pairs, although the placement of the pairs in the cable can actually provide some degree of shielding (for example, pair 1-2 and pair 7-8).

There is also a type of flat cable that consists of paired wires positioned side to side. This cable, which is used for undercarpet applications, has similar characteristics to other paired cables covered earlier in this chapter.

Nonpaired Wire Sizes

There are many wire sizes associated with nonpaired wire. Theoretically, any conductor size may be used, but you will probably find that sizes between 18 and 26 gauge (AWG) are the most common for this type of wire. Telephone 4-wire and silver satin are generally 22 and 26 gauge, respectively. Ribbon cable can really be any wire gauge, although the usual range is 26 to 30 gauge.

Nonpaired Electrical Characteristics

Nonpaired wire is generally characterized in terms of resistance per linear foot. For example, 24 gauge annealed copper wire has a resistance of

25.7 ohms per 1000 ft. Complete specifications for copper wire are available in several reference handbooks. This is very useful information if you are using the wire to supply power or signaling voltages, such as door bells or fire alarms.

Nonpaired wire is really not intended for transmission at any frequencies above voice, and even there, the transmission characteristics are poor to the extent they are consistent. Of course capacitance may be specified for ribbon cable, since it is used to interconnect electronic circuits.

Nonpaired Insulation

The insulation types for nonpaired wire are identical to those for twisted pair wire. The same insulation types are used and care must be taken in plenum or riser spaces, in accordance with the NEC and local standards.

Nonpaired Color Coding and Marking

A very wide variety of wire insulation and jacket colors are available for nonpaired wire. However, for our purposes in telecommunications, there are some common color-coding standards that are useful.

The round 4-wire nonpaired cable that is used in some residential wiring used the colors green, red, yellow, and black. This cable was originally designed for single-circuit use—that is, one phone line. The green and red wires supplied the voice signal (tip and ring), and at one time (a long, time ago), the yellow supplied ringing voltage and the black, ground. The 4-wire cable continued to be used long after the ringing voltage was moved to the tip and ring and the ground was deleted. For a time in the 1960s, the yellow and black were used to supply power from a wall-mounted transformer to the lamp that lit the dial of the famous "Princess" phone. This was later eliminated as the lamp power was also supplied on the tip and ring wires. When a second phone line was to be added, the yellow and black wires were generally available, and that's where it went. Unfortunately, the nonpaired 4-wire cable can generate serious crosstalk between the pairs, and you may faintly hear the other person when you are on a different line. This may be an advantage for parents, but is a problem for LAN circuits and must be avoided.

Nonpaired Shielding

Nonpaired cables may be shielded with foil or braid shields, as with twisted pair cable. A typical use for a shielded, nonpaired cable is for microphone cable. Even flat cable may be shielded. Such cables are useful in meeting electromagnetic interference (EMI) requirements of regulating agencies.

Fiber Optic Cable

Fiber optic cable may also be used in LAN wiring. It is generally specified for backbone wiring, such as vertical wiring between wiring closets and is not yet as common as twisted pair for horizontal wiring. However, fiber optic cabling is one of the acceptable types of cable in TIA/EIA-568-A/B and it is quite possible to run most LAN topologies entirely on fiber optic cable.

Fiber optic cable offers several unique advantages over conventional metallic wiring. It neither generates nor is susceptible to electrical or magnetic fields. High bandwidth signals can travel very long distance with little attenuation. The nonconductive fiber can eliminate ground and bonding problems, with the associated electrical safety hazards. Outdoor-rated fiber optic cables often are completely nonmetallic, which makes them resistant to lightning strikes that plague interbuilding metallic cables.

Fiber optic cable still has a few disadvantages, though. For example, when compared to conventional copper twisted pair wiring, fiber optic cable is more expensive, requires more expensive workstation outlet configurations, cannot transmit DC power to remote devices, and might be more difficult and time-consuming to terminate, in addition to requiring special termination equipment. It is also more difficult to test than lower-performance Category 3 copper, where a simple continuity check almost guarantees operation for low-speed LANs. Optical TDR equipment costs, however, are on a par with advanced Category 5 test sets.

The lack of any metallic conductors presents a surprising problem if a defective fiber optic cable must be located. The lack of a conductor means that inductive cable tracers cannot be used to find the suspect cable in a bundle of other fiber cables. In addition, direct, physical length measurements may be needed to pinpoint the location of the break, even though the distance is easily identified on optical test equipment. It may be less expensive to replace an indoor fiber cable rather than attempt to locate a break or other discontinuity. Outside plant cables will need to have faulty sections spliced or replaced, as it would be cost-prohibitive to replace an entire run. However, a longer than necessary section may need to be replaced to allow for measurement uncertainty if the location of the break is not obvious.

Fiber optic cable is much maligned as it is surprisingly fragile. The truth is that the thin glass fiber (about twice the width of a human hair if the cladding is included) is incredibly flexible and resilient given the brittle expectation for something made of silica. Even so, additional care must be taken in handling the fiber, particularly near the point of termination, where the protective jackets and strength members are removed and the thin buffered fiber is exposed.

The use of fiber optic media for LAN and multimedia communication is becoming more popular as time goes on. However, for most installations, fiber optic cable will primarily be used in applications that exploit its advantages over conventional metallic wiring.

Fiber Optic General Construction

Fiber optic cable has a construction surprisingly similar to some multiconductor wire. One or more buffer-coated optical fibers are enclosed in a protective outer jacket with a fibrous strength member in between. This construction is shown in Fig. 3.9. The buffer is rather like the insulation that surrounds a copper wire in a conventional cable. The strength fibers are added to prevent stretching of the cable, which would fracture the optical fiber.

Optical fiber is made of extruded glass (silica) or plastic, specially formulated to pass light of specific wavelengths with very little loss. Most fiber optic LAN wiring is used for backbone wiring and uses glass fiber. Plastic step-index fiber is available from a few sources, but has traditionally suffered from high attenuation and limited bandwidth, factors that are critical to the deployment of fiber technology. Recent developments in graded-index plastic optical fiber (GIPOF) may offer increased bandwidths in the future, perhaps with lower cost and suitable attenuation when compared to glass fiber. Plastic fiber construction is not yet blessed by the standards for LAN wiring, so the rest of this section will concentrate on the characteristics of standard glass fiber.

The Fiber Debate

The choice between installing copper wiring and fiber optic cable to the workstation is the subject of much debate—an important issue, particularly in new cabling installations where the cable plant is expected to have a useful life of 15 to 20 years. Few would suggest that anything but fiber should be used for campus wiring or for longer high-speed cable runs between wiring closets. The controversy centers on whether fiber should be used from the wiring closet to the workstation—so-called "fiber to the desktop." Current estimates for horizontal fiber range from 5 to 15% of all installations. Copper wiring installations still dwarf fiber installations, yet there is much disagreement on the issue.

What are the two sides to the debate? The proponents of fiber argue that it offers greater signal bandwidth, runs longer distances, offers future growth, costs nearly the same, and is now easy to install and more rugged than generally thought. Proponents of copper over fiber argue that expanded bandwidth is being implemented for copper, that network switching equipment often does not support fiber, that the installed cost of fiber is greater than copper, and that fiber is subject to breaks that are expensive to repair. In general, the most avid fiber supporters are either those companies that offer fiber optic cable, connectors, and related hardware, or they are installers and users who have made a significant investment in fiber optic technology and very naturally defend their decision. Please do not construe this statement to mean that either of these two groups is wrong, but simply that they have a certain bias that leads them to the conclusion that fiber is always the best choice. In some circumstances, fiber cabling may clearly be the best, if not the only reasonable choice. But, in many other situations, the issue is less clear. What should be your choice?

The fiber-versus-copper issue can be reduced to three questions:

- Should fiber be used because of its greater bandwidth and for future migration to higher-speed networking?
- What are the cost implications of installing fiber over copper?
- Are there any special applications that favor fiber over cooper?

The bandwidth question is a moving target. Technology simply will not stay still, so we might as well give up trying to say with certainty what the future will bring. We can say one thing for certain: future networking applications will use much more data bandwidth than in general use today. Copper twisted pair technologies, particularly with the use of enhanced Category 5e products, easily support 100 to 1000 Mbps now. Of course, fiber can certainly take that signal over longer distances, but 100 meters is all that is required in horizontal cabling.

The question of relative cost is easier to approach. Fiber proponents have stated that installed fiber costs only about 20 percent more than Category 5e copper. Fiber optic cable is indeed much less expensive than it used to be. However, users may wish to use protective innerduct or use more expensive strengthened cable. In addition, no one would argue that the interface adapters for network equipment and hubs cost more for fiber optic capability. How much do the installation components relative to copper and cable cost?

Let's look at fiber versus copper termination components first. The traditional drawback to fiber installation has been the amount of time it takes to prepare, cleave, epoxy, and polish each fiber connection. New, quick-termination fiber connector assemblies are available that cut termination time to about one minute per connector. However, these connectors may cost (the end user) about $10 each, so you are trading installation labor for component cost. Adding the cost of the fiber optic outlet box with one duplex 568SC adapter (total cost about $35) to the cost of the two quick-termination connectors yields a fiber termination components cost of about $55. In contrast, a good-quality Category 5e modular connector and plate would cost about $15 (to the end user). Installation labor, at best, would be approximately the same, but the fiber termination requires the purchase of an installation kit, at a cost of approximately $500 to $1000.

Cable prices bounce around a lot because of periodic shortages in plenum-rated materials, but the ratios remain about the same. If the installation calls for plenum-rated cable, then plenum-rated jacketing and inner materials must be used even for fiber cable. The recent plenum price for two-strand plenum 62.5/125 fiber cordage was 30% higher than for the same manufacturer's Category 5e plenum copper cable. You can draw your own conclusions from these cost comparisons.

The third question asks whether your application requires some of the special characteristics that make fiber shine. For example, in an industrial plan with long distances, lots of electromagnetic field interference, and potential grounding problems, fiber cabling makes lots of sense. Fiber is immune to electrical interference, eliminates ground loops, and can transmit much longer distances than copper. If your network equipment has existing fiber interfaces or if you require very large data bandwidths, fiber is again the obvious choice for your use. And fiber is an appropriate choice for backbone and campus cabling.

But what about the future? Will technology make copper cabling facilities obsolete in a few years? Should you install an all-fiber facility now so that doesn't happen? What about your current equipment that may not have fiber interfaces? Can you afford new network adapters and hubs that support fiber, or can you even upgrade your existing equipment? Wow, what a mess! How can you decide between the two types of cabling?

If relative expense is not an issue, must one really choose fiber *or* copper? Why not install both? Install Category 5e copper for today's applications and fiber for the future. In fact, most telecommunications outlet manufacturers have outlet modules that can easily accommodate both a fiber duplex connection and two or more twisted pair cable jacks. It is certainly less expensive to install both technologies in a new facility than it would be to go back in five or ten years and add fiber. The best advice, if the budget allows, is to put in twin fiber and copper facilities.

Several variations of cable construction are available, as shown in Fig. 3.10. The cable may use either a tight or loose buffer construction. A tight buffer closely surrounds the fiber and adheres to it. The tight buffer must be removed with a tool similar to a wire insulation stripper before the fiber can be terminated in a connector. Loose buffer cables use buffer tubes to surround and protect each fiber, which is loose in

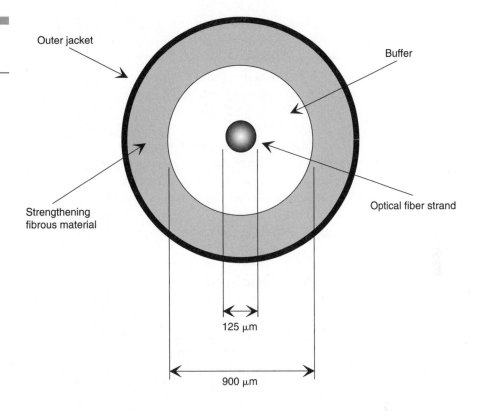

Figure 3.9
Fiber optic cable, cross section.

Outer jacket

Buffer

Strengthening fibrous material

Optical fiber strand

125 μm

900 μm

Figure 3.10
Several variations of fiber optic cable construction are available.

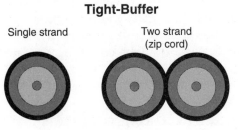

Tight-Buffer

Single strand

Two strand
(zip cord)

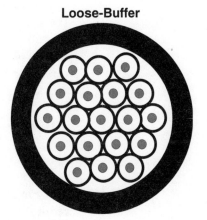

Loose-Buffer

the tube. A variation of tight buffer cable is intended for breakout of the fibers without requiring a protection breakout box to safeguard the exposed fibers. The type of cable encloses each fiber in a breakout jacket that contains its own strength fibers. The entire assembly is covered with a protective outer jacket. Filler cords may be used to make the overall cable round. A variation of the breakout cable is the two-fiber duplex, or *zip cord,* fiber cable that is often used for jumper or patch cords.

Fiber optic cables are commonly available with 1 to 36 fibers. At two fibers and above, even numbers of fibers are the most common (e.g., 2, 4, 6, 12) because fibers are often used in pairs for full-duplex circuits. Cables are also available with a combination of optical fibers and metallic pairs. These special-use cables combine the properties of each type of cable, including the advantages and disadvantages.

Fiber Optic Sizes

The most common fiber size that is used in LAN wiring is multimode 62.5/125 μm fiber. This number describes the diameter in micrometers, of the fiber core and the cladding. The tight buffer surrounds the cladding and brings the overall diameter to about 900 μm. One of the fibers that are specified in TIA/EIA-568-B is 62.5/125 μm. It can be used for many applications, including Ethernet 10BaseF, 100BaseFX, 1000BaseF, FOIRL, optical Token-Ring, FDDI, and ATM.

Other fiber sizes that are commonly available include 50/125, 85/125, and 100/140 μm fiber. The 50/125 μm fiber was added in TIA/EIA-568-B because it offers better performance in some applications, such as Gigabit Ethernet. It is important to use the same type and size of fiber in an installation, since transitions between sizes can cause excessive loss. Also, the light-emitting transmitters and receivers are optimized for fiber of a certain size. Using the wrong size fiber will cause some of the light signal to be lost because of poor coupling. This may result in marginal operation or even circuit failure.

Fiber Optical Characteristics

Unlike copper wire, which can carry electrical signals of any frequency from DC to many megahertz, fiber optic cable is designed to carry light in a range of optical wavelengths. These wavelengths are typically from around 800 to over 1500 nanometers (nm). Operating characteristics of a

Figure 3.11
Fiber optic cables.
Top: a tight-buffer
single-strand cable.
Bottom: a loose-
buffer multistrand
cable.

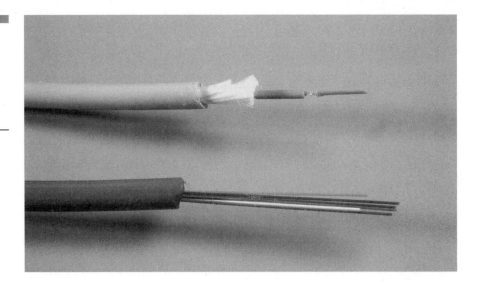

particular fiber are normally specified at discrete wavelengths which correspond to the output wavelengths of common light-emitting diode (LED) and semiconductor laser emitters/transmitters. Multimode fiber is normally used at 850 nm or at 1300 nm, while single mode fiber is used at 1310 nm or 1550 nm. For any given installation, only one mode and one wavelength are used. The fiber cable should be characterized for operation at that mode and wavelength.

Optical fiber is rather like a waveguide for light. While RF waveguides are hollow with metal sides, optical fiber is solid. During manufacture, the refractive index of the glass is made to vary with the diameter of the fiber core. A difference in refractive index is what makes light actually bend between air and water, for example. The light is sent down the fiber and literally bounces off the "walls" formed by the step in the refractive index. In a single mode fiber, the transmitting emitter is a laser diode, chosen so that the light entering the fiber will be coherent and straight along the axis of the fiber. A multimode fiber, on the other hand, often uses a graded refractive index. Light entering from a light-emitting diode (not a laser) enters at all angles, but is coached down the fiber by the gradient of refraction. Multimode fiber is also available in stepped index, but the step occurs much nearer the outer diameter than with single mode fiber. Single and multimode operations are shown in Fig. 3.12.

Optical fiber can carry very large bandwidths of information at low attenuation. Multimode fiber can be used at distances exceeding 3000 m, although other constraints, such as transmitted bandwidth, may limit

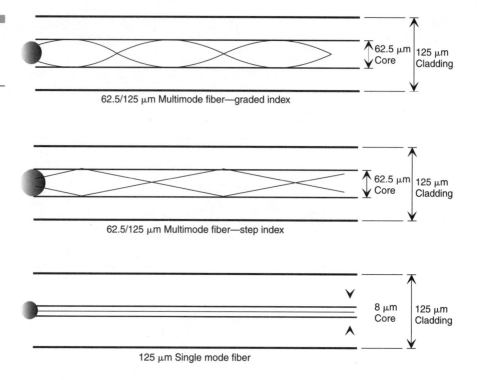

Figure 3.12
Single and multi-
mode fiber optics
operation.

operation to less distance. Single mode fiber can be used at many times that distance and is frequently used in long distance distribution of T-carrier and Sonet signals. Typical attenuation for 62.5/125 multimode fiber is 3.5 dB/km at 850 nm and 1.0 dB/km at 1300 nm. Typical numbers for single mode 8 μm fiber are 0.5 dB/km at 1310 nm and 0.4 dB/km at 1550 nm. Table 3.9 lists representative values.

Transmission bandwidths of fiber optic links are primarily limited by the bandwidths of the transmit and receiving devices, rather than the fiber core's transmissivity. However, as you can see from the attenuation characteristics above, the attenuation does vary somewhat with wavelength of the transmitted light. Bandwidth is also limited by chromatic dispersion of the transmitted signal, which increases linearly with distance along the fiber. Short distance transmission, such as in the horizontal link, greatly minimizes this effect.

When a fiber link is used for low-bandwidth LANs, such as Ethernet 10BaseF, its bandwidth is not significant, relative to the allowed transmission distances. However, beginning at 100 Mbps, the bandwidth of the fiber starts to have a limit on transmission that is independent of the needs of the LAN technology. And at Gigabit Ethernet

TABLE 3.9

Typical Fiber Losses

	Multimode fiber (62.5/125 μm)		Single mode fiber	
Wavelength, nm	850	1300	1310	1550
Loss, dB/km	3.5	1.0	0.5	0.4

speeds, the fiber bandwidth cuts the usable distance of some types of multimode fiber to only a couple hundred meters.

So-called fiber bandwidth is actually a bandwidth-distance product that can be roughly translated into usable distance, simply by dividing the specified number by the distance of the link. Thus the bandwidth-distance product of 500 MHz-km yields an estimated usable bandwidth of 500 MHz at 1 km, 1000 MHz at 500 m, and 2000 MHz at 250 m.

Fiber Optic Insulation/Jacketing

Insulation/jacketing material types are roughly the same for fiber optic cable as for metallic cable, but the rating codes are different. The ratings include plenum use, general purpose, and conductive cable types. Conductive fiber optic cables contain non-current-carrying metallic strength or vapor barrier components. A listing of the fiber optic cable types is shown in Table 3.10. For more information, see Section 770 of the NEC.

From a flame-resistance standpoint, the requirements for the use of fiber optic cable are identical to conventional cable. Plenum-rated fiber cable must be used in air plenums and riser-rated cable must be used in vertical riser shafts. In general, a plenum-rated cable may be substituted for riser use or general use. Likewise, a riser-rated cable may be used for general use. Ports and pathways for fiber optic cables should meet the appropriate flammability and fire-stop requirements, as with metallic cable.

Fiber optic cables that are intended for outside use or for direct burial should be appropriately rated by the manufacturer. Because fiber optic cables do not carry voltages, the cables technically do not need to meet the NEC requirements for direct burial; however, the resistance to moisture, chemicals, and abrasion should be the same. Remember that a cable that runs any length inside a plenum must be appropriately rated, even if the cable originates outside the plenum space. An exception to this rule is an outside cable at an entrance facility. See the NEC for details. Fiber optic cable with UV resistant jackets and armoring is also available.

TABLE 3.10

National Electrical Code (NEC) Cable-Use Codes for Fiber Optic Cable

NEC article	Code	Meaning	Allowable Substitutions*
770	OFNP	Optical Fiber Plenum Nonconductive	none
	OFNR	Optical Fiber Riser Nonconductive	OFNP
	OFNG	Optical Fiber	OFNR
	OFN	Nonconductive General purpose	
	OFCP	Optical Fiber Plenum Conductive	OFNP
	OFCR	Optical Fiber Riser Conductive	OFCP OFNR
	OFCG	Optical Fiber	OFNR OFCR OFN
	OFC	Conductive General purpose	

*In general, a cable with a more strict usage code may be substituted in an application that allows a less strict usage code. For example, a cable rated to the stricter Plenum code may be substituted in a Riser application. The chart does not list all possible substitutions, only those to the next level. A nonconductive cable is one that contains no metallic elements, including conductors, metallic sheaths, or strength members.

SOURCE: *1999 NEC.*

Fiber Optic Color Coding and Marking

Fiber optic cables are generally colored black for outdoor use. For indoor use, a variety of colors are available. You may wish to choose a bright color to distinguish the fiber optic cable from other cable. Fiber optic jumpers are often bright orange or yellow. The glass fiber may be damaged by tight bending or sudden impacts and is easier to avoid when it is brightly colored.

Marking is the same for fiber optic cable as for conventional metallic cable. The marking will generally include the name of the manufacturer, the manufacturer's part number, the NEC or UL rating (or similar rating for the country of use), and perhaps the fiber size (such as 62.5/125) and mode.

Fiber Optic Shielding and Armoring

Fiber optic cables transmit light and do not need to be shielded from electromagnetic fields. Some cables are armored or include a wire strength member for aerial installation. You should be aware that the metallic armoring or wire reduces the cable's relative immunity to lightning. In addition, under some circumstances, high levels of static electrical charge can build up on outdoor fiber optic cables, and workers may need to use special precautions including grounding of the cable to discharge built-up voltages.

Combination cables are available that contain both optical fibers and metallic conductors. These cables may be offered with shielding for the wires. See the section on UTP and STP cable for a complete discussion of shielding types for these conductors.

Special Purpose Wire

If you intend to place LAN cable outside of ordinary office locations, you may need to specify special purpose cable types. Most normal locations for LAN cable are benign environments that maintain the cable at mild temperatures, away from moisture, contaminants, harsh or dangerous chemicals, and the effects of the sun. However, many locations exist that require special wire insulation, jacketing, armoring, or ratings. Most wire types are available in special versions to accommodate all of these needs. Several of these special-use ratings for cables are described here.

Plenum and Riser Rating

Cables that are placed in plenums must be plenum rated (Fig. 3.13). A plenum is a space used to move air to workspaces for the purpose of ventilation, heating, or cooling (HVAC systems). The informal words for plenums are "air duct" and "air return."

In small buildings, plenums can be individually ducted for forced air and return air. Such plenums rarely contain cabling, and you may be able to use cheaper non-plenum-rated cable (if you stay out of the plenums). Large buildings often have individual forced-air ducts to deliver conditioned air to the workspace, but use vents into the over-ceiling area to

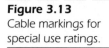

Figure 3.13
Cable markings for special use ratings.

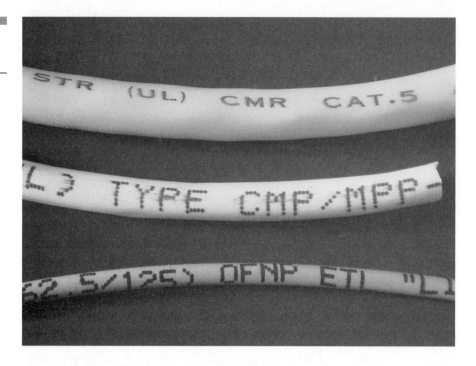

return "used" air to the air handler. In such cases, the entire area above the ceiling grid is classified as plenum space.

The whole purpose of plenum rating cable is to reduce or eliminate the transport of hazardous gases to uninvolved workspaces, in the event of a fire. Cable placed in the plenum spaces must be tested for flammability and smoke generation to minimize this hazard.

Riser cable is intended for use in vertical shafts that run between floors. Many buildings have a series of equipment rooms that are placed vertically in a reinforced shaft for the purpose of enclosing power distribution equipment, HVAC units, telephone distribution, and other utility services throughout the building. Large buildings may even have multiple utility shafts. Cable placed in these shafts must not contribute to the spreading of fire from floor to floor. However, the smoke generation requirements are not as important, because the shafts are not plenums. For that reason, the riser rating is less stringent than plenum. Plenum rated cable may generally be substituted for uses requiring riser rating.

These special plenum and riser cable requirements are covered in detail in the NFPA's (National Fire Protection Association's) National Electric Code (NEC). NEC Sections 770 and 800 explain these ratings for fiber optic, LAN, and communications cabling.

UV Light Rating

The ultraviolet light from the sun can badly damage cable jackets that are not specially formulated. This damage will not be immediate but will take place over a period of years. The cable jacket will become discolored, then cracked and brittle. It will fail to keep out moisture and contaminants. Moisture penetration may cause the LAN connection to fail.

UV-rated cable may be obtained from most cable manufacturers. In fact, many cables already include sunlight resistance as a standard feature. You should check with your wire manufacturer if you are not certain.

Outside Locations

Cables must be specially designed for use outside a buildings (Fig. 3.14). The harsh outside environment of temperature, moisture, light, and stress place unusually stringent requirements on cable of any type. Outside cable is suspended long aerial distances, run in underground cable ducts, and directly buried. Cable may need to be gel-filled or pressurized to exclude moisture. It may need special strength members of an integral messenger cable to span aerial distances. Metallic armoring may be needed to resist damage from animals.

If you must run outside cables into buildings, you may want to consider a cable that is also rated for indoor use. Such cable meets the NEC

Figure 3.14
Outside plant cable.

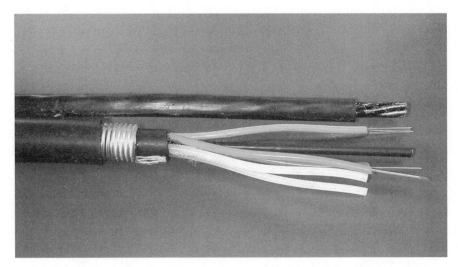

code requirements for indoor use as well as being designed for the harsher outside environment. The NEC specifies that typical outdoor cable may only be extended 50 feet inside a building, unless that cable is rated for the proposed indoor use. The alternative is to place a junction or splice point inside the building, near the point of cable entry, to convert to inside-rated cable.

Hazardous Locations

Special cable, conduit, and junction boxes may be required for cable that is run in so-called hazardous locations, including places where flammable or explosive gases or liquids might be present. Surprisingly, hydrogen and oxygen gas fit into this description, as do volatile liquids such as gasoline and solvents. The NEC addresses the installation of cabling in hazardous areas.

4

LAN Topologies

Chapter 4 Highlights

- Ethernet10 Mbps—Gigabit
- Ethernet UTP pinouts
- Ethernet fiber
- Legacy Ethernet coax
- Token-Ring
- 100VG-AnyLAN
- Isochronous Ethernet
- FDDI, ATM, Arcnet, etc.

An understanding of LAN topology—how the wires are hooked up—will be very useful in our discussion of LAN wiring. In this book, we cover wiring suitable for many of the popular types of LAN topologies used today. Most of the other chapters are concerned with wiring devices and techniques that are not specific to any particular LAN type. This chapter will detail several LAN types and their variations, with reference to specific wiring techniques, where appropriate.

We will explain the wiring patterns and connections, called the *topology,* that are needed to support each type of network. We will not be covering the intimate details of a LAN's protocols, as this is not our topic.

If you are familiar with LAN protocols or the OSI (Open Systems Interface) model, you will recognize the LAN topology as Layer 1, the physical layer. Layer 1 is subdivided into additional layers. Part of Layer 1 comprises the actual wire, connectors, and hubs that must be connected properly for the rest of the protocol model to function. This wiring layer is critical to the functioning of the network. If the wiring fails, the network fails. Nothing else in the protocol stack matters. This may be held over the head of the nearest cable installer (perhaps your own head), if it gives you any comfort.

Many types of LAN topologies may be supported on the twisted pair wiring described elsewhere in this book. However, many of these LAN topologies function on other types of wiring. For completeness, we will include LAN wiring methods that use coaxial cable, as well as the more modern shielded and unshielded twisted pair and optical fiber.

The three basic types of network topologies are bus, ring, and star. Some LANs, such as Ethernet, use more than one basic type. The three topologies are illustrated in Fig. 4.1. Each topology has its advantages and disadvantages. In some cases, such as Token-Ring, one topology can be wired very much like another. (The token "ring" is normally cabled like a "star," but the legs are interconnected at the hub in the telecommunications room so as to form the ring.) As you can imagine, each topology and LAN standard has its fans, as well as its detractors. We will stay away from the contest, except to point out when a wiring method can be used for several topologies.

We will cover, in order: Ethernet coax, twisted pair, and fiber; and then Token-Ring, Arcnet, 100VG-AnyLAN, Isochronous Ethernet, and others.

Ethernet Coax

Ethernet topology is the granddaddy of LAN systems. It was the first widely accepted, nonproprietary, standardized multiple-access network.

Figure 4.1
The three basic LAN
topologies.

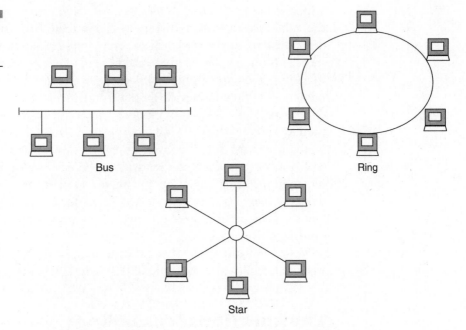

Bus

Ring

Star

Intended for the routine connection of computers and workstations, it allowed many devices to share a high-speed communications line without regard to the customary hierarchy of permanent computer-to-terminal connections.

Ethernet was originally developed by the Xerox Corporation at its Palo Alto Research Center. In the late 1970s, Xerox Corporation, Digital Equipment Corporation (DEC), and Intel Corporation agreed to jointly promote Ethernet as an open standard for computing. This standard eventually became the IEEE 802.3 standard, administered by the Institute of Electrical and Electronics Engineers. IEEE 802.3 has been revised and refined several times and now includes coax, twisted pair, and fiber optics.[1] Modern Ethernet, as embodied in IEEE 802.3, actually differs in some ways from the original Xerox/DEC/Intel standard (sometimes called DIX). This older standard has been revised and lives on as Ethernet Version 2. However, most installations have shifted to the IEEE standard and we will refer to that simply as "Ethernet" in accordance with common usage, unless there is a significant difference.

Ethernet uses a method of signaling called *Carrier Sense Multiple Access with Collision-Detect* (CSMA/CD). At the signaling rate of 10 Mbps it uses

---

[1]Wireless multiple-access connection methods are standardized in IEEE 802.11, which is covered in Chap. 14.

a special method, called *Manchester coding,* to preserve a null DC level on the cable; this allows collisions to be detected when they occur. Each device listens to the network for transmissions from other devices. If a transmission is received that matches the device's address, it is processed in accordance with higher-level protocols. If not, the transmission is simply ignored. When a device wishes to transmit, it first "listens" for another existing transmission. If none is present, it transmits its message, and then notes the voltage level on the line to see if another device happened to transmit at the same time. This would be called a *collision.* If a collision occurs, each device waits a short, random time period and retransmits, after checking for an existing transmission. Thus, many devices have access to the network (multiple access), they each listen before transmitting (carrier sense), and they retransmit if a collision occurs (collision detect).

Newer 100 and 1000 Mbps Ethernet technologies use variations of this basic scheme to encode data and detect carrier and/or collisions.

Thicknet (10Base5) Cabling

The original implementation of Ethernet used a large, 50-ohm coax trunk cable that is now referred to as thicknet. This cable is specified in the 10Base5[2] standard of IEEE 802.3, and is used at the 10 Mbps signaling rate.

A 10Base5 network consists of a thicknet backbone cable that is tapped with a series of *transceivers* or *media attachment units* (MAUs). Each workstation or server is connected to a single transceiver with a transceiver cable. Thicknet is a *Tapped Bus* topology. A typical thicknet installation is shown in Fig 4.2. The transceiver literally taps directly into the Ethernet cable by means of a drilled hole and a probe contact. The transceiver has a 15-pin connector, which carries the Ethernet 10 Mbps LAN signal, error and status signals, and power. This interface is called the *Attachment Unit Interface* (AUI). The transceiver is connected by a transceiver cable, often called an AUI cable, to the workstation or server. This AUI cable is a stranded, 20 gauge, 4 twisted pair, shielded cable terminated at each end by a 15 pin d-shell connector with a special locking mount. The AUI connections are shown in Fig. 4.3.

---

[2]The "5" in 10Base5 indicates the segment maximum length of 500 m. The 10 indicates the symbol rate of 10 Mbps. In Manchester coding, the symbol rate is the same as the signaling rate (in MHz).

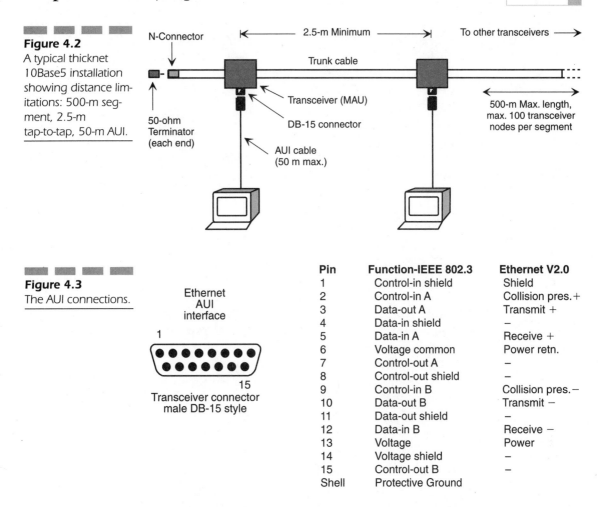

Figure 4.2
A typical thicknet 10Base5 installation showing distance limitations: 500-m segment, 2.5-m tap-to-tap, 50-m AUI.

N-Connector

2.5-m Minimum

To other transceivers

Trunk cable

Transceiver (MAU)

DB-15 connector

50-ohm Terminator (each end)

AUI cable (50 m max.)

500-m Max. length, max. 100 transceiver nodes per segment

Figure 4.3
The AUI connections.

Ethernet AUI interface

1

15

Transceiver connector male DB-15 style

Pin	Function-IEEE 802.3	Ethernet V2.0
1	Control-in shield	Shield
2	Control-in A	Collision pres.+
3	Data-out A	Transmit +
4	Data-in shield	–
5	Data-in A	Receive +
6	Voltage common	Power retn.
7	Control-out A	–
8	Control-out shield	–
9	Control-in B	Collision pres.–
10	Data-out B	Transmit –
11	Data-out shield	–
12	Data-in B	Receive –
13	Voltage	Power
14	Voltage shield	–
15	Control-out B	–
Shell	Protective Ground	

The thicknet cable is run in a continuous fashion, from a starting point to an area near each workstation or server that is to be connected. If a cable is cut, or if another length of cable is to be attached, both cable ends are connectorized and a barrel connector is used to join the two sections. Type-N connectors are used with thicknet. The backbone cable must be terminated at each end by a 50-ohm terminating resistor, which is usually incorporated into a coax connector. The terminating resistors minimize reflections of the Mbps signal that would otherwise occur. Removing a terminator or cutting a backbone cable will cause the network to fail, as the reflections are inverted and cause a failure in the carrier sense and collision detection mechanism. Transceivers may generally be connected to a live network cable, although there is always a possibility that the cable shield and center conductor may short briefly during the process. Such a short will temporarily disable the network, but should be cleared quickly when the tap is installed. Some network managers schedule transceiver installations only during scheduled downtime.

Transceivers are mounted directly to the thick cable and secured in place. A stepped hole is drilled into the cable with a special drill and guide. When the transceiver mounts to the cable, a probe makes pressure contact with the cable's center conductor and the ground of the transceiver is placed in contact with the shield of the cable. (As mentioned, this connection method is sometimes called a *vampire tap*.)

Thicknet is most often placed into ceilings or walls near the workstations. Prewire methods may place the cable into wall outlets with an integral transceiver tap. Power for the transceiver is provided from the workstation when the transceiver cable is connected to the wall outlet. The cable may also be placed above the ceiling, suspended from structural members or lying directly on the ceiling grid. The transceivers are tapped into the cable as needed and the transceiver cables are run to the workstations by dropping directly or running through a wall opening.

The backbone trunk cable may be run a maximum of 500 m (about 1640 ft) without using repeaters or bridges, with each of these runs being called a *segment*. Segments may be joined with a repeater or a bridge. A repeater simply receives, reconstitutes, and retransmits the Ethernet signal (packet) to the next segment. A bridge or router, on the other hand, reads the packet and determines if the packet's destination is in another segment. If so, the bridge or router recreates the packet and retransmits it on the next segment. Ethernet 802.3 has a maximum limit of two repeaters between segments (four repeaters, if two of the segments are repeater-to-repeater links). Ethernet version 2 defines things a little differently to determine the repeater limit. Bridges and routers may be used to extend the network beyond this repeater limit or to off-premises locations via wide-area networking (WAN).

The transceiver cable is limited to a length of 50 m (164 ft). While this distance might mean that it would be possible to run all the way from a wiring closet to a workstation, the usual procedure is to run the cable to a nearby section of the trunk cable and install a transceiver. Each transceiver must be separated from others by a minimum of 2.5 m (about 8 ft, 2.5 in). This minimum distance makes it impractical to place many transceiver taps in a single location. In instances where several AUI cables must be run, a multiport transceiver with several AUI interfaces may be used. The 10Base5 standard allows a maximum of 100 transceiver nodes per segment, including repeaters.

Thicknet installations are most often found in large buildings and are frequently older network installations. The problems that may occur are often due to corrosion of the connectors, barrels, or even transceiver probes. Additional problems may occur because of improper grounding. More information on grounding is given at the end of this section.

Thinnet (10Base2) Cabling

A newer implementation of Ethernet is the 10Base2 *thinnet* standard. This cabling uses a less expensive cable type that is basically RG-58A/U 50-ohm coax cable. Thinnet was developed during the early 1980s as a cheaper, easier-to-install version of traditional Ethernet. Sometimes the names *thin Ethernet* and *cheapernet* are used to describe 10Base2. Thinnet is also a Tapped Bus topology. A typical thinnet installation is shown in Fig. 4.4. This cable is specified in the 10Base2 standard of IEEE 802.3.

The main topology difference between thick and thin Ethernet is the disappearance of the discrete transceiver and its associated AUI cable. In thinnet, the transceiver is an integral part of the network adapter of each workstation or server. The thinnet coax is simply routed from one workstation or server to another in a daisy-chain fashion. At each workstation's network adapter, a T connector is used to effectively "tap" the coax. In general, a coax cable runs to each workstation where it is connected to another coax cable that runs on to the next workstation, in turn, until all have been connected. At the two ends of the run, a 50 ohm terminator is placed to minimize reflections of the LAN signal. As

Figure 4.4
A typical thinnet 10Base2 installation showing distance limitations.

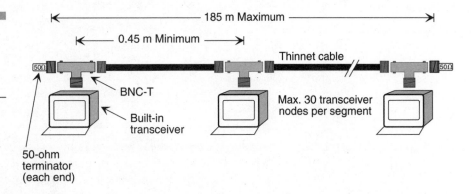

with thicknet, removing a terminator or disconnecting a cable will cause the network to fail.

The connectors that are used with thinnet are called *BNC connectors* (see Fig. 4.5). The cable-end BNC connectors are a male, bayonet-style (push on and twist to lock) connector, originally used for RF connections. The BNC-T has two female BNC connectors at the top of the T, which are connected to each incoming cable or a terminator, and one male BNC at the bottom of the T, for connection to the workstation adapter.

Thinnet is often placed into the walls or simply run along the floor and behind furniture to connect to workstations. If a wall plate is used, two BNC bulkhead connectors must be provided, since the cable must be extended all the way to the workstation and then back to the wall to complete the network path. For grounding reasons, isolated-bulkhead connectors should be used (see the section on grounding, below). If no workstation is present, a short length of coax, with connectors at both ends, can be used to jumper the two bulkhead connectors and maintain the network path. Sometimes, instead of placing an outlet at the wall, the two lengths of coax cable are run unbroken through an opening in the wall, or simply dropped down from the ceiling. In this case, the two cables are simply joined with a BNC-T, whether or not a workstation is present.

An entire thinnet segment is limited to 185 m (607 ft), which is approximately 200 m, thus the "2" in 10Base2. This includes every leg as

Figure 4.5
The connectors used with thinnet are called BNC connectors—BNC male cable connectors and BNC-Ts.

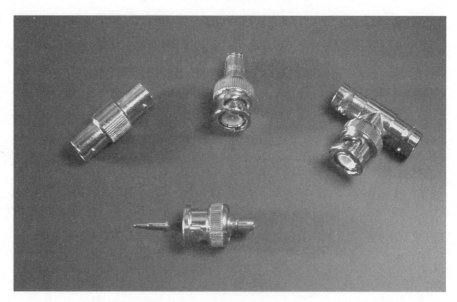

well as the small contribution of the connectors and T's. Although each consecutive length of cable to each workstation is added to the total, thinnet still makes conservative use of cable. Adjacent offices can be wired together with only a small run of cable. Workstations in the same room can be served with very short runs of cable (0.45 m or about 1.5 ft minimum). Thinnet may have as many as 30 nodes on a segment. As with thicknet, only two repeaters may be used between any two workstations in a network, without using bridges or routers (four repeaters, if two of the segments are repeater-to-repeater links). This means that no more than three 185-m workstation segments can be linked with repeaters. Bridges and routers may be used to extend the network beyond this limit or to off-premise locations via wide-area networking.

Problems to watch for in thinnet installations are intermittent connectors and T's, connectors which may have pulled loose from the cable, and easily disconnected BNC-T's. The T may be disconnected from the workstation without disrupting the network, as long as both cables to the next adjacent workstations remain connected to the T. In large installations, ground loop currents between areas of separate electrical service may cause network errors and even safety problems.

Grounding

Proper grounding of coaxial LAN cables is very important, both for safety and operational reasons. Both types of Ethernet coax use unbalanced "ground-referenced" signaling. That is, the LAN signal is applied to the center conductor and is referenced to a signal ground, connected to the shield. Most people assume that a ground is a ground. However, it is intended that this signal ground be independent of the "earth" ground (sometimes called *chassis* or *safety ground*). Unfortunately, the independent signal ground of the network and workstations can sometimes get interconnected with safety grounds. If the signal ground on a cable and the chassis ground are not at the same electrical potential, significant signal distortion and even potentially lethal voltage differences can result. Grounding is illustrated in Fig. 4.6.

The Ethernet specification calls for the trunk cable to be grounded at each wiring closet or thinnet cable to be connected to earth ground at one end (not at both ends). In addition, you should ensure that the building electrical system is properly bonded and grounded together, preferably at the one point where the power entry to the building is located. All electrical distribution rooms in the building should be grounded back to the

Figure 4.6
Typical grounding
recommendations for
coaxial Ethernet
cabling.

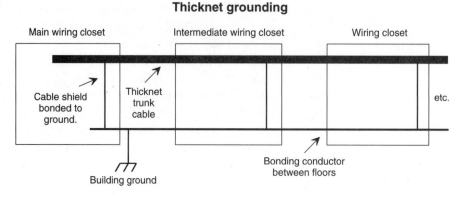

Thicknet grounding

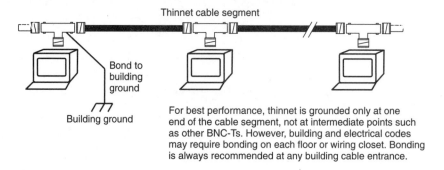

Thinnet grounding

For best performance, thinnet is grounded only at one end of the cable segment, not at intermediate points such as other BNC-Ts. However, building and electrical codes may require bonding on each floor or wiring closet. Bonding is always recommended at any building cable entrance.

same entry point, as should all power transformers. This is clearly an area for an experienced electrician, but as the network manager or contractor, you should know how to assess a proper building ground system. EIA/TIA 607 gives extensive information on providing proper building grounding and bonding.

Ethernet Twisted Pair

The introduction of twisted pair wiring into standard Ethernet networking ushered in a new age of network connectivity. For some time, frustrated users had been experimenting with thinnet-to-twisted-pair baluns in order to use existing telephone wire for network links. Several manufacturers even introduced proprietary network adapters and hubs for twisted pair Ethernet. Eventually a standard was fashioned, under the umbrella of the IEEE 802.3, to deploy a new twisted pair Ethernet topology called 10BaseT.

This new standard has been well accepted and robust, indeed. It has been so successful that it has virtually eliminated the installation of new 10Base5 and 10Base2 networks. The need to specify a system of universal telecommunications cabling that would allow the proper operation of 10BaseT gave rise to a series of new cable, wiring, and component standards that eventually resulted in the EIA/TIA-568 standard.

Now that 10 Mbps Ethernet is beginning to seem very slow, Ethernet networking has taken a jump to 100 Mbps, and even 1000 Mbps, over twisted pair cable. Fortunately, the standards committees have kept pace with these developments and have released 100 Mbps and 1000 Mbps standards for network adapters, hubs, switches, cable, and wiring components.

We will cover classic 10BaseT topology first, because it is essentially identical to 100/1000BaseT. Then we will point out the things that make 100 and 1000BaseT different.

10BaseT Networking

The 10BaseT (and 100/1000BaseT) Ethernet topology is an Active Star, rather than the Tapped Bus topology of 10Base5 and 10Base2. The operation of 100BaseT is identical to 10BaseT.[3] This star topology is quite compatible with the standard home-run method of commercial telephone-style wiring. Although the use of existing telephone wiring was originally the goal of a twisted pair Ethernet, the standards now recommended that LAN twisted pair wire be used exclusively for the LAN network connection. Telephone wiring should be done using a separate cable to be totally in step with the standard.

A typical 10BaseT installation is shown in Fig. 4.7. The center of the star topology is a 10BaseT hub. Each workstation (or server) has a 10BaseT network adapter port that is connected to the hub over a twisted pair cable. The standard specifies a maximum 90-m distance for each cable leg, plus a total of 10 m for interconnection, both at the workstation and at the hub. Modular cords, jacks, punchdowns, cross-connects, and patches are all allowed. While these wiring devices resemble their telephone counterparts, there are important differences that make them

---

[3]In fact, most medium to large hub/switches today are dual 10/100 speed ports, often with a high-speed (Gigabit) uplink.

Figure 4.7
In a 10 or 100BaseT shared hub, transmission by a station first goes to the hub, which then repeats (retransmits) the signal to all of the other connected stations.

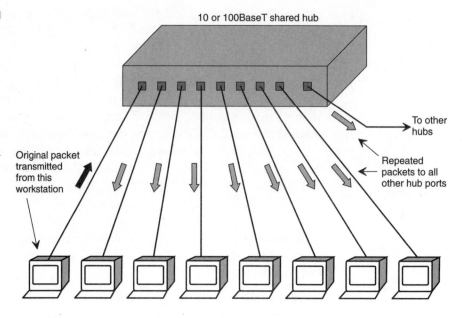

data grade." Some of these differences are evident by the performance certifications needed to meet the standards for LAN wiring.

A significant difference between 10BaseT and Ethernet coax topologies is the addition of an active hub device. 10BaseT still uses the CSMA/CD signaling method. With the coax topologies, any transmission by a station is passively distributed by the coax cable to all the other connected stations. In 10BaseT, however, a transmission by a workstation first goes to the hub, which then repeats (retransmits) the signal to all of the other connected stations. Each port thus acts as a transceiver (repeater hop).

A 10BaseT hub (or switch) typically has 8, 12, 24, or even more 10BaseT ports. It may be a standalone hub, or part of a chassis with plug-in hub cards. Standalone hubs usually have one port each for thicknet and thinnet connections, in addition to the 10BaseT ports. The thicknet port has an AUI interface and the thinnet port has a BNC connector. Hubs may be used in concert to build bigger networks by interconnection through the thicknet port, the thinnet port, or one of the 10BaseT ports. The 10BaseT interfaces can be converted to AUI or thinnet interfaces or even fiber optic links with appropriate transceivers or repeaters.

The active hub gives 10BaseT some unique advantages over the coax topologies. Because each hub port is repeated to the rest of the network, each 10BaseT port is independent of the others. This means that the cable length of a given port is not affected by the cable length of any other port. It is much simpler to plan and test a 10BaseT network, because the cable length for a port must merely be less than the allowable maximum of 100 m (90 m for the horizontal cable, and 10 m for the patch and user cords at either end). As an added benefit, the hub can automatically isolate any port that misbehaves.

A variation of the 10/100BaseT hub device is called a *switched hub*. Standard hubs send all packets received from any port to any other port, as you can see from the figure. In this way, they share the twisted pair media among all ports. Undesired collisions can sometimes occur, because there are so many devices sharing the same *collision domain*. In order to reduce this problem, it is possible to use a simple technique called Layer 2 switching.

A switched hub is a very significant improvement in network interconnection. In most (nonbroadcast, nonmulticast) situations, a network intends to send a data packet to one other device on the network. Our switch/hub can identify these two stations through their unique Layer 2 MAC (media access control) addresses and send the data packet only to the one hub port that has that device attached. This technique greatly minimizes traffic going to individual ports, allows two (or more) transmissions to/from two (or more) stations on independent ports to occur simultaneously, and can allow a high-speed uplink to a backbone server to handle many simultaneous data exchanges with workstations. The basic operation of a switched hub is shown in Fig. 4.8.

Workstations may be connected to or disconnected from 10BaseT hubs without interrupting other stations on the network. Hubs usually have status lights that indicate proper connection on the 10BaseT ports and collisions or other error conditions. Standard 10BaseT hubs will automatically isolate (or partition) a port for a wiring reversal, short, or open connection (including no device attached). More sophisticated hubs may also detect more subtle errors and allow for automatic or manual port isolation. Hubs may also be connected to a variety of management tools, from Simple Network Management Protocol (SNMP) to sophisticated network hardware managers, such as Hewlett-Packard Openview™ and Sun NetManager™.

The basic interface for 10BaseT is the *medium-dependent interface* (MDI), a special wiring configuration of the common 8-pin modular jack that is used by some telephone equipment. Two pairs of wires are used, one for the transmit data, and one for the receive data. The connections are polarity sensitive. That means that the connection will not

Figure 4.8
In switched
10/100/1000BaseT a
transmission by a sta-
tion first goes to the
hub, where it is
switched to a single
port attached to the
destination station.

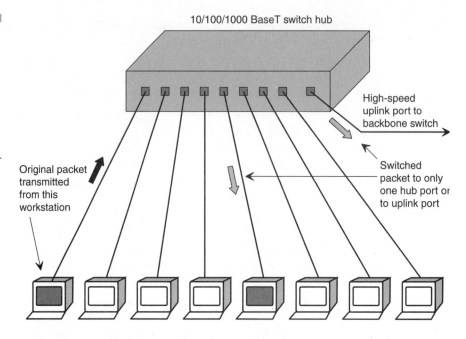

operate if the two wires of either pair are reversed. The wiring of the 10BaseT interface is shown in Fig. 4.9.

Although 10/100BaseT uses only two wire pairs, it is customary to use 4 pair cable to make the station drops. Several of the 100 Mbps network schemes require all four pairs. In addition, a jack that is wired for all four pairs can support many other types of data and voice connections, which may use some other combination of the eight connector pins. Some installations use the other two pairs for a telephone connection or for another 10/100BaseT connection. Be wary of telephone installers "stealing" your network wiring pairs for their use. Telephone wiring is often at cross purposes from network wiring. The standards are different, as are the installation practices.

The TIA/EIA-568-B wiring standard supports 10BaseT wiring. The wiring standards of Category 3 are normally adequate, although you may wish to use Category 5e in a new installation to allow for future enhancements. Cabling for 10BaseT is simply a wiring star of hub and nodes. Generally, a hub is located centrally, perhaps in a telecommunications room, and station cables are run to each workstation location. The cables may be terminated in a patch panel or on a punchdown block adjacent to the hub. A patch cord or octopus (fan-out) cable connects each station cable to a hub port. These terminations are covered in detail in Chap. 8.

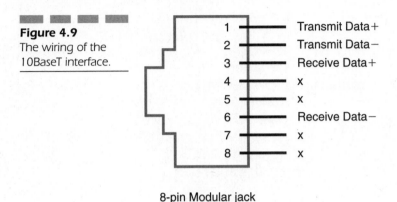

Figure 4.9
The wiring of the
10BaseT interface.

1	Transmit Data+
2	Transmit Data−
3	Receive Data+
4	x
5	x
6	Receive Data−
7	x
8	x

10BaseT
Medium-dependent
interface
(x = unused)

8-pin Modular jack
(Front view)

At the workstation location, station cables are normally terminated in a wall jack and a user cord connected to the workstation. The wiring can be the standard two pair 10BaseT wiring shown in Fig. 4.9; however, the more robust 4 pair wiring pattern of TIA/EIA-568-B offers effectively the same connections. A connector jack's internal wiring may present some confusing variations in wire colors and numbering; some of these variations are shown in Fig. 4.10. Note that the wiring colors may be completely different from those of the station cable and the plates sometimes have numeric markings that do not correspond to the pin numbering of the jack itself. Fortunately, 10BaseT has the ability to sense misconnections, and the hub status light for the port is a good diagnostic tool. Inexpensive modular test indicators are also available to help you make this connection.

All wiring in 10BaseT is straight through (pin 1 to pin 1, pin 2 to pin 2, and so on). Be careful to use only adapters, patch cords, adapter cables, and accessories that are wired this way. Many similar cables and adapters that are intended for telephone use are wired "crossover," which means pin 1 to pin 8, etc. In addition, all 10BaseT wiring is twisted pair. Avoid using the flat telephone wire, such as silver satin, that is sometimes available with 8-pin modular connectors.

Each leg of a 10BaseT network is limited to 100 m (328 ft). Of this 100 m, 90 m typically may be station wire and an additional 10 m may be the total for the patch cord, cross-connect (if any), and workstation cable (user cord). Cable, component, and installation standards should be in accordance with TIA/EIA-568-B Category 3 or higher. Note that either T568A or T568B modular jack wiring patterns will support 10BaseT. Just be sure that you use the same patterns at each end of the cable,

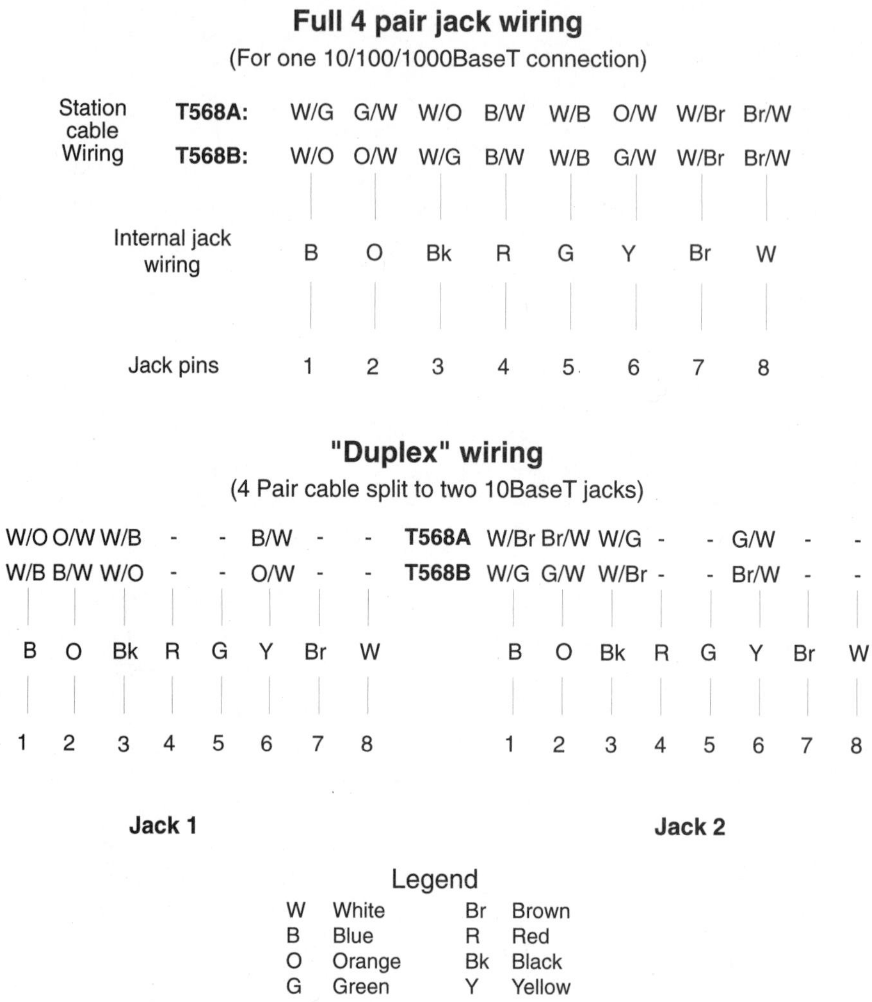

Figure 4.10
Typical 10/100/1000 BaseT jack wiring. Some connector jack internal wiring may present some confusing variations in wire colors and numbering. Note that the wiring colors may be completely different than those of the station cable and the plates sometimes have numeric markings that do not correspond to the pin numbering of the jack itself.

Full 4 pair jack wiring
(For one 10/100/1000BaseT connection)

Station cable Wiring								
T568A:	W/G	G/W	W/O	B/W	W/B	O/W	W/Br	Br/W
T568B:	W/O	O/W	W/G	B/W	W/B	G/W	W/Br	Br/W
Internal jack wiring	B	O	Bk	R	G	Y	Br	W
Jack pins	1	2	3	4	5	6	7	8

"Duplex" wiring
(4 Pair cable split to two 10BaseT jacks)

W/O O/W W/B	-	-	B/W	-	-	**T568A**	W/Br Br/W W/G	-	-	G/W	-	-
W/B B/W W/O	-	-	O/W	-	-	**T568B**	W/G G/W W/Br	-	-	Br/W	-	-
B	O	Bk	R	G	Y	Br	W		B	O	Bk	R	G	Y	Br	W
1	2	3	4	5	6	7	8		1	2	3	4	5	6	7	8

Jack 1   **Jack 2**

Legend

W	White	Br	Brown
B	Blue	R	Red
O	Orange	Bk	Black
G	Green	Y	Yellow

whether in the wall or as a user/patch cord. Termination of the wires is covered in detail in Part 2.

Problems in 10BaseT usually relate to individual workstations. If you are having a problem, the hub status lights can diagnose total failures of the cable to a workstation or server. Common cable failures include damaged wires, connectors pulled loose or not properly seated, and poor or improper connections to the modular plug. Use only the expensive, tool and die type, crimp tools for 8-pin plugs. The cheaper tools will not properly seat the center contacts on the plug and the cable may fail. In addition, you should be very cautious if you use solid wire with modular plugs (see Chap. 10 for more information). If the cable tests good, suspect the workstation's network adapter, if the problem is with a single workstation. If all workstations are affected, suspect the hub or any thinnet or thicknet connections between hubs or to other devices.

100BaseT Cabling

Standards for 100 Mbps Fast Ethernet networks have rapidly developed. Many proprietary methods and interim standards were put forward since 10BaseT was first introduced. The IEEE 802 standards committee resolved the competing technologies with a supplement that details the implementation of 100BaseT. There are two copper wire 100BaseT variations under what has become known as the IEEE 802.3u Supplement standard. They are 100BaseTX and 100BaseT4. A companion 100BaseFx (covered later in this chapter) rounds out the 100 Mbps CSMA/CD standard offering. The non-CSMA/CD 100VG-AnyLAN (which we cover in another section) was relegated to a new committee, IEEE 802.12, after much debate.

In some ways, the new standards encompass several formerly proprietary technologies under the same standards umbrella. The variations are not cross-compatible, although some vendors may offer adapters or hubs that can meet more than one standard. You should still be cautious in what you get, as the product nomenclature can be quite ambiguous.

These 100 Mbps Ethernet topologies are sometimes referred to as Fast Ethernet. The implementations of the two main standards, TX and T4, differ in the minimum link performance category and the number of pairs required. The important point is that either of the copper Fast Ethernet standards can be supported on an unmodified TIA/EIA-568-B cabling system, although the required category of wire is different. And implementations of both fortunately offer backward compatibility with 10BaseT signals and networking equipment. Because of the differences, though, let's cover them one at a time.

100BaseTX   The simplest of the Fast Ethernet standards is 100BaseTX. This standard is quite similar to 10BaseT, except that it runs at 100 Mbps, 10 times faster. Hubs, switches, and NICs often offer dual 10/100 Mbps speeds with automatic sensing. Also, it requires only two pairs, one for transmit and one for receive. The pairs are wired exactly the same as the 10 Mbps version. The main difference from the slower standard is that the signal itself is at a full 100 Mbps data rate. A Category 5 or higher link is required. The signal is simply too fast for Category 3 or 4 cable. This may be the primary justification for placing Category 5/5e standards into any new cable installation. Remember that all components (not just the wire) must meet Category 5/5e and proper installation techniques must be used.

The wiring pattern for 100BaseTX is shown in Fig. 4.11. Although only two pairs are required for this topology, it is still recommended that all four pairs be connected in the T568A pattern given by TIA/EIA-568-B.

An additional feature may allow negotiation between NIC and hub to enable a fallback to 10BaseT if both devices are not 100BaseT compatible. This is a real advantage if you are gradually upgrading to 100 Mbps. You can install dual 10/100 Mbps NIC cards in all your new workstations at relatively little extra cost. When you change your hubs to 100BaseTX, the dual-speed cards will automatically make the 10 times speed shift.

A 100BaseTX network has some rather severe distance restrictions in a shared hub environment. Because of the timing constraints of CSMA/CD, the total "radius" of the network is one-tenth the size of a Mbps Ethernet network. Hubs may be linked, but only on a very short backbone of 5 m. The hubs (which are repeaters) can be linked to stations by 100 m of cable, which fits into the TIA 568-A limit of 90 m + 10 m nicely. This computes to a maximum station-to-station distance of 205 m (100 + 100 + 5). A fiber link from a hub to a bridge, router, or switch can be up to 225 m, and a nonhub fiber link can be 450 m with CSMA/CD enabled and 2 km, disabled. These limits encourage the use of Fast Ethernet switches.

An interesting problem exists if the environment has both 10 and 100 Mbps stations. Any simple hub must be only one speed, either 10 or 100,

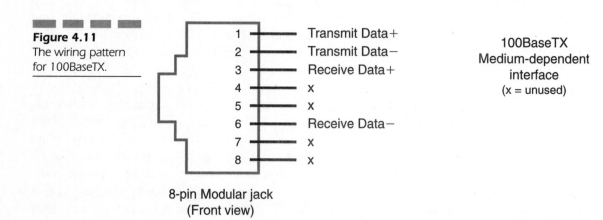

Figure 4.11
The wiring pattern for 100BaseTX.

8-pin Modular jack
(Front view)

Pin	Signal
1	Transmit Data+
2	Transmit Data−
3	Receive Data+
4	x
5	x
6	Receive Data−
7	x
8	x

100BaseTX
Medium-dependent
interface
(x = unused)

because a 100 Mbps station obviously could not transmit to a 10 Mbps station. The switch must buffer the high-speed data packets and restrain the higher-speed port to allow time to send the packets out at the slower data rate.

100BaseT4 For networks built to accommodate 10BaseT wire speeds, the move to 100 Mbps is a tough one. Many twisted pair networks were installed to Category 3 wiring standards that will not accommodate the 100BaseTX line speed. For those installations that have installed 4 pair Category 3 cable, the 100BaseT4 standard allows the operation of a 100 Mbps CSMA/CD connection over Category 3 cable.

The implications of this are enormous, if you have a large installed base of 10 Mbps (or 16 Mbps) users. You can selectively increase your LAN connections to 100 Mbps without pulling new Category 5 cable. This can save a lot of money if you have a large number of terminals. Better yet, many of the suppliers of 100BaseT4 workstation adapters offer dual speed operation at 10 and 100 Mbps. The adapter automatically detects which speed is in operation and makes its connection at that speed.

The main drawback to 100BaseT4 is its low acceptance in the market. The choices for hubs, switches, and NICs are limited and the prices are higher. The 100BaseTX version is the dominant technology.

However, if you are installing a new cabling system, you should not use the existence of this type of high-speed connection as an excuse to put in Category 3 links. Full Category 5e installations are not that much more expensive than new Category 3 ones, and you can move to Gigabit technologies that require Category 5e performance.

The wiring pattern for 100BaseT4 is shown in Fig. 4.12. All four pairs are required for this topology, so you cannot use it if you have split your cable into 2 pair jacks. However, it will run fine on any standard TIA/EIA-568-B cabling system.

The four pairs of a 100BaseT4 connection are used in a clever manner. The traditional pairs for transmit and receive are maintained on the same pins as the lower speed 10BaseT network, but the other two pairs are used alternately for transmit or for receive when sending data. Thus at any one time, three pairs are available for either transmit or receive. This technique is coupled with an advanced line coding to achieve the high bit rate at low symbol rates.

The 100BaseT4 technology suffers from the same distance and repeater limitations as 100BaseTX and has the same speed buffering problem if mixed with slower networks.

Figure 4.12
The wiring pattern
for 100BaseT4.

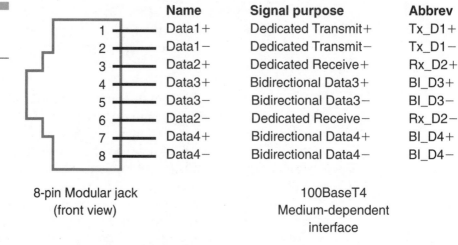

	Name	Signal purpose	Abbrev
1	Data1+	Dedicated Transmit+	Tx_D1+
2	Data1−	Dedicated Transmit−	Tx_D1−
3	Data2+	Dedicated Receive+	Rx_D2+
4	Data3+	Bidirectional Data3+	BI_D3+
5	Data3−	Bidirectional Data3−	BI_D3−
6	Data2−	Dedicated Receive−	Rx_D2−
7	Data4+	Bidirectional Data4+	BI_D4+
8	Data4−	Bidirectional Data4−	BI_D4−

8-pin Modular jack
(front view)

100BaseT4
Medium-dependent
interface

Ethernet Fiber, 10/100/1000 Mbps

Ethernet standards allow the use of fiber optic cable at 10, 100, and 1000 Mbps. Emerging standards will even allow the speed to grow to 10 Gbps. This section will briefly describe the Ethernet fiber standards. More information is contained in Chap. 12.

10BaseFx Wiring

The increasing use of fiber optics to transport Ethernet signals over extended distances or between buildings eventually led to the standardization of this method by the IEEE committees. The 802.3 standard now recognizes several variations of the fiber optic link that started out as FOIRL (Fiber Optic Inter-Repeater Link). The replacement for FOIRL is known as 10BaseFL. This standard, as the original name implies, is used to interconnect repeaters, although several manufacturers have used it to link stations as well. Additionally, supplements to the 802.3 standard have added 10BaseFB and 10BaseFP.

The 10BaseFB specification is a "synchronous Ethernet" link between repeaters that extends the limit for repeaters and segments in a single unbridged network. Normal repeaters use a portion of the preamble of the Ethernet package to distinguish incoming data from noise and to synchronize with the data. Each repeater, in effect, gobbles up some of these "sync" bits, which limits the number of repeaters a packet can traverse. Remember, a repeater does not store and forward a packet, so the lost bits cannot be replaced in the outgoing signal. Synchronizing the links means that the repeater must only distinguish data from noise and the entire preamble may be preserved.

The 10BaseFP specification is a passive star configuration for fiber optics. The signal is shared with other fiber arms of the star via a unique system of optical distribution.

All of these 10BaseF standards use graded-index 62.5/125 µm fiber optic cable with two fibers per link (one for transmit and one for receive). Distance limits are 2 km for 10BaseFL and 10BaseFB, and 500 m for passive star 10BaseFP. The 10BaseFx data rates, of course, are the same 10 Mbps as copper 10BaseT and consequently offer distance and electrical isolation as the primary advantages over copper.

Active point-to-point fiber optic links offer few problems that are not quickly revealed by the link status indicators. The 10BaseFP installation, however, does present the same problems as the coax copper star of Arcnet, since the hub is passive and has no indicators. Fiber optic links offer unique advantages such as extended transmission distances and immunity to interference. However, fiber is not indestructible and may require some special handling during and after installation. For example, fiber optic cable is susceptible to moderate to severe performance degradation from a one-time bending of the fiber below the bending minimum radius. While light may still transfer past the microfracture, transmission loss may soar. Fiber cables may also fracture when they receive a sharp impact. Fiber optic light sources, meters, and even OTDRs (optical time-domain reflectometers) may be required to troubleshoot a fiber optic installation.

Fast Ethernet 100BaseFX and SX Wiring

The 100BaseFX standard allows Ethernet transmission at 100 Mbps using multimode fiber at the long wavelength, 1300 nm. The 100BaseFX standard allows links of up to 2000 m and is widely used both in high-speed fiber backbone uplinks and in campus links between buildings. The less expensive LED transmit sources are generally used. As FX uses a different light wavelength than 10BaseFL, it is not compatible or upgradable.

The 100BaseSX standard, on the other hand, uses the same short, 850 nm, wavelength that 10BaseFL uses. This allows a transceiver to be designed that can handle either speed and allows for a gradual migration from 10 to 100 Mbps (or even 1000 Mbps). However, the SX distance is limited to about 500 m because of the characteristics of multimode fiber at that wavelength.

Both 100BaseFX and 100BaseSX support the relatively short 100-m distances of the horizontal cabling in TIA/EIA-568-B. The 100-m link is no problem to fiber, with its inherently greater signaling distances. The moderate backbone connections of 300 m are also easily accomplished by 100 Mbps Ethernet fiber.

We cover much more information on fiber optic cabling in Chap. 12.

Gigabit Ethernet 1000BaseSX and LX Wiring

The majority of Gigabit Ethernet cabling is done on either 1000BaseSX or 1000BaseLX standards. As fiber was the first available technology for Gigabit, it naturally has a significant installed base. However, both these gigabit fiber standards are relatively limited in distance capability (at fiber norms). They will both easily handle the 100-m horizontal limit of TIA-EIA-568-B, but when used for backbone or centralized fiber, range in distance from as little as 220 m to about 550 m. Of course, these are the minimum limit for IEEE 802.3z, so most manufacturers exceed these distances by moderate amounts.

As with the other fiber standards, the "S" stands for the short wavelength of 850 nm and is used with multimode fiber. The "L" indicates the longer wavelength, which is 1300 nm for multimode and 1310 nm for single mode fiber. As it turns out, the distance ranges for each technology also mock the S (short distance) and L (longer distance). Much more technical detail is shown in Chap. 13.

Several longer-range offerings are emerging for the extension of Gigabit Ethernet into large campus and metropolitan area networks. Among these are proprietary operating modes called SLX, ELX, and ZX that extend the fiber distances to 10, 70, and 80 km, respectively. These longer-distance Gigabit Ethernet connections mean that many networking applications can use the commonly associated IP (Internet protocol) networking within a metro area.

Token-Ring

Token-Ring networks were introduced in the mid 1980s by IBM Corporation and others. The network topology is now embodied as the IEEE 802.5 standard. Token-Ring was originally implemented on shielded twisted pair (STP) cable using a unique hermaphroditic connector, commonly called the IBM Data Connector. This cable type and connector is now specified in the TIA/EIA-568-B standard. More recently, Token-Ring has been migrated to conventional unshielded twisted pair (UTP) cable (what we simply call twisted pair in most of this book).

Token-Ring has a clever topology that allows an electrically continuous ring to be implemented with wiring that is installed in a hubbed-star configuration. Each arm of the star is called a *lobe*. As shown in Fig. 4.13, a special type of wiring hub, called a MSAU (Multistation Attachment Unit), allows signals from workstations (or from servers or bridges) to be looped through to the next workstation, in turn, until the signal is ultimately looped back to the beginning workstation. This effectively turns the star wiring into a loop or ring, from which the Token-Ring gets the "ring" part of its name.

The "token" part of the name comes from the fact that a so-called token is passed from station to station, along with data, commands, and acknowledgments. The LAN signal thus proceeds in an orderly fashion around the ring in a loop. Here is a very simplified explanation of the process. A station may transmit only when it "has the token." All transmitted frames are passed from station to station around the ring. All stations test each passing frame for messages addressed to them, process the message if it is theirs, and pass a marked token back around the ring. The original transmitting station "releases the token" when it returns around the loop.

The Token-Ring signaling structure is called a Differential Manchester Code, and has no DC voltage component. It can thus be directly, inductively, or capacitively coupled to networking components. This lends itself to the "self-healing" aspect of the physical topology. The

Figure 4.13
A special type of wiring hub, called an MSAU (Multistation Attachment Unit), is used for Token-Ring networks.

MSAU is the key to this architecture. (The MSAU is also commonly called a MAU and pronounced as a word by Token-Ring users, although it is quite different from the Ethernet component by that same name.)

Each port of the MSAU contains a small relay that connects the ring signal to the next port, in turn, when no cable is plugged in. If no cables are plugged in, the MSAU is a small ring within itself. As each cable is plugged into a port on the MSAU, a phantom voltage from the associated workstation opens the relay and the ring signal is diverted down that cable to the workstation. The workstation monitors the ring signal and repeats it back to the MSAU, where it may be diverted to the next active workstation. If a workstation is not active (powered up or inserted), the relay remains closed and bypasses the ring signal to the next port, ignoring the attached cable.

The data rate of Token-Ring is either 4 Mbps or 16 Mbps. A particular ring must operate at just one speed. A ring may be extended to additional MSAUs through the Ring-in/Ring-out ports of each MSAU. Copper repeaters and fiber optic repeaters (or converters) may be used to extend the distances of a ring. Rings are joined together by means of bridges or routers, as with the other network topologies.

Token-Ring uses two pairs of wires to connect each workstation to the MSAU. The pairs may be incorporated in a shielded or unshielded cable that typically comprises four pairs. The popular IBM Cabling System uses STP cabling in conjunction with special Data Connector jacks and cables to support the traditional Token-Ring installations. The IBM Cabling System is described in Chap. 5. Unshielded twisted pair wiring normally uses an 8-pin, modular (RJ-45 style) plug and jack. This is called the *medium dependent interface connection* in the 802.5 terminology. The wiring patterns of the Data Connector and the 8-pin modular jack are shown in Fig. 4.14.

Station wire for the Token-Ring requires two pairs and the cable may be either STP or UTP. Cables are always run in a star pattern from each

Figure 4.14
The wiring patterns of the Data Connector and the modular jack for Token-Ring networks.

Station signal	Data connector colors	Modular jack 8-pin	Pinouts 6-pin
Transmit −	Black	3	2
Receive +	Red	4	3
Receive −	Green	5	4
Transmit +	Orange	6	5

SHORTING BARS

workstation outlet to a central wiring closet. The TIA/EIA-568-B wiring standard will support Token-Ring networking since it contains four pairs in a compatible pattern. Category 3, 4, or 5 wire will carry either the 4 Mbps or 16 Mbps data rates, although the Category 5e standards offer future growth to 100 Mbps networking. The standard distance limits for station cable in TIA/EIA-568-B will take care of Token-Ring networks in most instances, because each workstation acts as a repeater to the next workstation. However, if you wish to calculate the actual working distances according to IBM's recommendations, you must include a complex series of derating factors that take into account the number of wiring closets and MSAUs. These calculations yield the maximum allowable station cable lengths and the allowable cable lengths between wiring closets. Table 4.1 shows a typical planning table for Type 1 STP Cable at 16 Mbps. For a relatively simple network that involves only one or two wiring closets, or if you use repeaters or bridges between closets, you may be able to bypass the calculations. However, if your network is complex, you should consult one of the network planning guides from IBM and other Token-Ring MSAU component vendors.

Because Token-Ring may be run on either 150-ohm STP or 100-ohm UTP, potential impedance-matching problems will exist if you have a mixed-media network. For example, the Token-Ring Network Interface Card (NIC) generally has a 9-pin d-shell connector with a 150-ohm interface. If your station cable (from the wall jack to the wiring closet) is 100-ohm UTP, you will have to use a "media filter" cable from the NIC to the wall jack to translate from the 150-ohm NIC interface to the 100-ohm cable. Figure 4.15 shows a typical media filter. The media filter actually contains an impedance transformer (balun) that compensates for the impedance difference. Operating without a media filter can cause serious problems that will limit lobe distances and cause unwanted signal reflections.

Wiring-related problems that may be encountered with Token-Ring networks include excessive distance between workstations and MSAUs or between wiring closets, and problems that occur when an insufficient number of active workstations are on a large ring. Extended distance MSAUs or repeaters may be required to solve these problems. In addition, the normal problems of bad or intermittent connectors, cables, and jacks may affect the network. Token-Ring does exhibit a delay if an improperly wired workstation/cable tries to insert itself into the ring. After a period of time, the offending workstation adapter should deactivate, thus restoring the ring. Some nonpowered MSAUs may get a relay stuck in the open position when no cable is connected. The only way to cure this

TABLE 4.1

Typical Token-Ring Planning Calculation in 16 Mbps Networks

The maximum lobe distances are calculated from the adjusted ring length and the maximum transmission distance (which varies with the numbers of MSAUs and telecommunications rooms).

Number of MSAUs	MTD, m (ft) Number of telecommunications rooms				
	2	3	4	5	6
1	130 (430)				
2	130 (420)	125 (410)			
3	125 (410)	120 (400)	120 (390)		
4	120 (400)	120 (390)	115 (380)	115 (380)	
5	120 (390)	115 (380)	110 (370)	110 (360)	110 (360)
6	115 (380)	110 (370)	110 (360)	105 (340)	105 (340)
7	110 (360)	110 (360)	105 (350)	105 (340)	105 (340)
8	105 (350)	105 (350)	105 (340)	100 (330)	100 (330)

NOTE: The maximum lobe distance is calculated for Type 1 and 2 cable as follows:

$$L_{max} = MTD - ARL$$

where MTD = maximum transmission distance from the table
ARL = adjusted ring length (the total cable length connecting wiring closets on the same ring, minus the shortest closet-to-closet length and excluding all patch cords between MSAUs in the same closet)
For Type 3 (unshielded twisted pair) cable, divide the Type 1 L_{max} distance by a factor of 2. This factor does not apply to Category 3 and above UTP, which has better transmission characteristics than Type 3/Category 2 cable. Category 3 lobes should observe the 90-m limit of TIA-568/EIA-B or the recommendations of the MSAU and network adapter manufacturers, whichever is less.

SOURCE: Andrew Corporation

Figure 4.15
A typical Token-Ring media filter.

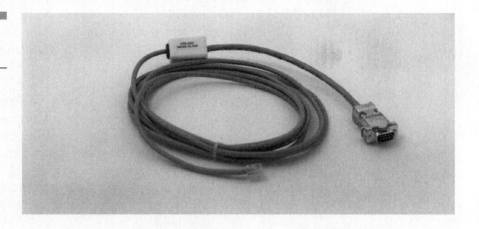

is to connect a cable from an active station or use a battery-powered test plug to reset the relay.

Arcnet

The Arcnet network type actually comprises a variety of physical wiring topologies that share a common signaling protocol. Arcnet protocol is implemented as a token-passing RF bus that combines some of the features of other network methods, although it was independently developed by Datapoint Corporation in the mid 1970s. Arcnet can be implemented on coaxial cable, twisted pair cable, fiber optic cable, or a combination of the three. The defining factor is the particular type of Network Adapter Cards (NICs) you choose.

Arcnet operates at 2.5 Mbps, although a newer ArcnetPlus at 20 Mbps has been introduced. There are also proprietary implementations at 100 Mbps. The method of protocol operation is similar to Token-Ring, but without a physical ring. All stations are connected to a common bus. When any station transmits, it is simultaneously (more or less) received by all other stations. A token is passed to each station, in a logical sequence, and a station must have the token to transmit. When a transmission is made, the receiving station makes an acknowledgment reply.

The topology of Arcnet can be a star, a bus, or a daisy-chain bus. The star topology may be used with coax, fiber, or twisted pair. The center of the star is either a passive hub or an active hub. A passive hub simply combines all signal inputs (so they may be shared) and provides impedance matching (to minimize reflections). Some of the signal is lost in this process, depending on the number of hub connections, so passive hubs are limited to a small number of ports. Active hubs amplify each incoming signal and distribute it to the other ports. Passive hubs are used with coax cable. Active hubs come in varieties for coax, twisted pair, or fiber. An example of Arcnet star network is shown in Fig. 4.16.

The Arcnet coaxial bus topology is shown in Fig. 4.17. It is quite similar in structure to Ethernet thinnet, using coax cables with BNC-Ts and terminators. However, because it does not provide for any active signal amplification, it is generally limited to about eight stations.

Several types of coax cable may be used for Arcnet, depending upon the particular NICs you are using. Arcnet may be used with RG-62/U coax at distances up to 2000 ft, which might be useful as a method to upgrade from IMB 3270 terminals to PC networking without extensive

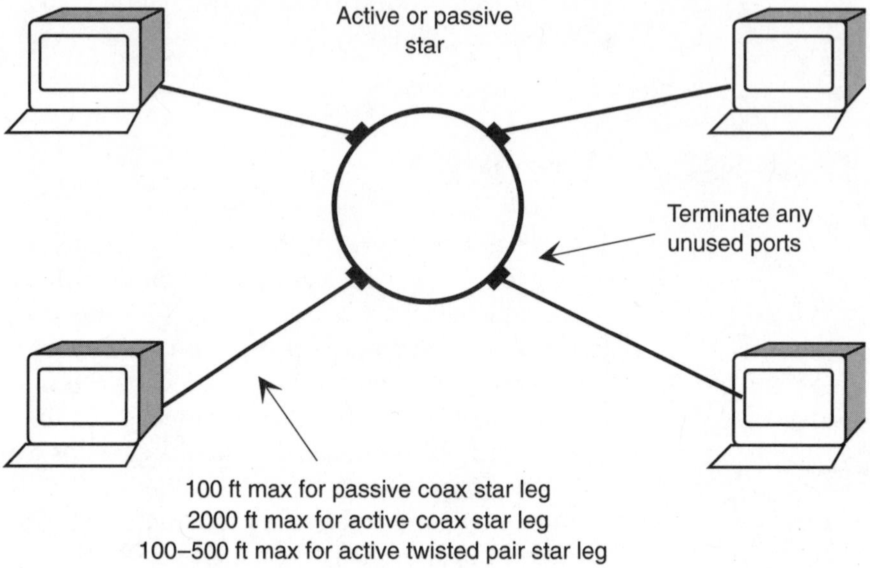

Figure 4.16
The Arcnet star topology.

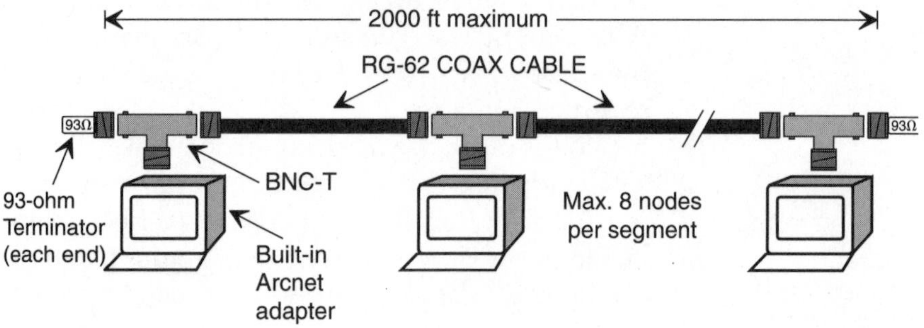

Active or passive star

Terminate any unused ports

100 ft max for passive coax star leg
2000 ft max for active coax star leg
100–500 ft max for active twisted pair star leg

Figure 4.17
The Arcnet coaxial bus topology.

2000 ft maximum

RG-62 COAX CABLE

93Ω

93Ω

BNC-T

Max. 8 nodes
per segment

93-ohm
Terminator
(each end)

Built-in
Arcnet
adapter

rewiring. Arcnet is also used with 75-ohm RG-59/U and the larger RG-11/U coax. A table of cabling alternatives and some of their properties is shown in Table 4.2.

A popular alternative to Arcnet coax is to use twisted pair wire for the transmission medium. An Arcnet twisted pair network can be wired as an active star or as a daisy-chain bus. Technically, a passive star could be used for a small number of workstations, although the active star is more common. A twisted pair star is wired in the same fashion as with other networks. Workstation cables are run outward from a wiring closet or other central location to each workstation outlet. Each cable is terminated at the workstation with a 100-ohm resistor plug. The active star hub includes built-in terminators.

TABLE 4.2

Arcnet Cabling
Alternatives

Cable type	Nominal impedance	Typical connector
RG-11/U Coax	75 ohms	Type N or BNC
RG-59/U	75 ohms	BNC
RG-62 Coax	93 ohms	BNC
IBM Type 1	150 ohms	IBM Data Connector Adapter/Balun
IBM Type 3	100 ohms	IBM Data Connector Adapter to 6- or 8-pin modular plug (RJ-11 or RJ-45 style)
Category 2 (or higher) twisted pair—2 pairs used	100 ohms	6- or 8-pin modular plug/ jack (RJ-11 or RJ-45 style)

The workstation outlets are sometimes 6-pin RJ-11 jacks with matching cables, since only two pins are used (see Fig. 4.18). Only two pins are required since both transmit and receive occur on the same pair. This is possible, of course, because an Arcnet station may transmit only when it has the token. If you should use telephone flat cable (not really recommended), be careful to get straight-through cords, rather than the "flipped" cords that are standard for telephone use. A flipped cord will reverse the polarity of the Arcnet wires and cause the connection to fail.

Standard TIA/EIA-568-B wiring may be used with the Arcnet twisted pair star (or with the daisy-chain bus described below). However, most Arcnet twisted pair active hubs and workstation NICs use the 6-pin RJ-11 style connector and will require an adapter. An adapter cable may be built for the workstation outlet and the patch panel with a 6-pin modular plug at one end and an 8-pin modular plug at the other.

User and patch cords should be twisted pair wire, as should the station cable. Short lengths of flat telephone-type cord may work for the lower Arcnet speeds, though, since only one pair is used and the problem of crosstalk does not exist.

The daisy-chain twisted pair bus is a variation of the coax bus using twisted pair. It is useful for small networks with existing cable and in situations where the workstations are all in one room. The one-room design is the easiest to visualize. As with the coax bus, a twisted pair cable is run from one workstation to the next in line, where another cable is run on to the next workstation, forming a chain of workstations. Most twisted pair Arcnet adapter cards have two common connectors, so this is easily done. A 100-ohm terminator plug must be used in the two workstations

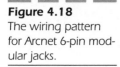

Figure 4.18
The wiring pattern for Arcnet 6-pin modular jacks.

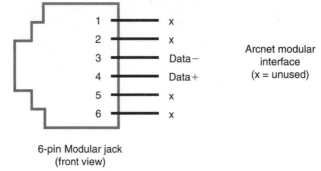

at the ends of the chain. Of course, the wire can also be run between workstations in different rooms, in the same daisy-chain fashion.

Things are a little more complex if this is attempted with installed wiring. The loops of the chain must be in the same physical station cable. One way to accomplish this is to use a special adapter cable from the wall plate that is split out to two plugs at the NIC. The cable has two active pairs, say Pair I and Pair 2, that are split to the two plugs, both wired for Pair 1 so they will connect to the NIC's pins 4 and 5. Then at the punch block in the wiring closet, Pair 1 on the first drop is cross-connected to Pair 2 on the second drop, and so forth through the chain. The first and last workstations in the chain will use only one of the plugs, the Pair 1 plug for the first and the Pair 2 plug for the last, and will need a 100-ohm terminator plugged into the other port on their NICs.

Problems in Arcnet cabling are as varied as the many types of cable topologies that are used. The coaxial bus is subject to the same potential problems with BNC-Ts and terminators as thinnet. If a terminator is missing, the network will at least partially fail. As with the other LAN types, any break in a bus cable will also cause the network to fail, although the twisted pair bus may be somewhat more tolerant of this. Of course, no signals may pass across the break! If a break occurs, look for damaged cable, connectors, Ts, and so on.

Rarely will a bad adapter card cause a total network failure, since the cards are not active until they transmit. An adapter failure will usually affect only that workstation.

Star-type Arcnet networks will usually have a problem only on one leg, although that will take your network down if that leg is the server. Try plugging into another port on the hub. Most twisted pair hubs and some coax active hubs have status lights that will indicate good cable. An active hub can have a total electrical failure, or in some cases have damage to only one port. A passive hub must have the proper value terminator connected to all open ports. Many passive hubs have terminators that are permanently tethered to the hub with a ball and chain, so the terminator won't disappear.

Other Network Types

Many other types of networks can operate in the office environment. Some of these other technologies have similarities to standard Ethernet and Token-Ring, but are not truly a part of those networking methods. We will cover a few of the other networks here, including some of the fiber optic topologies. Do not presume that these topologies are out of favor merely by being called "other." In some cases, many technologies, such as Isochronous Ethernet, 100VG-AnyLAN, and ATM, are emerging, and it is simply premature to say which will dominate.

100VG-AnyLAN

A new technology called 100VG-AnyLAN overcomes some of the distance and repeater limitations of the 100BaseT and allows operation on 4 pair Category 3 cable. The standard specification for 100VG-AnyLAN is contained in IEEE 802.12.

The 100VG-AnyLAN technology is an interesting combination of asynchronous communication, such as that employed by Ethernet, and round-robin message scheduling, such as that of Token-Ring. The 100VG-AnyLAN topology is an arbitrated star arrangement, that is, a star with transmission control in the hub. Workstations and servers are on the legs of the star, with an intelligent hub at the center. The hub performs the functions of an arbitrated gateway and a switch. A workstation that wants to transmit a message makes a request to the hub. If the network is idle, the request is granted. The workstation transmits its message and the hub switches it to the proper outgoing port. If more than one station needs to transmit, the hub acknowledges each one in turn, according to a predetermined priority scheme.

The wiring systems supported by a 100VG-AnyLAN include 4 pair Category 3 or higher UTP, 2 pair STP-A, and single or multimode optical fiber. The signaling rate on 4 pair UTP is kept low by splitting the signal into four data streams, one per cable pair. The bandwidth requirement of approximately 15 MHz is within the capability of standard Category 3 cabling. The wiring pattern for 100VG-AnyLAN is the same as for any standard TIA/EIA-568-B cabling system.

Because 100VG-AnyLAN is not dependent upon a collision detection mechanism, it is not subject to the distance limitations of unswitched Fast Ethernet. Some 100VG-AnyLAN adapter implementations offer dual speed capabilities, as with the 100BaseT products.

Isochronous Ethernet

One of the greatest challenges in integrating voice, data, and even video, is that the technologies heretofore have used entirely different means of encoding and transmitting information. The only practical solution has been to separate those signals in separate cables, each routed to its respective switching or distribution equipment. Isochronous Ethernet proposes to end that divergent situation by combining multiple signals onto the same datastream, so that all three technologies can be simultaneously carried on the same physical medium. Indeed, the technology actually allows quite a number of channels to be carried over the standard 100-m 4 pair cable link. Isochronous Ethernet is standardized in IEEE 802.9, and the information here reflects the proposed draft revisions to become IEEE 802.9a.

Isochronous Ethernet uses a clever scheme of time-division multiplexing (TDM) and signal encoding to place a full 10 Mbps Ethernet channel plus an additional 6 Mbps or so into the same composite signal. The total data rate in the signal is 16.384 Mbps. A 4B/5B encoding scheme is used instead of the Manchester coding of standard Ethernet. This yields a 10.24 MHz line transmission rate, thus making the signal well within the capabilities of Category 3 cable or higher cable.

Two modes of operation are available. In the multiservice mode, the signal is divided into three user channels, the 10 Mbps (actually 10.016) Ethernet P-channel, a 6.144 Mbps bearer C-channel, and a 64 kbps signaling D-channel. The C-channel is easily further divided into 96 channels of 64 kbps subchannels. Of course, the subchannels are exactly the same size as an ISDN bearer channel, so the network can easily support data, digital voice, or compressed video. An alternative isochronous-only mode divides the total signal into 248 bearer subchannels, each of 64 kbps.

A third mode of operation allows the devices to operate as standard 10BaseT, Manchester encoded network interfaces, in full accordance with IEEE 802.3 and with no support for the TDM channels.

AppleTalk/LocalTalk

AppleTalk has turned into a somewhat generic term that is generally used to describe Apple LocalTalk networks. AppleTalk is actually a higher level protocol that can be transported over several types of network topologies. The wiring for classic LocalTalk is a low-speed two-wire bus that is usually run in a daisy-chain fashion from computer to computer. The unique feature of most Apple computers since the introduction of the Macintosh is the built-in LocalTalk networking port. The provision of a built-in networking capability has made printer sharing and file transfer a normal part of operation with Apple computers.

The ease of networking, even though it is a slow, peer-to-peer method, has both helped and hurt Apple users. It has helped by making the network a quick and natural solution to data and print sharing that must have made IBM-style PC users envious (or at least confused). On the other hand, the relatively low speed and proprietary nature of the native network, without expensive upgrade modules for Ethernet or Token-Ring, initially put AppleTalk users at a comparative disadvantage.

The original LocalTalk connection used a cumbersome shielded cable, but then Farallon Computing made a compatible adaptation called PhoneNet on twisted pair wiring. This is the simple LocalTalk connection that most users are familiar with. Many users employ standard telephone cord to connect devices. This telephone cord is typically the flat silver-satin type 4-wire cable, although twisted pair cable will work even better. The network runs at only 230.4 kbps, so the normal crosstalk and attenuation problems are not as severe. LocalTalk can operate well on TIA/EIA-568-B.

The same comments apply to potential wiring problems with LocalTalk that were stated for any of the twisted pair networks. Because LocalTalk can use ordinary flat phone wire, many users will simply purchase preassembled cables, which have few faults. However, this wiring method sometimes leads to long runs of wire being routed around furniture and across traffic areas. Any abrasions or cuts in the cable will have network consequences. Often a network must be split at some point and terminated to diagnose such cable problems.

FDDI

FDDI (Fiber Distributed Data Interface) is a fiber optic networking technology that operates at 100 Mbps. It employs dual, counterrotating, token-passing rings. The network concept is actually fairly complex and is different from either Ethernet or Token-Ring.

Network nodes that sit on the primary fiber ring are called *dual-attach nodes* and employ a sophisticated automatic wraparound of the FDDI signal in the event of a cut in the ring. This is somewhat akin to the backup path in Token-Ring, but the fiber in an FDDI ring is expected to take a diverse path to the next node in line. In the event of a fiber cut or other failure, the dual-attach nodes at each end automatically loop back and the network continues without the missing component.

Each dual-attach node has two dual fiber connectors for a total of four individual fiber optic connections. Each connector has a fiber input and output, but the FDDI signal actually passes from one input, through the node processor, and out the opposite connector's output fiber on to the next node. This produces dual signals in opposite directions.

Another type of node, called a *single-attach node*, implements the basic FDDI protocol, but attaches point-to-point in a star fashion to a concentrating node. This concentrating node is customarily on the actual ring (and is dual-attach), and benefits from its fault-tolerant design. The single-attach node lacks the loop back path to its concentrator node, and so is normally used within the same building where the fiber cuts would be rare.

The FDDI topology is an ANSI (American National Standards Institute) standard. The fiber is standard 62.5/125 μm fiber terminated in the unique duplex FDDI connector. Both multimode and single mode fiber are supported, although the interface module is different, since single mode requires a laser-diode transmitter. An FDDI single-attach connection could be supported on a TIA/EIA-568-B fiber optic link, but an adapter cable would have to be used to connect to the 568-SC connectors in the workstation outlet.

Because FDDI is neither an Ethernet nor a Token-Ring protocol, those and other protocols are usually encapsulated into an FDDI packet and decoded at the receiving station. This method simply uses the FDDI link as a transport mechanism that frees one from the point-to-point nature of most WAN links.

ATM

Asynchronous Transmission Mode (ATM) is a very high speed communications protocol that does not really imply a particular LAN topology. As a matter of fact, the ATM protocol can be used over long distances in what is often called the WAN (Wide Area Network). The ATM protocol specifies a 53-byte "cell" that is somewhat analogous to a packet or frame of conventional variable-length MAC-layer protocols. Unlike those protocols, the ATM cell does not always contain source and destination addresses, nor does it contain the higher-level addressing and control of longer packets.

ATM is actually a very abbreviated point-to-point, switched protocol. An ATM message typically sets up a virtual connection between a source and destination, transmits cell-size chunks of the message, and ends the connection. The cells may go through many switching points before reaching the destination. The protocol allows cells to be received at the destination out of order and reordered before being presented to the receiving application.

ATM can be transported at many rates and over a wide variety of media. ATM protocol is implemented over copper and fiber at rates from about 25 Mbps to over 2 Gbps (gigabits per second). Because the protocol is not dependent upon a particular connection scheme, many ATM-based technologies have emerged under the ATM umbrella. They are not the same, and should be treated as separate topologies that share a common link basis.

Fiber implementations of ATM often run at Sonet physical parameters, such as 155 Mbps OC3 and 2.4 Gbps OC48. Copper implementations have been introduced at 25, 52, 100, and 155 Mbps. The 25 and 52 Mbps speeds are intended to operate over Category 3 twisted pair cabling. The current ATM UNI-155 standard requires a full Category 5 link for operation. You should be aware that the ATM speeds shown are rounded off from the actual data rates. For example, UNI-155 and OC3 actually run at 155.52 Mbps. To get the exact data rate, multiply 51.84 Mbps by the multiple. For example, OC12 would be 51.84 Mbps $\times$ 12, or 622.08 Mbps.

Not to be outdone by fiber optics, copper-based 622 Mbps technologies are under development at this very moment. Watch your mailbox for details! Copper proponents claim that 1 GHz bandwidths are possible for twisted pair, although it is doubtful that this could be achieved with the present complement of connectors and cabling.

Unlike the standard LAN topologies, ATM is a switched protocol. It is not really suitable for standard bus or shared star (repeater-coupled) topologies. It does not imply a collision detection or a token passing mechanism. When used for a LAN, it is switched, with the physical switch being at the hub of a star of workstation legs. Often, LAN protocols such as Ethernet or Token-Ring are transported over ATM links via a method of LAN emulation.

Structured Cabling Systems

Chapter 5 Highlights

- TIA/EIA-568-B structured cable
- Lucent (AT&T) Systimax Premises Distribution System
- IBM cabling system
- NORDX/CDT IBDN
- DECconnect

We will describe several LAN cabling systems in this chapter. These systems define the characteristics of cable, connectors, and general wiring schemes without regard to any of the specific LAN topologies that were described in Chap. 4, and all of them also use some form of twisted pair wire as their basis. The general purpose TIA/EIA-568-B cabling system will be described, as will several of the proprietary cabling systems that are in wide use.

Many proprietary cabling plans have been offered over the history of LAN wiring technology. Early on, the direction of LAN cabling was uncertain and several major equipment vendors devised detailed cabling systems that met the needs of their equipment. Any customer who followed their guidelines could be reasonably assured of a LAN cable plant that would operate well with that vendor's products. Of course, use of a particular vendor's products also frequently implied use of a particular LAN topology.

Over time, many LAN technologists have learned that the use of a particular vendor's cabling plan might actually limit the utility of their cable plant for other uses. In some cases, adherence to a cabling system might significantly increase system cost over other methods. However, a few cable plans offered wide connectivity of both voice and data. Out of those plans, a universal cabling system finally emerged, EIA/TIA 568, now revised as TIA/EIA-568-B.

It is widely agreed that the flexibility and utility of the standard TIA/EIA-568-B system offers the best alternative for new installations. The standard can be used in lieu of the proprietary cabling systems, with little difference in performance. It can easily support a wide variety of LAN topologies and offers growth to 100 Mbps and beyond.

TIA/EIA-568-B Structured Cabling

The TIA/EIA-568-B Commercial Building Telecommunications Wiring Standard describes a generic cabling system that can support many types of LANs, as well as many other telecommunications applications. In this section, we will cover the hardware and wiring issues presented by the standard. And of course, the details of many of the wiring components are covered elsewhere in this book. We will rather loosely group the requirements of TIA/EIA-568-B and its predecessor documents, EIA/TIA 568, TSB 36, TSB 40, TSB 53, TSB-67, TSB-72, and TSB-95 under the label "TIA/EIA-568-B." The installed link testing requirements

TIA/EIA-568-B Highlights

The latest revision to the famous TIA/EIA-568 standard is the "B" revision. The standard is split into three parts, designated "B.1," "B.2," and "B.3" that cover general requirements, 100-ohm twisted pair cabling components, and optical fiber cabling components, respectively. Here are the formal standard names and their predecessor documents (in parentheses).

ANSI/TIA/EIA-568-B *Commercial Building Telecommunications Cabling Standard* (2000)

This new standard replaces ANSI/TIA/EIA-568-A; incorporates and refines the technical content of TSB-67, TSB-72, TSB-75, ANSI/TIA/EIA-568-A Addenda A-1, A-2, A-3, A-4, and A-5.

TIA/EIA-568-B.1 *Commercial Building Telecommunications Cabling Standard, Part 1: General Requirements* (SP 4426)

TIA/EIA-568-B.2 *Commercial Building Telecommunications Cabling Standard, Part 2: 100-Ohm Balanced Twisted-Pair Cabling Components Standard* (PN 4425)

TIA/EIA-568-B.3 *Commercial Building Telecommunications Cabling Standard, Part 3: Optical Fiber Cabling Components Standard* (SP 3894)

The B revision basically incorporates the technology within 568-A, plus all of the addenda and the applicable telecommunications standards bulletins (TSBs). The new structure allows the specifications for the wiring components to be completely separate from the general structured guidelines. It also allows each section to be separately revised and additional sections to be added as necessary. Although the standard basically incorporates changes to the technology and practices since the previous standard was published, there are some significant differences. Here is a summary of the changes in the B revision:

New specifications and changes in 568-B:

- Replaces "telecommunications closet" with "telecommunications room."
- Defines performance specifications for Category 5e balanced 100-ohm cabling (keeps Category 3 and drops reference to Categories 4 and 5, in favor of Category 5e).
- Defines performance specifications for 50/125 μm optical fiber cables (in addition to the 62.5/125 μm cables).
- Allows alternative fiber connector designs in addition to 568-SC (such as the new SFF, small-form factor, fiber connectors).
- Divides the standard into three parts:
    B.1—General requirements
    B.2—100-ohm balanced twisted-pair cabling components
    B.3—Optical fiber cabling components

The change in specifications to Category 5e is significant for early Category 5 users, as additional parameters are added for such things as attenuation-to-crosstalk ratio (ACR), delay-skew, power-sum near-end crosstalk (PSNEXT), and far-end crosstalk (FEXT and ELFEXT). These additional specifications were found to be necessary for proper operation of gigabit-speed technologies, which often use bidirectional, full-duplex techniques to multiply the effective bandwidth of the cable.

As the additional cable characteristics became known, many manufacturers marketed advanced "Category 5—Plus" or "enhanced Category 5" cables which met the new requirements. Now cable constructions that greatly exceed Category 5e are being marketed as "Category 6," although those performance levels are not yet approved and may well change before they are finalized. All this makes the marketing claims of advanced cable performance much more credible. Let the buyer plan ahead!

of TSB 67, now a part of TIA/EIA-568-B, also provide guidelines for this cabling system.

The TIA/EIA-568-B cabling system actually includes four types of cable options: unshielded twisted pair (UTP), shielded twisted pair (STP), multimode 50/125 and 62.5/125 optical fiber, and single mode fiber (50-ohm coax is grandfathered for existing installations only). See Chap. 12 for a discussion of fiber optic cabling.

The TIA/EIA-568-B cabling combines the low cost of modular wiring connections and twisted pair cable with a system of performance classifications and installation procedures. The use of modular connectors and twisted pair wire for LANs was made popular by the availability of inexpensive, existing components that were originally intended for telephone system use. Unfortunately, users soon learned that the cable needed to have specific characteristics, such as minimum twists per foot, to work properly. The cable needed to meet certain performance levels that varied with the application. A cable vendor's rating system, supplemented by Underwriters Laboratories testing, eventually was incorporated into the TIA/EIA-568-B cabling system. The rating system divides wire and components into numbered categories that are associated with increasing levels of performance needed for LAN operation.

Six categories of TIA/EIA-568-B cabling are now specified, as shown in Table 5.1, although low-performance Categories 1 and 2 are specifically excluded from consideration in the standard. Two of these categories, Category 3 and Category 5e, are the dominant ones used in most LAN cable installations today. Category 3 can support 10 and 16 MHz networks, such as traditional Ethernet and Token-Ring, while Category 5e supports 100 MHz (100 to 1000 Mbps) networks, as well as the lower speeds. Categories are rated in terms of bandwidth, in MHz, rather than data rate, in Mbps, because data encoding techniques often obscure the actual cable signaling rate. The popular 10 and 16 Mbps networks require matching cable bandwidths of 10 and 16 MHz because of their signal structure, but 100 Mbps network techniques can range from 15 MHz (per pair) to 33 MHz because of encoding. Since Category 3 is rated to only 16 MHz and Category 4 only to 20 MHz, some of these 100 Mbps networks require the higher bandwidth of Category 5. Category 5e is also often mentioned as a requirement for some of the 155 and 1000 Mbps data rates. Category 6, the newest of the standards categories, extends operating frequencies to 200 MHz (tested to 250 MHz) and increases operating margins at lower speeds. Components, such as jacks and patch panels, are also rated by category of performance.

TABLE 5.1

UTP Wiring Performance Categories

Performance	Wire gauge	Nominal impedance	Transmission	Typical uses
Category 1	18–26	Not specified	Audio, DC	Speaker wire, door bells :-)
Category 2	22–26	Not specified	Up to 1.5 MHz	Analog telephone
Category 3	22–24	100 ohms ± 10%	Up to 16 MHz	10BaseT, 4/16 Token-Ring
Category 4	22–24	100 ohms ± 10%	Up to 20 MHz	10BaseT, 4/16 Token-Ring
Category 5/5e	22–24	100 ohms ± 10%	Up to 100 MHz	100BaseTX, ATM, 1000BaseT (4)
Category 6	22–24	100 ohms ± 10%	Up to 200 MHz	1000BaseTX

Standard 8-pin (RJ-45 style) modular connectors are used for the TIA/EIA-568-A. Four pair, 24 AWG, unshielded twisted pair provides the station cable. All 8 wires are connected to the jack. A system of optional punchdown termination blocks and patch panels completes the connection system. Wiring connection is *straight through,* which means simply that Pin 1 at one end of a cable will correspond to pin 1 at the other end. This is true whether we are talking about a user cord, a patch cord, or the in-wall station cable. This is a departure from normal 6-wire telephone cords, which reverse the connections from end to end.

The TIA/EIA-568-B system allows two different color code wiring patterns for the specified 8-pin modular connectors. Figure 5.1 shows the two patterns, called T568A and T568B (not to be confused with the "A" or "B" revisions of the original standard). The two wiring patterns are electrically equivalent, but the positions of Pair 2 and Pair 3 are reversed. Practically, this means that the color-to-pin correspondence varies, and it makes little difference which you use, as long as both ends of the cable are pinned the same. In other words, if you use the T568A pattern at the workstation outlet (called the telecommunications outlet in the standard), you must also use a T568A patch panel or fan-out cable in the telecommunications room. Pairs 2 and 3 just happen to be the two pairs used by 10/100BaseT Ethernet, so ignoring this wiring convention will cause the connection to fail. Token-Ring and 4-wire communications circuits, including T1, will also have a failure. Many prewired components are not marked as to wiring pattern, particularly nonrated octopus fan-out cables, so be aware of what you install.

Figure 5.1
The T568A and
T568B wiring
patterns.

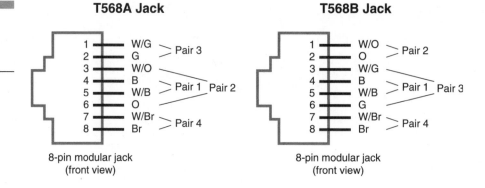

The TIA/EIA-568-B standard defines many wiring infrastructures that apply to all telecommunications wiring, including traditional telephone system wiring. The hierarchy of these wiring structures is shown in Chap. 15. Although the standard creates a very serviceable telephone wiring system, the real need lies in LAN applications which are much pickier about transmission characteristics. The portions of the standard that apply particularly to LAN wiring restrict such things as the length of cable and cross-connects, the characteristics of the cable, and installation techniques.

If you use the TIA/EIA-568-B system for a LAN, you may need to restrict the wiring to a particular category, such as Category 3 or 5e. This will place additional requirements on your wiring, such as the maximum cable run, the type of cable, the amount of untwist, and number of cross-connect (punchdown) points that are acceptable.

Figure 5.2 shows a summary of the characteristics for TIA/EIA-568-B cabling. Standard wiring patterns for the plugs, jacks, punchdown blocks, and patch panels are used for the TIA/EIA-568-B system. Refer to Chaps. 7, 8, and 9 for these details.

Lucent (AT&T) Systimax PDS

The Lucent Technologies (AT&T) Premises Distribution System (PDS) is a wiring system that bears great similarity to the TIA/EIA-568-B system. The original PDS system, however, is basically a multipurpose telecommunications wiring system that can also be used for LANs. In contrast, the 568/UTP system is primarily designed with LANs in mind, but can also be used for telecommunications.

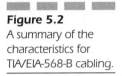

Figure 5.2
A summary of the characteristics for TIA/EIA-568-B cabling.

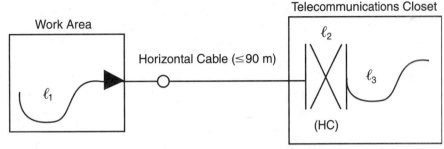

```
                  LEGEND

4 Pair UTP Cable:                    ——

Telecommunications                   ▶
    Outlet/Connector:

Transition Point (optional):        -O-

Mechanical Termination:              —|

Cross-connect Jumper                 ✕
    or Patch Cord:

WA Equipment Cable:                  ℓ₁

Patch Cord or Jumper:                ℓ₂

TC Equipment Cable:                  ℓ₃

Horizontal Cross-connect:          (HC)
```

```
                CHARACTERISTICS

Horizontal Cabling:
        4 twisted pairs
        24 AWG solid conductors
        Thermoplastic insulated
        Unshielded
        100-ohm ±15%
            characteristic impedance
        Other performance
            characteristics per
            Cat. 3, 4, or 5
Horizontal length:
        ≤90 m (295 ft.)
Patch Cords:
        Same as horizontal cable,
        except stranded conductors
        and up to 20% higher atten.

Connecting cord and
    Jumper length:

        ℓ₁ + ℓ₂ + ℓ₃  ≤ 10 m (33 ft)
```

The original Systimax PDS system as been updated to reflect all of the TIA standards. Original PDS is a comprehensive system that covers telecommunications wiring from the building cable entry, through intermediate cable distribution or wire centers, to the telecommunications outlet. It also encompasses campus distribution. PDS uses either unshielded twisted pair wire or fiber optic cable. Copper is used for horizontal wiring to the workstations and fiber is used for vertical wiring between wiring closets. The copper wire is 4 pair, 24 AWG cable, while the fiber is 62.5/125 μm cable. Eight-pin modular connectors are used for the copper, and ST connectors are used for the fiber.

The basic layout of a telephone-style wiring plan assumes that there will be hierarchical cross-connect points where large-pair count cables are terminated and fanned out into smaller cables or to station cables. The length of individual runs and the number of cross-connect points are not critical because traditional telephone equipment can run much longer distances than LAN cable.

PDS uses the 110-block as the basic cross-connect component. Station outlet modules use insulation-displacement connectors that snap into a modular faceplate. All connectors are normally specified as 8-pin RJ-45 style connectors, although 6-pin RJ-11/12s are also available. The connector modules are available with markings for "Data" or "Voice" and the rears of the modules are color coded to the wires of the 4 pair cable. The Lucent (AT&T) color code and pin mapping are shown in Table 5.2. Note that the standard wiring pattern for the PDS 8-pin modular jack is identical to the T568B pattern of TIA/EIA-568-B. This is not a coincidence, as the influence of standard telecommunications wiring plans, such as PDS, carried great influence in the design of the EIA-TIA standards. An earlier Western Electric plan, referred to as Distributed Inside Wire (DIW), was a logical precursor to both PDS and the EIA/TIA systems.

IBM Cabling System

In the 1980s, IBM Corporation designed a multiuse cabling system to support the rapid introduction of twisted pair wiring for computer networks. The cabling system uses a unique hermaphroditic connector (called the Data Connector) and several types of cable identified by a simple numbering system. Figure 5.3 shows a picture of a Data Connector, and Table 5.3 lists the cable types and their characteristics. The cable types, designated Type 1 through Type 9, include shielded twisted pair (STP) and unshielded twisted pair (UTP) cables as well as combinations of both.

The concept behind the cabling system is that most workstation locations actually need at least one data circuit and one telephone circuit. Both needs could be provided by one cable, such as a Type 2 cable, with shielded pairs for the data and unshielded pairs for the telephone.

The cabling system uses a clever genderless connector that eliminates the need for two complementary connector styles (such as the plug and

TABLE 5.2

Lucent/AT&T Systimax Structured Cabling System (SCS) Wiring Scheme

Modular twisted pair cord sets:

4-wire cord		6-wire cord		8-wire cord	
Plug pin	**Wire color**	**Plug pin**	**Wire color**	**Plug pin**	**Wire color**
1		1	White-blue	1	White-orange
2	Black	2	White-green	2	Orange-white
3	Red	3	White-orange	3	White-green
4	Green	4	Orange-white	4	Blue-white
5	Yellow	5	Green-white	5	White-blue
6		6	Blue-white	6	Green-white
				7	White-brown
				8	Brown-white

Outlet jacks:

102A-style information outlet (110-type cable terminations), using T568B wiring:

Modular jack pin	Internal lead color	110 Block position	Horizontal wire color
5	White-blue	1	White-blue
4	Blue	2	Blue-white
1	White	3	White-orange
2	Yellow	4	Orange-white
3	Red	5	White-green
6	Black	6	Green-white
7	White-brown	7	White-brown
8	Brown	8	Brown-white

M11BH-style information outlet (H-pattern IDC-type cable terminations), using T568B wiring:

Modular jack pin	Outlet position	Horizontal wire color
1	1	White-orange
2	2	Orange-white
3	3	White-green
4	4	Blue-white
5	5	White-blue
6	6	Green-white
7	7	White-brown
8	8	Brown-white

SOURCE: Lucent/AT&T

Figure 5.3
The Data Connector.
(*Courtesy of Hubbell
Premise Wiring, Inc.*)

TABLE 5.3

IBM Cable System
Cable Types

31BM type	Pairs	AWG	Impedance	Shielding*	Rating
Type 1	2	22	150 ohms	e/f + o/b	Plenum
Type 2	2	22	150 ohms	e/f + o/b	Plenum
	4	22	600 ohms†	No	Plenum
Type 3	4	24	150 ohms	No	Plenum or non-plenum
Type 4	Not used				
Type 5	2 fibers	100/140 μm	850 nm	No	Plenum
Type 6	2	26	105 ohms	e/f + o/b	Nonplenum
Type 7	Not used				
Type 8	2-flat	23	150 ohms	e/f + o/b	Nonplenum
Type 9	2	26	105 ohms	e/f + o/b	Plenum

*e/f = foil shield, each pair; o/b = overall braid shield
†Impedance at 1 kHz, rather than 0.250 to 2.3 MHz.

NOTE: A-suffix cables are rated at higher frequencies than original 20 MHz rating cables (e.g., Type 1A is tested to 300 MHz).

jack of modular wiring). Data Connectors simply plug together to make a connection. This requires that the identical style of connectors be used at the cable end, wall plate, and patch panel. The unique Data Connector was probably designed with Token-Ring in mind, since that has long been IBM's network of choice. It contains a mechanical shorting-bar mechanism that can electrically loop back a cable or jack when disconnected, an operation very similar to that of the Token-Ring MSAU port.

IBM cabling system installations consist of a series of interconnected wiring closets with star wiring to each workstation location. As we mentioned, either STP or UTP pairs or a combination are contained in each station cable. The STP station pairs are terminated in a Data Connector at each end. The Data Connectors have no gender such as the male/female or jack/plug arrangement of most cabling systems. These Data Connectors are mounted in a flush-mount plate at the workstation outlet and in a patch panel (called a Distribution Panel) at the wiring closet, but are otherwise identical in function to those used on cable ends. The Distribution Panel typically contains 64 mounting positions for Data Connectors that terminate the end of the station cables. It is customarily mounted on freestanding 19-inch rails. The Distribution Panel also contains strain reliefs for the station cables. Data Connector style cables, made using Type 6 cable, are used for patch cords in the wiring closet.

The connection at the workstation is a cable with a Data Connector at one end and, at the other end, an appropriate connector for the particular equipment to be connected. By simply terminating the cable with an appropriate connector, the IBM cabling system can actually support quite a variety of data communications equipment, including coax, twinax, RS-232, and Token-Ring or Ethernet. In the case of Token-Ring, a 9-pin d-shell connector is the appropriate connector at the workstation. The workstation's Token-Ring interface is 150 ohms, as is the STP station cable, so no matching balun is needed. At the wiring closet end, many MSAUs are equipped with Data Connector ports and use an ordinary Data Connector patch cable.

At the workstation location, the telecommunications outlet is composed of a faceplate and a regular Data Connector, which terminates the station wire. Type 1 wire terminates in a single Data Connector, while Type 2 wire terminates in one Data Connector and one telephone jack connector. Both connectors are designed to clip into the faceplate. The IBM system offers two basic styles of faceplates. A Type 1 Faceplate has a single Data Connector position, while Type 2 also includes a position for an 8-pin RJ-45 style modular connector for telephone use. Conve-

niently, the Type 1 outlet is used with Type 1 cable, and likewise, the Type 2 outlet with Type 2 cable. Both faceplates are available in surface mount as well as flush mount options. Figure 5.4 shows a typical workstation outlet for use with Type 2 cable. Other vendors offer Data Connector outlets in several other connector combinations.

The cable and connectors have a matching color code to make wiring simple. However, the cable has a thick insulation and shielding that complicates the termination process in comparison to other types of twisted pair wire. This is undoubtedly why even the station cables are often terminated directly into a Data Connector, rather than a punchdown arrangement as is conventional twisted pair wire. Cable is available in both plenum and nonplenum ratings and must be placed appropriately. Termination of the Data Connector is described in Chap. 10.

The cabling system also includes Type 5 fiber optic cable. The fiber and connectors are basically the same as with other wiring methods. Termination and descriptions of fiber optic cable are described elsewhere in this book.

Several other variations of IBM cabling system wire may be found, the most common being Type 3, Type 6, and Type 9 cable. Type 6 and Type 9 are 26 gauge nonplenum and plenum versions of Type 1 cable. Type 3 wire is 105 ohm, 24 gauge, unshielded, twisted 4 pair cable that meets IBM's specifications. It is available in plenum and nonplenum insulation and is similar but not identical to Category 3 cable. When Type 3 wire is used as station cable, a special adapter cable (called a *media filter*) must be used to match the 105-ohm cable to 150-ohm device interfaces, such as Token-Ring interfaces. Some newer Token-Ring interfaces also have a 105-ohm modular jack, which eliminates the need for the media filter. Similar media filters may be used with 100-ohm Category 3, 4, 5, cable. Note: Older media filters may support only 4 Mbps Token-Ring and will fail when used with 16 Mbps rings.

The IBM cabling system provides some important advantages in a mixed application environment that may use coax, twinax, and twisted pair terminations. The cable has been designed to support all these types of signals using the proper adapters or impedance matching baluns.

However, in a LAN-only environment, such as Token-Ring, the functions may all be satisfactorily accomplished by the TIA/EIA-568-B system at a lower cost. In fact, most estimates place the cost of an IBM cabling system installation at two to three times that of a standard TIA/EIA-568-B installation.

Other Standard Variations

Many other vendors have developed proprietary wiring patterns for cabling of data and voice. Most of these plans support specific features of that manufacturer's equipment and are not general purpose at all. Others incorporate variations of the cabling systems previously mentioned. Of these, two other systems are widely used and deserve mention.

NORDX/CDT IBDN

Another variant of twisted pair wiring is the Integrated Building Distribution System put forth by Northern Telecom, often referred to as Nortel. The IBDN components are now marketed by NORDX/CDT. This system is quite similar to the Lucent/AT&T PDS and primarily supports telecommunications wiring. As we saw with PDS, data networks can be run over such systems, but the distance limitations and other installation practices must be observed.

IBDN uses the common system of modular connectors with cross-connects based around the BIX block. IBDN also contains fiber components, as does the PDS. While the IBDN system is not specifically designed to match TIA/EIA-568-B requirements, it will allow common types of network connections if the distance limitations for the network are not exceeded. When used in accordance with the guidelines in the standard, an IBDN installation can be said to be TIA/EIA-568-B compliant.

All of the modular connectors, jacks, cable, and color codes of the AT&T PDS system are common for IBDN. The interconnection component, of course, is the BIX block, rather than the 110 block. Please refer to the PDS section in this chapter and the BIX block section of Chap. 8 for more information.

DECconnect

Created by Digital Equipment Corporation, the DECconnect system supports four communication technologies in an attempt to provide a global wiring system. The four technologies are voice, video, low-speed data, and network. DEC's approach uses five different types of cable that are run

	DECconnect type	Cable	Connector
TABLE 5.4	Voice	4 pair	RJ-45
DECconnect Cable Types	Video	75-ohm coax	F-type
	Low-speed data	2 pair	MMJ (modified RJ-45 type)
	Network	50-ohm coax	BNC
		62.5/125 micron fiber	ST or SMA

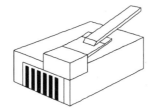

Figure 5.5
The DECconnect system uses a unique MMJ connector with an offset locking tab and keyway.

individually to the DEC faceplate at the communications outlet. The five cable types are listed in Table 5.4, along with the preferred connector.

The complete DECconnect system is rarely used because of its separable cable technologies and its unusual data connector. If you must run a different cable type to each workstation, you might as well keep each technology totally separate. The only advantage the DEC system offers to a mixed media user is a common faceplate mounting for the connectors. The system uses a unique MMJ connector with an offset locking tab and keyway, shown in Fig. 5.5. Unlike a normal keyed jack, with an extra keyway at the side of the opening, the MMJ will not accept a non-MMJ plug at all. It is totally incompatible with all other modular devices, including connectors, jacks, adapters, fan-out cables, patch panels, and test equipment. Special MMJ devices must be used, or an adapter cable must be fashioned. (The first M may stand for "maddening.")

Because of the shortcomings of the DECconnect system, DECconnect wiring is primarily used to connect legacy DEC asynchronous computer terminals using a variant of RS-232/RS-422 wiring. These async terminals use an MMJ connector for the interface. The DECconnect system specifies a clever symmetrical wiring pattern for the data connector that avoids the failure that would result from reversing the order of pin wiring, a common error. Ironically, the most frequent source of this type of error would come from the use of standard telephone cables that flip pins from end to end. These are the very cables that will not plug into the MMJ outlets.

LAN Wiring for the Future

Chapter 6 Highlights

- 100/1000 Mbps networking
- The need for speed
- The importance of category
- "Enhanced" Category X
- Fiber to the front
- The future of copper wiring
- Migrating to higher speeds
- Future (wiring) shock

It has been said that the only thing that remains constant is change itself, and that's certainly true in reference to changes in computer technology. Even so, it is possible to accurately predict the near-term future of LAN wiring technology.

One of the interesting features of technological change in the computer industry is the steady progression of a technology from experimental to leading-edge, and then to widely accepted. While experimental technologies are quite divergent, those that are feasible eventually become the hot new technologies that define the true state of the art. Of course, a new feature or capability does not immediately supplant existing products, but eventually products with useful new features become readily available at a reasonable cost. At that point, standardization emerges, whether formal or de facto, and the product becomes widely accepted.

Picking the best emerging technology for your network is an exciting and potentially rewarding game. The game is really not in trying to guess which experimental developments will become accepted, but rather in observing which leading-edge technologies have begun to be accepted, and then employing them in the future. The creation of recognized standards is the key to which technologies will be used in the future.

In local area networking, we are fortunate because we must operate in a multivendor environment. The need for interoperability means that we must have strong, internationally accepted standards for networking. Another fortunate circumstance is that no proprietary technology dominates the field. Many, if not most, of the technologies that have become widely used are a result of unprecedented industry cooperation through partnerships, consortia, and standards committees. In addition, because of the need for interoperability, standards tend not to change radically once they are implemented. This is particularly true of LAN wiring.

You can expect your LAN wiring system to remain totally usable over many years if your needs remain relatively constant. This is not to say that if you progress from simple printer and file sharing to high-bandwidth graphic or video applications that you might not need to upgrade some network components. However, most local area networks support organizations with relatively constant needs. For example, if you have a standard 100BaseT network in place today, you may reasonably expect it to remain viable for a number of years.

On the other hand, if you are installing new network wiring, you should plan for the future. For LAN wiring, the future is fairly clear. There will be a gradual move from the current 100 Mbps network speeds to 1000 Mbps and higher speeds. There may be no immediate need (or budget) to convert initially to 1000 Mbps, but you should probably not install any new cable that will not serve that speed.

So, how do you plan for the future? How do you put in a 1000 Mbps network? Should you use copper wire or consider fiber? Can you hedge your bets? Let's see.

100/1000 Mbps Networking Compatibility

If you currently are using 10 Mbps Ethernet or 16 Mbps Token-Ring, you may not be using LAN wiring components that will move you to higher speeds such as 100 or 1000 Mbps. These lower speeds are much less demanding in their need for transmission media than the high-speed networks, such as Fast Ethernet.

For example, 10BaseT will run at a workstation-to-hub distance of 100 m on Category 3 cable. If you try to upgrade such a network to 10 times the speed, however, it will probably fail. Gigabit speeds would be impossible.

So, what can you do if you may need to upgrade to 1000 Mbps? The answer is to install a LAN wiring system that meets the performance specifications for the higher speed now; it will operate very satisfactorily at the lower speed for the present and will provide additional service as you upgrade to high speeds.

One sure way to provide such a wiring system is to use the TIA/EIA-568-B standard for a Category 5e or Category 6 cable system.[1] Category 5e wiring is designed to support full 1000 Mbps speeds. At the same time, Category 5e wiring will fully support lower network speeds, such as 10/100BaseT, Token-Ring, and CDDI as well as intermediate speeds for ATM and Fast Token-Ring. For that matter, it will also support voice, ISDN, and T1/E1 circuits very nicely. Table 6.1 shows a table that summarizes the supported uses of Category 3/6 installations.

"True" Category 5e, 6, and, proposed Cat. 7 wiring should be used for a single application, not shared between data and voice, for instance. If your needs are uncertain, you may wish to overwire your workstations by pulling more than one cable to each workstation location. There is a school of thought that holds that total separation of voice and data facilities—

---

[1]Although Category 6 is an emerging standard, most manufacturers already have components that meet the requirements.

TABLE 6.1

Common Applications for EIA/TIA 568 B Category 3, 5e, 6, and 7 Facilities

Category 3	Category 5e	Category 6	Category 7
1.5 Mbps IBM 3270*	All Category 3 applications	All Category 3 applications	All Category 3 applications
2.5 Mbps Arcnet	25 Mbps ATM	All Category 5e applications	All Category 5e applications
4.0 Mbps Token-Ring*	52 Mbps ATM	1000BaseTX	All Category 6 applications
10 Mbps 10BaseT	100BaseTX	1244 Mbps ATM	2.5 Gbps ATM
16 Mbps Token-Ring*	100BaseT4		10 Gbps Ethernet
25 Mbps UNI ATM	100 VG AnyLAN		Baseband video and CATV
52 Mbps UNI ATM	155 Mbps ATM		
100BaseT4	622 Mbps ATM		
100 VG AnyLAN	1000BaseT(4)		
Baseband voice			
ISDN Basic Rate Interface			
ISDN Primary Rate			
T1/E1 carrier (1.544 Mbps)			
RS-232D (partial interface)			
RS-422			
Baseband video*			

*Application may require baluns or special interfaces.

down to separate wiring closets and equipment rooms—is the safest course. This certainly might make sense if your telephone system and LAN system are maintained by separate entities. Standard practices for telephone wire installation are much looser than the strict requirements for a Category 5e/6 data installation. Many LAN managers have had bad experiences when the two types of installation methods are mixed. However, if all of your cabling is installed and/or maintained by the same people, you should be able to ensure that Category 5e/6 wiring stays Category 5e/6.

Ensuring the performance to Category 5e/6 standards requires much more than just Category 5e/6 components. Location, routing, and workmanship issues are at least as important. These three issues are difficult to judge because they are related to the installation process. You can buy the finest certified components and still have them installed improperly. That is why we have devoted an entire part of this book to the installation of the wiring devices.

Category 5e/6 Testing and Certification

The simple proof of 100/1000 Mbps network performance is to connect your 100/1000 Mbps switch-hubs and workstations and see if they work. Unfortunately, many of us must install cabling systems long before we buy the first 100 Mbps hardware. Even if high-speed hardware were to be installed immediately, the time to cure cable problems is during the build-out stage of construction, not after move-in.

The solution is to have the cable system tested and certified as part of the installation process. This testing and certification process is described in more detail in Chap. 16, but we can summarize it here.

The first step in certification is to use the proper materials. All cables, jacks, patch panels, patch and user cords, and connecting blocks should be certified to the proper category by the manufacturer. There is no major benefit to specifying all this by brand name unless you are using a particular manufacturer's connecting hardware system that offers unique features. For a smaller installation of 50 station drops or less, this will probably just raise the price and narrow the field of bidders.

The second step is to use properly trained installers, particularly for Category 5e/6 installations. Several certification programs exist for cable installers. The vendor-specific programs generally emphasize a common set of workmanship techniques and are adequate training in most situations. You should not insist that installers be certified by a particular vendor unless you are willing to put up with the extra expense and aggravation of such a policy; it is more important that each installer either be formally trained at a recent course, or be closely supervised during the installation.

A few industry-wide certification programs exist, including the Registered Communications Distribution Designer (RCDD) program from BICSI, Inc. (formerly the Building Industry Consulting Service International, Inc.). The RCDD requires experience, successful completion of a test, and membership in BICSI. A LAN Specialization program also exists. Plans are under way to introduce an installer certification program that does not require membership in the organization. See Chap. 19 for additional information on training and certification.

The third step is to individually test and certify each workstation link. The worst-case link performance criteria and testing guidelines of TIA/EIA-568-B (TSB-67) are an excellent basis for certification. Each link should pass all the tests specified for the category of the installation and you should get a printed report. Keep in mind that you will have to pay extra to have each link scanned. The equipment to perform the testing is expensive and the installer will expect to be compensated.

The Need for Speed

"Why do I need higher network speeds?" you might ask. The answer lies in understanding the increasing size of data transfers in the modern network. The common word processing tasks have modest file transfer requirements, even if files are maintained on a server. However, we are doing many more tasks than letter writing today, and these new applications have placed great demands on our network transfer rates.

For example, the average user today accesses graphics-rich Internet Web sites, often with audio and video multimedia content. We routinely attach fairly large documents, such as Word™ documents and Power-Point™ files in our e-mails. Businesses are increasingly putting voice traffic onto digital networks. And applications such as video teleconferencing and telemedicine have high constant-bandwidth requirements. In fact, digitized x-rays and CT scans are routinely being transmitted over the networks of medical providers. Many cities now offer real-time video of roadway traffic, and broadcast television and entertainment performances are close behind.

How much bandwidth is needed for efficient networking? That depends on the application and the perception of the user. The chart in Table 6.2 shows some comparative applications and the amount of bandwidth each would need to complete its transmission in one minute. Now imagine staring at your computer monitor for 60 seconds while you are waiting for this application to load. It seems a lot longer than you would normally tolerate, at least without some choice words under your breath.

You can also scale the numbers a little bit to bring them into the reality of your network. For example, if the magazine clip were only 10 pages, it would need 40 Mbps of network bandwidth to load in 1 minute. Or it would need 240 Mbps to load in 10 seconds, which might be much more tolerable. However, do you have a 100 Mbps network

TABLE 6.2

Network Speed
Required for File
Transfers in 1
Minute

File size	File application	1-minute transfer rate*
300 kbytes	Web page with graphics	40 kbps
3 Mbytes	PowerPoint™ file	400 kbps
30 Mbytes	MP3™ audio—30 min.	4 Mbps
300 Mbytes	DVD clip—9 min	40 Mbps
3 Gbytes	Magazine—100 pages (MPEG-2)	400 Mbps
30 Gbytes	CT scan, three views	4 Gbps

*Since 1 byte = 8 bits, 300 kbytes × 8 bits/byte = 2400 kbits, and 2400 kbits ÷ 60 seconds = 40 kbps (kilobits per second).

now? What impact does devoting 40% of your network's speed to one transfer have? The generally accepted utilization limit for Ethernet networks is 60 to 70% utilization. So, the 40 Mbps transfer would eat up about two-thirds of your network capacity, and the 240 Mbps would chew out about a 30% hunk of a Gigabit Ethernet pipe.

The Importance of Category

Installing cabling and components of the proper category is important for future upgrades in the data rate of your network. However, network data rates and cable bandwidth do not directly correspond to each other. Data rates, of course, are given in bits per second, typically in megabits per second (Mbps) or even gigabits per second (Gbps). Cabling components (including cable and connectors) are specified in terms of useful bandwidth in hertz (cycles per second), typically in megahertz (MHz) or in gigahertz (GHz). Data bits are usually encoded in a way that reduces the actual bandwidth requirement for a given data rate. For example, 100 Mbps Fast Ethernet actually requires about 30 MHz of bandwidth, because of the encoding. Similarly, ATM 155 uses an encoding technique called CAP-64 to place 155.52 Mbps on the cable at less than 30 MHz.

This is the reason that some data networking standards can operate at much higher data rates than would seem to be possible, given the cable's nominal bandwidth. Thus, we can put a 1 Gbps (1000 Mbps) data rate over cable that is rated for a 100 MHz bandwidth on each of its four pairs. Now, this is no easy feat. The interface and digital signal processing circuitry to accomplish this with copper wire is very sophisticated. Any minor disturbance in the cabling components can cause such a data link to fail. That is why it is so important to use the proper category of components for the application.

Table 6.3 shows the improvements that have been made to cable specifications to enhance the data-carrying capacity. The introduction of Category 5 expanded the useful operation frequency (bandwidth) to 100 MHz from 20 MHz. TSB-95 added testing requirements for older Category 5 links to near-5e performance in order to predict possible operation for Gigabit Ethernet, although the base bandwidth stayed at 100 MHz. Category 5e increased the NEXT isolation by 3 dB, incorporated the power-sum requirements for multiple disturbers (transmission on more than one pair), added far-end crosstalk requirements (FEXT), added propagation delay and skew, and added return loss to the more modest Category 5 parameters.

Category 6 and Category 7 increase working bandwidth to 200 MHz and 450 MHz, respectively. You may have heard these numbers as 250 and 600 MHz, but those are the required test limits. The attenuation-to-crosstalk ratio (ACR) must be positive to the lower frequency limit, and that is the generally accepted measure of the "useful bandwidth" of cables and connecting hardware. Some Category 7 hardware actually exceeds even the 600 MHz specification, but it is unclear whether that standard will be split or the higher limit will be renamed.

TABLE 6.3

Comparison of Category Parameters

	Cat 3	Cat 5	TSB-95	Cat 5e	Cat 6	Cat 7
Max. operating frequency, MHz	20	100	100	100	200	475
Test frequency, MHz	20	100	100	100	250	600
Attenuation	←			Yes		→
NEXT	←		Yes →	←	+3db	→
PS-NEXT				←	Yes	→
PS-ELFEXT			←		Yes	→
Propagation delay			←		Yes	→
Delay skew			←		Yes	→
Return loss			Yes	←	+2dB	→

"Enhanced" Category X

The categories and classes of operation of the international cabling standards organizations specify minimum performance parameters for cabling, components, and links used in structured LAN wiring systems. It is the purpose of the standard to set electrical characteristics of the wiring system, without regard to the particular type of LAN network technology that is used with the cabling.

Each of the category specifications, in reality, has been targeted at the operating requirements of a current or anticipated LAN topology. Consequently, Category 5 conveniently handles Fast Ethernet, Category 5e handles Gigabit Ethernet (4 pair simultaneous transmit/receive), and Category 6 handles the 2 pair transmit/2 pair receive version of Gigabit Ethernet, as well as ATM-622/1244 (OC-12 and OC-24 rates). With those technical guidelines, each category/class allows a wide range of applications in addition to the classic Ethernet ones.

However, it is quite possible and probably desirable for a manufacturer to build cabling components that actually exceed the minimum specifications of a particular targeted category. For example, at one time Category 5 was considered the ultimate and was a manufacturing challenge. Soon the leading manufacturers discovered that they could use innovative designs and production methods not only to comply with Category 5, but to actually exceed the parameter requirements, often by a significant margin. These better quality cables and connectors were initially called **Enhanced Category 5** and ultimately were found to be all but necessary to achieve Gigabit Ethernet operation. One forward-looking component distributor (Anixter) even promulgated an advanced specification that defined "levels" above minimal Category 5. Eventually, the standard was also enhanced, and Category 5e was not only born, but shoved conventional Category 5 aside.

The lesson here is that good engineering can create components that have better performance than the standards require. Can you benefit from those enhancements? Yes you can, particularly if you operate near the limits of distance or data rate. Better still, you may be able to plug in new technologies that were not even dreamed of when the cable standard was defined. Those of us who specified Enhanced Category 5 were rewarded with the ability to jump to Gigabit Ethernet over copper as soon as it became available, because our cable already met the TSB-95 and/or Category 5e performance levels. By the same token, enhanced cable and components for Category 6 already exist, and may well prove to offer more long-term stability as well.

You should buy the best cabling components you can afford, keeping in mind that the network technologies of today are transient and will be obsolete in a few years. Stepping beyond the envelope may be a little risky, as some predicted technologies never reach critical market mass. So, try to plan for one or two steps in increased network speed and leave the rocket science for the experimenters. In that way, you will have a network infrastructure that will provide optimum performance at minimum cost with a reasonable useful life.

Fiber to the Front

The importance of fiber optic cable cannot be overemphasized. Fiber cable has unique characteristics that make it suitable in many types of high-bandwidth, high-interference, and interbuilding applications. As recent very high-speed technologies have emerged, fiber interfaces have been developed and marketed considerably in advance of their copper counterparts. Whereas copper cables are limited to 100 m for almost all applications above 100 Mbps, such a short distance is trivial for fiber. Although structured standards still spec the 100-m limit, it would be just as practical to double or triple this distance limit for even the most rigorous high-speed network requirements. Not only has multimode fiber been able to handle 100 and 1000 Mbps networking, but it can also support 10 Gbps at over 300 m!

Fiber connector technology has been refined to greatly reduce interconnection losses, and to minimize reflections at the connections. This means that systems that use full-duplex transmission/reception on each fiber are feasible, and that the fiber may be used in critical applications such as CATV where minor reflections can cause serious problems. In addition, quick-connect fiber connectors are being offered that decrease costs, assembly labor, and required training for fiber termination.

The popular TIA/EIA-568-B standard still requires only two cables to each workstation. One must be Category 3 or above, and is presumably targeted for conventional telephone connection (either analog or digital ISDN). The other cable may be Category 5e or better, or it may be an accepted fiber. Why not choose both? Until IP telephony is in widespread use, you will probably still need to provide the Category 3 or better cable for telephone use. But why choose between copper and fiber for your data?

Here is an ideal scenario for LAN wiring for your future. Run two Category 6 (or 5e) cables and a 2 pair fiber optic cable to each work area outlet. You could actually use a fiber cable with one multimode pair and one single mode pair if you really wanted to push the envelope. Or you could use one 62.5/125 pair and one 50/125 pair, both multimode. Use a modular, four-jack connector plate. Now you have two copper cables, either of which will run virtually all voice, and high-speed data technologies, and you have the flexibility of using fiber for high-speed data or even video.

Chapter 12 contains a lot of additional information on fiber optic cabling options. Fiber optic cabling is clearly in our networking future.

The Future of Copper Wiring

In a paraphrase of Mark Twain, the report of the demise of copper wire is greatly exaggerated. There has been a great deal of press for several years that copper wire cabling should be abandoned in favor of either fiber optic cabling or wireless networking. While the claims of the proponents are not without merit, they are also not unbiased. The loudest supporters of copperless networking are the companies that produce those competing technologies. It is only natural that they would see their products as being technologically superior to older methods.

Truth, as often is the case, probably lies between the extremes. There is a certain philosophical attractiveness to alternatives for copper wire. The fiber alternative offers a considerable increase in bandwidth, which translates to an increase in LAN data rates. Unfortunately, it may come at greater cost and more limited flexibility. No one would argue that it would be difficult to adapt the hub, patch panel, and workstation outlet to fiber. Fiber has its own set of installation and testing difficulties, in addition to a higher component expense. Most of your existing networking equipment would have to be scrapped and replaced with optically interfaced equipment and adapters. Advantages in one area may be disadvantages in another. For example, the voltage isolation inherent in fiber optic cabling would make it impossible to centrally power telephone instruments, so communications could not be maintained in a power outage.

The solution, for the time being, is that fiber has an important place in telecommunications systems, but it will not immediately supplant copper wire as the cable of choice within buildings, if one must choose between the two. Fiber will continue to provide interbuilding links, links between far-flung hubs, and links needing greater bandwidth/distance performance than copper can provide.

The wireless alternative is an intriguing one. It eliminates the need for any direct placement of cable, whether copper or fiber. It is heavily promoted as the future of networking. However, it has some inherent disadvantages. The first disadvantage is that it knows no bounds. That means that your network connection signal can travel far beyond your immediate office (even outside your building). Not only is the signal transmitted where it is not wanted, but the wireless LAN adapter is also susceptible to outside receive interference. Second, as more and more wireless devices are placed in close proximity ("close" can be hundreds of meters in distance), more interfering signal clashes occur. Do you remember the early days of cellular radio, when the service actually met its claims of clear, interference-free communication from almost anywhere? The addition of thousands of users has greatly degraded the quality of reception. The same degradation can happen to crowded wireless LANs, except that the interference cannot be heard; it simply slows down the network in silence.

The third disadvantage is troubleshooting and management. A direct cable network is easy to manage and repair. With modern hubs, problems in the network cabling are easily isolated to one station, and monitoring of the process is straightforward. Finding a radio-frequency problem in a random, deterministic, virtual network is a much more challenging task. Fourth, wireless networking is more expensive in relation to direct cabling. Wireless network adapters are still usually 2 to 4 times the cost of copper cabling systems, including the copper adapters.

Wireless networking, however, has a clear advantage in places where running cable is impractical, impossible, or prohibitively expensive. It may be used as a very cost-effective link between buildings, especially across public rights-of-way. Practical or aesthetic reasons may tilt the scales toward wireless. For instance, a connection in a large arena or auditorium might be much easier to accomplish with a wireless link. Finally, if the workstation is mobile, wireless is crucial. (Picture a forklift-mounted workstation with a network cable trailing along behind!) Check Chap. 14 for more wireless information.

No, the demise of copper wiring is not imminent, but neither does it meet all possible needs. Each technology should be applied where it is the most appropriate and cost-effective.

Future (Wiring) Shock

Alvin Toffler, in his famous work, *Future Shock,* spoke of the pace of technological advancement as a shock to the individual and to society. In some ways, we in the computer industry have become accustomed to the rapid obsolescence of our equipment and methods. Computers, printers, and even network switches are relatively easy to upgrade or replace, as they are movable objects, attached only by their wire connections. LAN wiring is more of an immovable object, fixed into place within walls, ceilings, and cubicles. What can be the result when the irresistible force of change meets your wiring? How can you prevent this from happening?

Migrating to Higher Speeds

The Queen kept crying "Faster! Faster!"
..."Are we nearly there?" Alice managed to pant out at last.
"Nearly there!" the Queen repeated. "Why, we passed it ten minutes ago! Faster!"

Higher network speeds are upon us whether we like it or not. Actually, many data-transfer applications have begun to overload some networks, so the change may be welcome indeed. However, few network managers have the luxury of being able to totally upgrade their networks to the higher speeds of 1000 Mbps or higher. These networks will have to be gradually migrated to higher data rates, as money and the availability of technology allow.

Migrating to higher speed networks is a much broader subject than the LAN wiring issues being covered in this book. So, for simplicity, we will outline only the wiring implications of high-speed networking.

How can you plan a migration to higher-speed network cabling? First, unless you are designing a completely new facility, or planning to replace all of your existing network cabling, you should find where you are right now. For example, if you currently have a Category 5e facility, you can upgrade from 100 to 1000 Mbps easily. You will not be able to use 1000BaseTX or ATM-1244 at all, without replacing your cable with new Category 6 cable. On the other hand, if you already have Category 6 cable, you can upgrade to any of the 1000 Mbps technologies—that is, if that cable was installed to Category 6 standards.

If you already have fiber installed, you are probably set for any of the currently envisioned higher speeds. At the worst case, you may have to adapt the connectors or reterminate the fibers, replacing SC connectors with newer SFF connectors. You can also get to the higher speeds by installing new fiber, or by including a fiber run along with conventional twisted pair to each work-station area.

New installations that will initially employ lower-speed networking have a special problem. How can the network manager plan for the future implementation of these higher speeds when the exact technology choice is not yet known? The best bet is to put in the highest grade facility you can afford. It would be ideal to pull both fiber and copper to each work area. However, if your budget limits you to twisted pair, you should at least put in a full Category 6 cabling system, so you will get all the bandwidth possible. And remember, any decisions you make now may still not keep up with the future, but perhaps they will at least last through the 5- to 15-year useful life projected for most cabling systems.

"A slow sort of country!" said the Queen. "Now, here, you see, it takes all the running you can do, to keep in the same place. If you want to get somewhere else, you must run at least twice as fast as that!"

...and Alice began to remember that she was a Pawn, and that it would soon be time for her to move.

Lewis Carroll
(Through the Looking Glass)

Well, one thing you can do is to plan for the future. Design new cable systems so they meet tomorrow's needs as well as today's. If you are installing a 100BaseT network, that means you should install a cable facility that can go to the next step, 1000 Mbps. If you have an existing facility, add Category 6 cabling and hardware when you add workstation drops. Plan the additions so the old cable can eventually be converted to the new system. A graceful migration such as this can be accomplished over a period of time to minimize the budget impact.

Proper planning can give a cable system an extended life. Providing for the future may cost a little more now, but it will simplify change when it eventually comes.

The Tubing Pull

In some ways, it is impossible to predict every twist and turn technology will make. It would be nice to have some way of replacing the wiring periodically if needed. You might need to upgrade to fiber optic cable. Maybe each workstation will need another cable and outlet to add an application such as video. What if you could simply go to the telecommunications room and pull in some new cable?

It may be possible to do just that. A technique that works well in closed-ceiling buildings can also be extended to conventional grid-ceiling buildings. As the building is constructed, continuous plastic tubing home runs are made from the telecommunications room to a flush-mounted outlet box at each workstation. A pull string may be blown through the tube, or pulled in with a snake. The cable is then pulled through the tubing to the outlet and terminated normally on both ends. As needs change, new cables are pulled through the tube using the old cable as a pull string. One type of compatible tubing is called "3/4 Flex" in the electrical trades. Another thinner tubing is used in so-called blown-in fiber installations.

This method works very well in closed construction, which is very expensive to rewire. It also works in smaller offices where the cable runs are not too long. It may have a special advantage for Category 5e/6 runs, as it avoids the performance-robbing problems of kinks, minimum bend radius, tight tie-wraps, and metallic runs along pipes and structural steel.

2

LAN Wiring Technology

Work Area Outlets

Chapter 7 Highlights

- Outlet jack mountings
- Modular jacks for twisted pair
- Standard jack pin-outs
- Fiber optic outlet jacks
- Which fiber connector?
- Other outlet jack types
- Marking outlets
- Workmanship

In this chapter, we begin to go through the chain of wiring devices that make up a LAN wiring connection. The first item in our list is the outlet placed at the workstation area. The TIA/EIA-568-B standard requires that each work area be provided with two outlet jacks. One jack must be a 4 pair 100-ohm UTP or STP connection of Category 3 or better. The other jack may be Category 5e or higher 4 pair 100-ohm UTP/STP connection, or it may be either a 2 fiber 50/125 μm or a 2 fiber 62.5/125 μm fiber optic connection.

The work area outlet is referred to by several names. The TIA/EIA-568-B standard refers to the outlet as a telecommunications outlet/connector in the work area, to emphasize that it must include the actual connector jack itself. We refer to the outlet in this book by the name *work area outlet* or *station outlet*. This usage implies that the connector jack is not directly attached to the workstation or other network device. Such an attachment is made by an additional cable called a *user cord* or *equipment cord*. The modular or other jack typically attaches to a mounting plate or faceplate, which mounts in an enclosure. The entire assembly, the jack, plate, and enclosure, forms the station outlet.

We will cover the mountings and jack types first. We will also discuss mounting methods for the jack/faceplate assemblies, including surface and flush mountings. Finally, we will detail outlet identification and workmanship issues.

Outlet Jack Mountings

The connector jack at the work area outlet is mounted in several ways. We will cover the mounting first because it is useful to have an understanding of connector-mounting methods when we actually begin to discuss the jack types in the next section. In this discussion of jack mountings, we are actually talking about the faceplate/jack assembly in those instances where a separate faceplate is used.

Flush Mounts

Flush-mounted jacks are the familiar wall-mounted jacks we often associate with commercial telecommunications wiring. This type of jack arrangement places the jack faceplate flush (or even) with the wall. Typically, the jack extends into the wall and mounts via machine screws to an underlying outlet box or mounting ring.

Mounting Rings and Outlet Boxes The mounting ring or outlet box provides a secure location for mounting the work area outlet. Several types of each are available. Generally, either type of mounting is acceptable. Telecommunications wiring is a low-voltage, low-amperage wiring that is not required to be routed in conduit or mounted in an enclosure by the NEC. However, your local construction codes may require that the outlet be mounted in an enclosed box in some cases.

Traditional outlet boxes are the same design as those used for electrical circuit wiring. The boxes are available in metallic or plastic styles and different designs are available for prewiring, during initial construction, and for postwiring, when the building is complete. Figure 7.1 shows several types of outlet boxes that are suitable for telecommunications use. The 2-in-wide box is the most common for this style of wiring, although a 4-in-square box with a 2-in-wide mounting ring may also be used. In general, the telecommunications outlet plates that mount to these boxes are designed for the 2-in opening. The opening sizes are also referred to as "1 gang" (2 in wide) and "2 gang" (4 in wide) in the building trades.

Figure 7.1
Several types of outlet boxes that are suitable for telecommunications use.

Outlet boxes for prewire use will have nail plates for attachment to vertical supports or studs in the walls. Some of these boxes have captive nails already attached. The box should be securely attached to its support and positioned to allow for clearance through the wall sheathing. After the wall is complete, the mounting plates of the box should be even with the wall surface. The boxes should be positioned at the same height as required for electrical outlets. This might not necessarily be a construction code requirement, but it enhances the appearance of the final installation. The cable may be run through a bare knock-out hole in the box or through a conduit or flexible cable connector. Plastic sleeves are also available to minimize any chance that the cable might be damaged by the sharp hole edges.

Outlet boxes for postwiring use are also available. These boxes are sometimes called *old work* or *sheetrock* boxes. Because they mount in hollow walls that have already been constructed, they must be inserted from the front of the wall. An opening is cut in the wall big enough to insert the box body. Additional room may be required for the mounting device that secures the box to the wallboard. These mounting devices are usually screw-tightened to clamp the mounting flange of the box to the wall.

Another type of flush mounting device is the mounting ring. Mounting rings are available in several varieties, as shown in Fig. 7.2. Standard mounting rings provide a target location for telecommunications outlets during construction. These rings have mounting plates that allow them to be attached to the wall's vertical supports. The mounting ring normally has a 2-in by 4-in opening that is formed into a flange. The flange protrudes forward from the wall support and is the right depth to be flush with the front of the wall surface, when it is later attached. The top and bottom edges of the ring have mounting holes to which the mounting screws of the outlet attach.

Mounting rings for postwiring are also available. These rings have flexible metal fingers at the top and bottom of the opening that are part of the stamping. A rectangular hole of an appropriate size is cut into the wallboard. Then, the mounting ring is placed in the hole until it is flush with the wall. The metal fingers are bent around into the hole and back against the rear surface of the wallboard to secure the ring. An alternative design has a vertical screw slot incorporated into the metal fingers. After the fingers are bent into the mounting hole, a provided screw clamps the front flange of the ring tightly to the wall, using the slotted finger as a rear clamp. This prevents any undesired movement in the ring that would occur with the other type of ring.

Prewire and Postwire Methods Prewiring may be accomplished with actual cable runs or with a combination of pull strings and late-stage cable runs. If the actual cable is used for the prewire, it must be installed at the stage between wall support installation and wall sheathing attachment. This is often referred to as the *rough-in stage*. With the wall supports in place, the cable is run from each wiring closet to each served station outlet box or ring and either coiled inside the box or

secured to the ring. After the wall sheathing is in place, and usually after the wall has been finish-painted or papered, each cable is trimmed to working length and terminated in the jack. The assembled outlet is then mounted in the box or ring.

Using the actual cable in the rough-in stage can present some problems if you are planning a full Category 5e or Category 6 installation. The cable is left exposed and subject to damage by others during the construction. In addition, you must coil up the wire very carefully to avoid the sharp bends and kinks that can disturb cable performance at this category of operation.

Another technique is to leave a pull string in the wall during the rough-in stage—placed in a hollow wall or inside a riser that connects to the outlet box. Outlet boxes may have vertical risers (conduit) that run from the top of the box, through the wall's top plate, and half a foot or so into the plenum space above the ceiling grid. A pull string may be run through the conduit to assist in cable installation. If a mounting ring is used for the outlet, a pull string is particularly convenient. The opposite end of the string should be secured to anything above the top plate, where it will be visible to the cable installer. After the wall construction is finished, and preferably before the ceiling tiles are in place, the cables are run from the wiring closet to each outlet's pull string, attached, and pulled into the outlet where they may be terminated.

Pull strings and conduit are effective ways to run cable into difficult outlet locations, such as below windows or in columns. Cable may also be coiled and secured above a planned outlet location for later pulling into an outlet box or ring. Again, if you are planning a Category 5e/6 installation, be careful not to put sharp bends or kinks into the cable. Tie wraps may be used sparingly, but should not be overly tightened.

You should take care when running your cable into an outlet box so that you do not slice through the cable jacket. Metallic boxes often have sharp edges as a result of the way they are manufactured. These sharp places can cut into the cable and short some of the wires. Mounting rings may present other hazards to cable, such as the sharp screws that are used to secure some of the rings to the wall sheathing.

Surface Mounts

What do you do when you cannot mount the outlet jack inside a hollow wall? You can use an outlet box that is specially designed for mounting on the surface of a wall. These mountings are called *surface-mount* boxes and come in two varieties, self-contained and surface adapter boxes.

The self-contained surface-mount outlet contains one or more connector jacks for the type of telecommunications or LAN service you are installing. These boxes are often called *biscuit blocks* because of their appearance. A typical surface box, shown in Fig. 7.3, contains two modular jacks, one that may be used for data and one for phone. The

TIA/EIA-568-B standard calls for two circuit links to each workstation, but leaves the use of the links up to the user. Both should be fully wired, 8-position (RJ style) modular jacks to satisfy the standard, although you may wish to make one a 6-position (RJ-11 style) jack if your telephone system calls for that. This is particularly appropriate if you choose to totally separate the telephone and data wiring, as we discussed earlier.

Self-contained outlet boxes are appropriate for solid walls, modular furniture, and other difficult-to-reach areas. Care should be taken to protect the wire with a surface raceway if it is exposed. Providing strain relief to the cable is a significant problem with these boxes. Standard practice is to tightly secure the cable at the point of entry to the box, with a tie-wrap or a knot in the cable. This would obviously distort the cable jacket or break the minimum bend radius of the cable allowed by the standard for Category 5e/6. If the box can be secured to a surface, a better approach might be to use a surface raceway to protect the cable up to the point of entry to the box and leave the cable entry relatively loose. These surface raceways are described below.

The other variety of surface-mount box, shown in Fig. 7.4, is a rectangular plastic box with a 2-in by 4-in opening for a standard, flush-mount outlet plate. These surface adapter boxes are really part of an elaborate surface raceway system that is technically suitable for electrical

Figure 7.3
A typical surface box containing two modular jacks, one that may be used for data and one for phone. (*Courtesy of Leviton Telecom.*)

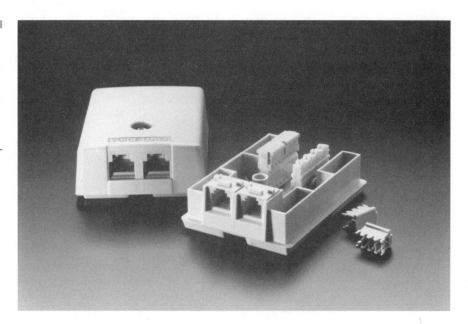

Figure 7.4
Rectangular plastic
box for a standard
flush-mount outlet
plate.

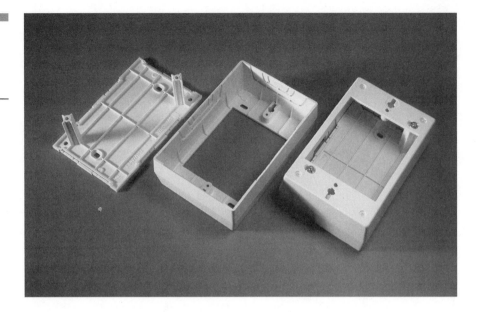

wiring in addition to telecommunications wiring. The system includes long, flat plastic raceways that may be joined together with a variety of couplings, including corners and straight couplers. Surface raceways are available from several manufacturers. An older system of metal raceways and boxes exists, but the plastic ones are much easier to use and are permitted in almost all jurisdictions for Class 2 or 3 wire or telecommunications wiring. Many of the plastic raceways have an adhesive foam tape backing for quick and easy mounting to the wall or other surface along which they run. The boxes may also come from foam mountings, but you are better advised to secure them to the wall with screws. The screws also help stabilize the entire raceway system.

Other plastic raceways are available that do not interlock with a surface box. These raceways use the same type of foam-tape mounting, but snap closed after installation so that the wires can be slipped in at the side of the raceway after it is stuck to the wall. These raceways are more appropriate for the self-contained surface outlets.

A note of caution should be stated regarding these raceways and matching boxes. The safety standards of the NEC and other agencies do not allow telecommunications wiring to share any raceway space or outlet box with Class 1 wiring. Class 1 wiring is wiring for lighting or electrical power circuits, the typical line power that supplies lights and electrical outlets. In some narrow cases, the wiring may be allowed in the same outer raceway cover, but it must be separated by a specific type of partition. See the code books for a detailed explanation of when and where this can be done. Bear in mind that the run-of-the-mill square power pole that drops from a suspended ceiling to a work area rarely meets this test, even though there is plenty of extra room inside for the telecommunications cable. Some modular furniture may have the same problem, as the wiring channel may be occupied by Class 1 wiring. In such cases, you may have to run the telecommunications wiring external to the wiring channel. Several varieties of multimedia (copper and fiber) outlet boxes are shown in Fig. 7.5.

Modular Jacks for Twisted Pair

Modular outlet jacks for twisted pair wiring have been around since modular connectors were first introduced several decades ago. The first of these jacks had loose wires from the connector to screw terminals mounted to the body of the connector assembly. Because these jacks were for analog telephone use, at very low frequencies, little consideration was given to the crosstalk and impedance issues that are very important at today's higher frequencies.

Jacks for LAN use are very different from these old telephone styles. In this section, we will describe the various types of jacks that are used for twisted pair LAN wiring. For our purposes here, the jack will include the modular connector itself, the means of terminating a twisted pair cable, and the wiring in between.

Jack Plate Types

There are three basic types of mounting plates for modular jacks: fixed plates, modular plates, and modular assemblies. Don't be confused by

Figure 7.5
Examples of flush-mounted and surface-mounted multimedia (copper and fiber) outlet boxes.

the word "modular" here, as we are referring to the fact that some of these devices snap together with different options.

Fixed jack plates are the traditional type of plate. These plates are made of molded plastic and include the opening for the modular connector as well as a mounting for the connector and termination components. The most common fixed jack plates have one or two connector positions and are often referred to as *simplex* or *duplex* jacks (respectively). The plate opening and its related molded structures form an integral part of the connector assembly. The connector cannot be separated from the plate and used alone. This is the reason the jack plate is called *fixed*. The supports for the connector terminations are also usually molded into the plate. The most common type of termination is a screw and washers that attach to a molded post on the back of the plate. An example of a fixed jack plate is shown in Fig. 7.6.

With the advent of structured wiring and the tough requirements of Category 5, modular jack plates and their associated connectors have become much more common than fixed jack plates. These modular plates have simple (usually square) molded openings into which are snapped a variety of connector modules. The advantage to this modular

Figure 7.6
An example of a fixed jack plate.

approach is that the connector style, terminations, and even color are independent of the mounting plate. Figure 7.7 shows the modular jack plate and connectors.

Plates are available to hold from one to six connector modules, so you can mix and match connector options. This means that you can decide to mix 6 and 8 position jacks on the same plate and in any order. Some of the modular systems even have modules for coax and fiber. You could have a phone jack, LAN jack, CATV jack, and fiber optic jack all on the same plate. Alternatively, you could have phone and LAN connections for two workstation positions all on the same jack plate. While the standards make no requirement for multiple module outlets, they certainly do not prohibit them. Just be careful to include a larger outlet box behind a plate with several connectors. You will need a space to coil up all that cable.

Figure 7.7
A modular jack plate and connector inserts. (*Courtesy of Leviton Telecom.*)

The connector module includes the modular connector jack and the terminal mechanism for connecting the cable. Termination is accomplished by several types of IDC connectors, as we will describe later. The construction of the connector/termination module generally produces much lower noise and crosstalk than with conventional fixed jack plates. For this reason, these jacks are preferred for LAN wiring, even in a Category 3 installation. They are also much easier and quicker to terminate than conventional screw terminals. Notice we did not say that they were less expensive.

Another variety of jack plate is a two-piece modular jack assembly. These assemblies are unique in that they separate the wire termination function completely from the connector function. A typical two-piece assembly will include a backplate module that the wire terminator snaps into after the cable is terminated. The terminator module is essentially a special type of connector that mates with a printed circuit board-edge contact area. This arrangement is commonly called an *edge connector.* The back module then mounts into the outlet box with the normal mounting screws. To finish the assembly, one or more connector modules, each with a PC-board edge contact, plug into the terminator module at the rear of the back module. A faceplate is then attached to secure the connector modules in place. This type of assembly has the unique feature that the front connectors may be changed without reterminating the cable. For example, you could change from a single 8-position connector to a dual 8-position one, or change to a 6-position or even coax module. Of course fiber connectors are out, because it is cable that is terminated, not fiber. Figure 7.8 shows an example of a two-piece modular jack assembly.

Modular jacks are often referred to as "RJ-45" jacks. This is not really the correct moniker, although it is in very common use. Because all modern LAN wiring uses the 8-pin (8 position, 8 conductor) modular plug and jack, we will call these conductors "8-pin modular" or just "modular." For a more complete discussion, see the box "RJ-45, What's in a Name" in Chap. 10.

Termination Types

Two basic types of wire termination are used with modular jacks for twisted pair wire. These are the screw terminal and the insulation-displacement connector (IDC). The screw termination is the original type that was used with telephone systems. It is still found in some LAN

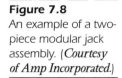

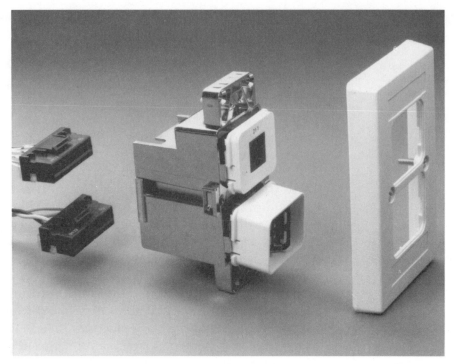

wiring installations, but is not recommended for Category 4 and 5 operation and certainly not Category 5e or 6. As a matter of fact, you probably will never see screw-terminal connectors certified above Category 3, if they are certified at all. You definitely do get a better connection, from the viewpoint of LAN transmission, with the other type of termination. For that reason, many installers do not recommend screw terminals for any LAN work. Of course, you do not need to rush around and pull out all of your old screw terminal plates that might be in your network now. Just remember that they cannot perform above Category 3 and that old installations may have "space-wired" station cable that is untwisted or crossed before it even gets to the screw terminals.

Nevertheless, there is still a right and wrong way to terminate wire on screw terminals. Proper wire termination for screw terminals is illustrated in Fig. 7.9. The screw terminal consists of a screw, two washers, and the plastic post of the jack plate. The jack's connector wires are normally terminated in an open-end spade lug that is positioned between the plastic post and the lower of the two washers. To terminate the station cable, the screw is first loosened. Each station wire's insulation should be stripped back about $3/_8$-inch and the bare copper portion wrapped between the two washers. The wire may be preformed before it is placed into position. The screw is tightened moderately and electrical contact is made in the sandwich made by the washers and the plastic post. The proper direction for the wrap is clockwise around the screw (and it must be between the two washers). No insulation should be between the washers and the wire should not wrap back across itself (to avoid breaking the wire when the screw is tightened). Some installers strip an inch or more and wrap the wire around the screw (with no overwrap). The excess wire should be trimmed after the screw is tightened. It is also a good idea for LAN circuits to try to maintain as much of the pair twist as possible, up to the point of termination. Some authorities recommend trying to limit the untwist of Category 3 cable to no more than 1.5 to 2 in.

The other type of wire termination is the IDC contact. This contact avoids stripping the insulation manually, by cutting through the insulation during the termination process. The wire, cut to length, is inserted into the contact area and snapped into place, either by a special tool or by snapping a plastic cap onto the IDC contact. Either action causes a small area of insulation to be pierced and electrical connection made with the underlying contact. There are three common types of IDC contact mechanisms: the 66 type, the 110 type, and the snap-in type. Other proprietary mechanisms include BIX and Krone types.

An older type of IDC jack contact field is fashioned from a shortened version of the traditional 66-type punchdown block. In this jack type, the wires of the modular connector are connected to positions on the block into which wires from the station cable are punched. The amount of untwisted wire from the block to the connector is usually

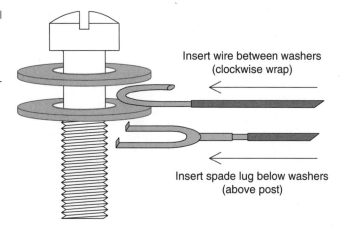

Figure 7.9
Proper wire termination for screw terminals.

Insert wire between washers
(clockwise wrap)

Insert spade lug below washers
(above post)

too great for this type of jack to be certified above Category 4, if indeed it is certified at all.

Another type of IDC jack contact incorporates the newer 110-type punchdown strip as shown in Fig. 7.10. The 110 strip may be separate from the modular connector and connected by wires, as with the 66-type jacks, or it may be an integral part of the connector module. If discrete wires connect the strip, the same untwisted-wire problem exists as with the 66 blocks, and certification above Category 3 or 4 is rare. However, if this block is incorporated into the connector module, the combination can be certified to Category 5e. The 100 strip has the advantage that it can be terminated with conventional punchdown tools and its connectivity and longevity parameters are well established. On the other hand, the strip is often wider than the connector module it is attached to, which limits placement on a mounting place. Variations of this scheme use the BIX or Krone blocks, both described in Chap. 8, in place of the 110 block.

An alternative to these two other IDC connectors is what we are calling the "snap-in" connector for lack of a better name (see Fig. 7.7). This connector type is made in a number of styles by many manufacturers.

Figure 7.10
A connector jack that uses 110-type blocks for the horizontal cable connections. (*Courtesy of Molex Fiber Optics, Inc.*)

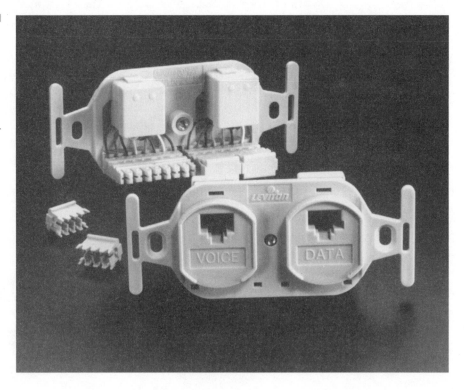

In general, these designs all have the common characteristic of compact construction and toolless termination. The connector contacts mount directly onto a printed circuit board, which also mounts the modular connector. The wires are each cut to size and placed in a channel above the contact edges. A plastic insert is used to push up to four individual wires into the jaws of the IDC contacts, where the insulation is pierced and the electrical connection is made between the wires and the contacts. The insert usually snaps into place with a click, which is why we are calling it a *snap-in connector.* The beauty of this connector style is that it terminates the wires without any tools. Its compact design leaves room on the connector housing for color codes to match the wire colors. Pin numbers may also be printed on the housing to help the installer.

Standard Jack Pin-outs

The standard pin connections for a variety of modular jack styles are shown in Fig. 7.11. Modular jacks are used for many different purposes, of which LAN wiring is only one.

The TIA/EIA-568-B standard describes two different wiring patterns for wiring the 8-pin modular jack. The two patterns are called T568A and T568B, and are shown in Fig. 7.11. The two patterns differ only in placement of Pairs 2 and 3 of the traditional 4 pair order. This order is defined as Pair 1 = Blue, Pair 2 = Orange, Pair 3 = Green, and Pair 4 = Brown. The T568A pattern places Pair 2 on Pins 3 and 6 of the jack, and Pair 3 on Pins 1 and 2. The T568B pattern places Pair 2 on Pins 1 and 2 of the jack, and Pair 3 on Pins 3 and 6. The A-pattern is actually unique to the TIA standard, but it is now the recommended pattern. The B-pattern is the traditional Lucent (AT&T) 258A pattern specified in their premises distribution system. Ironically, the A-pattern better follows the progression of wiring patterns from the 2 pair pattern for 6-pin modular jacks that is used with many telephone sets (the RJ-11/12 style jack). By the way, the "-" is used before the B in TIA/EIA-568-B specifically so there is less confusion with the wiring pattern called T568B.

There has been considerable controversy as to which pattern is the proper one to use. Depending upon whom you ask and where you ask, you can get strong opinions on either side of the question. The TIA standard, however, is quite clear. The "pin/pair assignments shall be..." T568A. Optionally, the T568B may be used, but only "if necessary to accommodate certain 8-pin cabling systems." These certain systems are

Figure 7.11
The modular jack
may be used for a
variety of purposes.
(See also Fig. 10-2.)

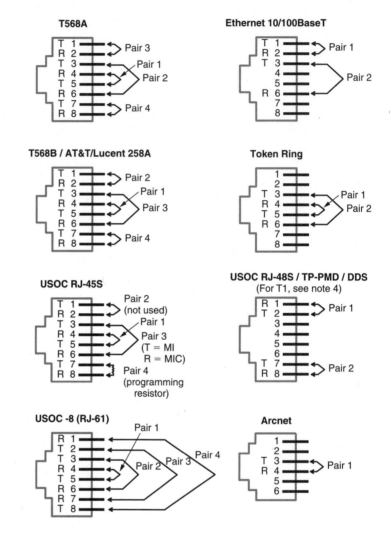

Pair Name	Conductor	Color Code (horiz. wire)	Solid Color (internal jack)
Pair 1	T1	White/Blue	Green
"	R1	Blue/White	Red
Pair 2	T2	White/Orange	Black
"	R2	Orange/White	Yellow
Pair 3	T3	White/Green	White
"	R3	Green/White	Blue
Pair 4	T4	White/Brown	Orange
"	R4	Brown/White	Brown

Notes:

1. T = Tip, R = Ring.
2. When polarities are used,
   Tip is +, Ring is −.
3. Jacks are viewed from the front.
4. T1 uses the RJ-48C and RJ-48X
   variations: Pair 1 is pins 4 & 5,
   pair 2 is pins 1 & 2).

ones that already use the optional pattern, for the most part. If you do not need to connect to an existing system, cable, patch, etc., that uses T568B, you should use the recommended T568A pattern.

What difference does it make which pattern you use—T568A or T568B? Electrically, it makes no difference, as long as exactly the same pattern is used at *both* ends. Let's say that again: The two patterns are electrically equivalent, as long as both ends of a cable run are wired with the same pattern. As has been said, "the electrons don't know the wire color." This means that as long as Pin 1 and Pin 2 on one end of a link connect to Pin 1 and Pin 2 on the other end, and the same is true of Pins 3 and 6, you can use either pattern. Both the outlet and the telecommunications room terminations (punchdown block or patch) of a horizontal cable run should be wired with the same pattern, as should both ends of any patch or user cord.

Technically, you can use either pattern; however, some patches, fan-out cables, equipment connectors, and adapters are prewired with the T568B pattern and you must match that pattern at the other end of the cable for the link to work. You can, however, mix patterns from one cable run to the next within the same link. For example, a patch cord could be wired with either pattern without regard to how the horizontal cable was terminated.

A common wiring pattern problem can occur in the horizontal cross-connect. If you terminate the horizontal cable into a punch block first, then cross-connect to a patch, you should be sure that the patch wiring matches that of the corresponding wall outlet, and that you have preserved the pair order during punchdown and cross-connect. See Chap. 9 on patch panels for more details on the internal wiring pattern of a patch panel.

The official stance of the TIA is that the A pattern is preferred. That statement certainly applies to new installations, since it would be problematic to switch the pattern in an older installation. You could also justifiably choose the B pattern if Lucent (AT&T) components are used, since that is their standard offering. There is great debate over which pattern is used the most. The truth is in the hearing. The customary pattern varies widely, depending upon whom you ask and what part of the country (or the world) you live in. However, it seems that the A pattern originally was not used as much because it was different from the pattern that had historically been required for telephone use. Now that most manufacturers offer both, you are free to follow the TIA's preference, or yours.

What happens if you use one pattern on one end of a cable run and another on the other end? Most LAN network connections will fail, and telephone connections that require two pairs will also fail. Pairs 2 or 3 are used by almost every device we connect to an 8-pin modular jack.

The TIA wiring patterns are considered universal in that they are applicable to several types of connections, because they cover all eight pins of the modular connector. For example, the TIA patterns can be used for any of the Ethernet connections, such as 10 and 100BaseT. The TX and T4 variations are both handled, as are 100VG-AnyLAN, 1000BaseT(4), and 1000BaseTX, since all four pairs are specified. In addition, the TIA patterns can support Token-Ring, Arcnet, ISDN, T1, ATM-PMD variations, and, yes of course, telephone wiring in its many variations.

You can make a very good argument that the TIA/EIA-568-B wiring system is as close to a universal wiring system as we will ever get. Of course, there will always be situations that do not fit this standard model, but more and more equipment is being designed to take advantage of its characteristics. If you choose to diverge from the standard, to implement some of the 2 pair technologies, for example, you may be left out in the cold in a very short time. For example, the only practical way to increase data rates on unshielded twisted pair wiring beyond the current 1000 Mbps data rates is to use additional pairs in a multiplexing or encoding scheme. The 100BaseT4 and 100VG-AnyLAN standards do this now with encoding such that the actual pair bandwidth requirements are under 32.5 and 15 MHz, respectively. This means that you can go to 1000 Mbps data rates without having to replace installed Category 5e cable links, which are rated to 100 MHz. However, these technologies need all four pairs. You cannot very well use four pairs if you decided to wire just two, just to save a little money in the short term. It is therefore strongly recommended that any new installations adhere very closely to the TIA standards, including the number of pairs to each outlet jack.

TIA Performance Category

It is very important to fully understand the categories of operation of the various modular jacks. The TIA/EIA-568-B standard contains strict performance requirements for the outlet connectors that are used for Category 3 or 5e operation. The requirements differ somewhat by category, although Category 4 and 5 parameters are often the same, and thus Category 4 has now been dropped. Although Categories 1 and 2 exist, they are not the subject of the standards because they really are not appropriate for any type of LAN cabling.

Components, as well as cable, should be marked as certified for a particular category of use. Thus, a connector that is intended for Category 5e use should be marked for Category 5e, as explained below. However, you may have occasion to need to determine the appropriate use of an unmarked jack. While it may not be possible to categorize an unmarked jack without some uncertainty, there are some guidelines that can help out. If you can identify the jack assembly style, you should be able to determine its probable category of rating. The jack styles and their appropriate categories for use are shown here:

TIA performance category	Termination style
1–3	Screw terminal or IDC
4	IDC, including 66- and 110-type
5e	IDC using PC board, only*
6	IDC using PC board, offset pins

*Category 5e jacks can also use 110, BIX, and Krone type connections.

Of course, any jack that is rated for a higher category than the intended use can be substituted. For example, it is perfectly acceptable to use a Category 5 or 5e jack for a Category 3 installation. However, you could possibly upgrade the entire wiring system for very little more cost. The IDC jacks are so easy to use, they are often used even when their higher category of operation is not required.

Keyed-Plug Entry

Some modular jacks have an extra slot at the side to supposedly limit connections to cable plugs with a matching slot. These side-keyed jacks are largely useless because they easily accommodate a nonkeyed plug. The original intent was to protect delicate digital equipment from the higher voltages that can sometimes be present on telephone circuits. For example, the common open circuit voltage for a telephone line in the United States is −48 DC, and the ring voltage can approach 90 V AC (both at low current). However, modern data circuits are fairly immune to these voltages and the fact that the keyed jacks are easily defeated has led to their disuse.

An alternate style of keyed jack is the MMJ style that was used by Digital Equipment Corp., described in Chap. 5. This jack has primarily been used for the connection of RS-232 and RS-422 low-speed devices and is rarely found in other uses. It is not appropriate for standards-compliant LAN wiring.

Fiber Optic Outlet Jacks

Fiber optic technology has become so important that we have devoted an entire chapter to it, Chap. 12. Outlet jacks for fiber optic connections have several unique requirements. The fiber optic outlet must allow the station cable fibers to be terminated and connected into a jack plate. In addition, the design of the jack plate must allow the easy attachment of an optical jumper to the workstation equipment. A special passive coupling device, called an adapter, is used to connect the two fiber optic cables at the outlet jack.

Unlike the termination of metallic conductors, the traditional termination of an optical fiber can be a delicate and time-consuming process, although the time may be reduced by using more expensive "quick-termination" connectors. The fiber must be stripped of its protective coverings, including the binder. It must then be precisely cleaved before being inserted into the mechanical coupling that is loosely called the *optical connector.* Some fiber connectors require the use of an epoxy or glue to secure the fiber and may use a crimp or clamp-in arrangement in addition (or in lieu of) the glue. Finally, the connectorized fiber end must be carefully polished to a smooth surface so that light entering or exiting the fiber is not bent off axis (the polishing step is not required for a quick-termination connector).

In order for light to couple properly between connectors, the internal fibers must be carefully aligned and placed virtually in contact by the coupling device. In fiber optics, this coupling device is called an *adapter* rather than a connector, because it does not actually terminate any fibers. It mounts in the fiber optic outlet plate much as a coax bulkhead barrel adapter does in coax systems. The fiber adapter provides a keyed location where the terminated station fiber connects on the back side of the outlet plate. The plate is then mounted in the fiber optic outlet box. The fiber optic user cord is connected to the section of the adapter that is on the front side of the outlet plate. Because of the nature of fiber optic transmission, two fibers and connectors are generally required for each LAN connection, one for transmit and one for receive. Keying these two connections, or differentiating them by color, is used to orient the transmit and receive fibers.

Fiber Optic Connectors

There are several types of fiber optic connectors. We will mention three of these that are often used in LAN wiring. The first is the optical SMA (SMA-905/906) connector shown in Fig. 7.12. This connector is quite similar to the miniature, screw-on SMA connector used for radio-frequency connections on small-diameter coax cables. The fiber is contained in a metal or ceramic ferrule and actually comes through to the very tip of the ferrule. The SMA connector allows very precise alignment of the fiber, but has the disadvantage of being awkward to use. To connect the SMA, the ferrule must first be aligned with the mating connector, and then the outer threaded retaining sleeve must be screwed down to hold the fiber in place. The 905-type connector has a straight 3-mm ferrule. The 906-type has a step-down ferrule that allows two ferrules to be connected by a coupler, with a plastic sleeve to align them.

SMA connectors are not designed to be easily paired together to support the standard dual fiber interface. You must mate each fiber connector

Figure 7.12
The optical SMA (SMA-905/906) connector. (*Courtesy of Molex Fiber Optics, Inc.*)

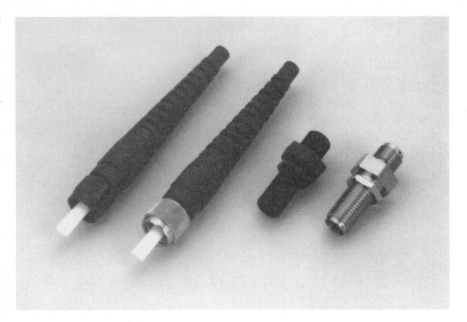

one at a time. Also, there is no easy way to key this connector so that the transmit and receive fibers do not get interchanged. Although this connector was once the most common fiber optic connector for routine LAN connections, it is not in as wide use today.

Another popular fiber optic connector is the ST connector, shown in Fig. 7.13. The design of this connector has a bayonet-style outer sleeve that allows the connector to be quickly inserted in its mate and secured with a quick quarter-turn. The connector looks somewhat like a miniature BNC connector. The ST connector had begun to supplant the SMA connector in most equipment and chassis connections. However, it still had the same single-connector problems that the SMA did. The connections must be made one at a time and there is no practical way to bind two fiber connections into one unit.

A newer connector is called the SC. It allows quick connections, good fiber alignment, and multifiber coupling so that two may be combined into a duplex connector for transmit and receive. The SC connector is an unusual looking square-tipped fiber optic connector that is slightly cone-shaped at the tip, as shown in Fig. 7.14. The SC is a push-on, pull-off connector that uses spring retention to hold the connector in place when mated. The square design and push/pull coupling mechanism make the connector usable in high-density applications. The SC connector may be coupled in duplex sets with a coupling receptacle or duplex clip. In a keyed duplex set, the SC connector easily implements a form of polarity matching with coupling adapters in fiber optic outlets or patch

Figure 7.13
Another popular fiber optic connector is the ST connector. (*Courtesy of Molex Fiber Optics, Inc.*)

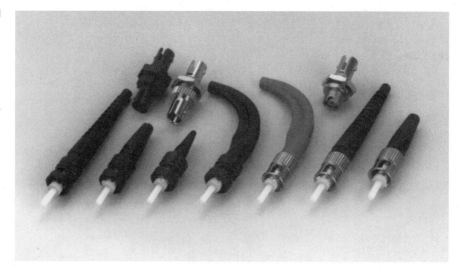

■■ ■■ ■■ ■■
Figure 7.14
The SC connector is
an unusual-looking,
square-tipped fiber
optic connector
slightly cone-shaped
at the very tip. (*Courtesy of Molex Fiber
Optics, Inc.*)

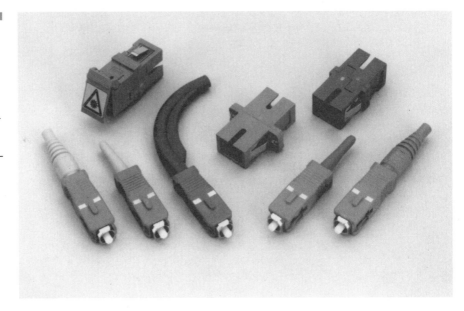

panels. This connector was adopted by the TIA as the officially recommended fiber optic connector in TIA/EIA 568-A. The connector is called the 568SC connector in the standard.

Small Form Factor (SFF) Connectors

The latest types of fiber optic connectors are generally called Small Form Factor (SFF) connectors. These connectors answer the most common objection to the SC style, namely that the connector cannot be contained in the space of a single modular (RJ-style) jack. In addition, the design of these SFF connectors allows two fibers to be terminated at once in a single plug or jack module. This simplifies the installation and use of duplex fiber connections and helps minimize polarity mistakes that come with improperly arranged fiber connectors of other types.

The SFF connectors are authorized as part of a general "deregulation" of fiber connectors in the new B revision of the TIA 568 standard. Specifically, TIA/EIA-568-B.3, *Optical Fiber Cabling Components Standard* allows the use of any connector that meets general performance requirements for fiber alignment, loss, and stability under physical loads. SFF-modular style connectors are not required, nor are they prohibited. In making this policy decision, the TIA declared its intention to foster inno-

Which Fiber Connector?

The advent of fiber optic cable communications systems has seen the progressive introduction of a wide variety of optical connectors. The dominant types that have been used for LAN connections are the SMA, the ST, and most recently the SC. Now, the landscape is littered with innovative new Small Form Factor (SFF) connectors, because of liberalized TIA/EIA-568-B rules. Which connector should you use?

All of the connectors mentioned provide close fiber alignment, reliability, and some measure of industry standardization. However, the SMA and the ST are not easily adaptable to a duplex arrangement so that two fiber connections can be made at the same time. Both of these connector styles must be used individually, because they require the twisting of a threaded or a bayonet-locking ring to secure the connectors to the mating jack. The SC connector, on the other hand, uses a push-on/pull-off mechanism that allows the two individual connectors to be attached together and connected or disconnected simultaneously. In addition, the new SFF connectors often use some sort of clip-in mechanism that retains the plug-jack mating under considerable physical loads.

The SC connector, mounted as a duplex unit, was originally selected as the recommended connector in TIA/EIA-568-A. However, this "568SC" has a 12.7-mm (0.5-in) ferrule spacing, which prevents it from mounting within the space of the RJ-style module plates that have achieved almost universal acceptance. The only alternative has been to redesign the plates, and often the modules, so that a dual-insert 568-SC jack could be accommodated. Even so, the SC takes up two module positions for one circuit connection, besides tying up a lot of real estate on router and switch option blades.

The answer to mating two fibers within the modular footprint is obviously to use one of the SFF connector designs that began to appear as soon as the SC problem was known. But there are now more than a half-dozen incompatible SFF connectors, each with its own advantages, available from multiple manufacturers. Moreover, the TIA now makes absolutely no recommendation, other than its general fiber connector requirements. For a solution to this dilemma, one could turn to the equipment manufacturers. Again, there is no unanimity, as switch and router manufacturers are split on SFF connector choices, and many continue to use the SC connectors. Because the TIA has decided not to decide, the market will eventually show a preference, but it will likely make no firm decision either. Just as there are many different types of computers, and many different makes of automobiles, there will continue to be many types of fiber connectors.

Over time, a few of the designs, perhaps two or three, will gain a majority of the market. Since the leading manufacturers have each licensed other strong producers to offer their designs, it is unlikely that only one will really emerge as a de facto standard. So, how should you choose which connector style to use for your network fiber?

The simple answer is to choose the connector that you believe will be the best for your environment. In reality, fiber optic transceivers are not actually connected into the patch terminations or the work area outlets. You must use a patch cord, user cord, or equipment cord to make the final connection. Also, it is unlikely that you will have the same SFF connector styles on all your equipment and workstation interfaces. Since different fiber connectors offer near identical optical performance, you can simply use a cord with the appropriate connectors at each end. Use a fixed connector and fiber cable combination that you feel comfortable with for your horizontal runs. Then get appropriate adapter cables for hub/switch/router equipment and for workstation network interface cards.

Most manufacturers offer adapter cords to transition their SFF connectors to virtually any other SFF design. It doesn't particularly matter if the cord ends are different, other than for arguing the fine points of connector performance. Fortunately, the TIA's performance guidelines assure that connector alignment and return loss will be well within that needed for almost all applications.

vation and let the market decide which connector style to use. Actually, the rush to non-SC RJ-style modular connectors had already begun, so problematic was the original SC implementation. The new standard simply allows this to continue, while specifying some very strict performance ground rules. A collection of the more popular SFF types is shown in Fig. 7.15. Additional information regarding fiber connectors, including several of the SFF types, is given in Chap. 12 in Table 12.2.

Other Outlet Jack Types

IBM Cabling System Jacks

A few other outlet jacks deserve mention. The first is the IBM Data Connector. The Data Connector, shown in Fig. 7.16, is a hermaphroditic (genderless) connector designed so that any connector mates with any other. It is a rather large, square-nosed connector with a rectangular body. The Data Connector is designed to be used with shielded twisted pair (STP) cable, although it can also be used with other connectors.

The Data Connector was specified in TIA-568-A as the standard connector for use with STP-A cable. The use of STP with the Data Connector was previously specified by an *EIA Interim Standard Omnibus Specification, NQ-EIA/IS-43.* The TIA-568-A standard detailed the Data Connector and its use with STP-A shielded cable. The Data Connector's complete electrical and mechanical specifications are contained in the first edition of IEC-807-8, *Rectangular Connectors for Frequencies Below 3 MHz, Part 8: Detail*

Figure 7.15
Examples of Small Form Factor (SFF) connectors: From Left to Right, the LC, the MT-RJ, the Opti-Flex, and the WF-45.

Figure 7.16
The Data Connector is a hermaphroditic (genderless) connector designed so that any connector mates with any other. (*Courtesy of Thomas & Betts Corporation.*)

Specification for Connectors, Four-Signal Contacts and Earthing Contacts for Cable Screen. Although this connector specification states that the connectors are for 3 MHz and below, the TIA has extended their characteristics to 300 MHz.

The color codes for horizontal 150-ohm STP-A cable are:

| Pair 1 | Red, Green |
| Pair 2 | Orange, Black |

This color code corresponds to the markings for the IDC contacts within the Data Connector and makes the cable conductors very easy to match with their proper contact.

The Data Connector outlet is a version of the standard Data Connector that is mounted in a cover plate. The cover plate is attached to a standard outlet box. The STP-A station cable is terminated directly in the outlet connector. There are versions of Data Connector faceplates available that contain both the Data Connector and one or more standard modular jacks.

The Data Connector and STP-A cabling system is a robust system that offers certification to 300 MHz, three times the current maximum frequency specification of standard UTP systems.

Coax Jacks

Coax jacks are another type of outlet jack. The BNC coax jack is commonly encountered in 10Base2 Ethernet cabling systems. This cabling system calls for the thinnet segment cable to be run from one workstation position to the next, and so on until you reach the end of the segment. At each workstation location, the two cable sections are joined together by a BNC-T and the center position of the T is connected to the workstation network adapter. The workstation essentially taps the coax line to become a part of the network. The method generally does not allow an extra stub of coax from the tap to the workstation; both cable sections must be extended all the way to the T at the workstation adapter.

Each fixed 10Base2 coax outlet, thus, needs two coax connectors, one for each cable section leading to the workstation. The customary outlet has two separate female BNC connectors on the faceplate, each connected to one of the two cable sections at the rear of the faceplate. There are two methods to connect the cable sections to the rear of the faceplate. One method connects bulkhead-mounted female connectors directly to each cable end. Each connector is inserted through a d-keyed hole in the plate and secured with a thin washer and nut. The alternative method places two bulkhead-mount barrel connector/adapters in the holes of the plate. The coax cable ends in the wall are terminated with male BNC connectors, which are then connected to the back side of the bulkhead adapters. From the front, the plates look the same, regardless of the method you use to connect the cable ends.

From the dual coax outlet, connection to the workstation is made with two coax jumper cables, with two BNC male connectors on each cable end. At the workstation, a BNC-T is connected to the network adapter and the two cables connected to the arms of the T. If the workstation is at the end of the coax segment, only one cable is used and a 50-ohm terminator is placed on the other T position. If there is no workstation in place, a short coax jumper cable is used to connect the two bulkhead connectors on the outlet's faceplate, in order to complete the cable circuit to the next workstation. If the workstation position would have been at the end of a coax segment, a single terminator may be used instead. The 10Base2 system requires end-to-end connectivity and dual-ended termination of the coax segment for proper operation. Any disconnected cables or terminators may cause the entire cable segment to cease to function.

Marking Outlets

Outlet jacks should be clearly identified on the faceplate. The identification should include a cable number and jack number, as well as an indication of whether the jack supports data (such as a LAN connection) or telephone. Some jack inserts and faceplates are marked to indicate "data" or "phone." Telephone extension numbers generally should not be marked on the plate, since many phone systems allow the number to be assigned to any physical outlet. The same applies for data jacks that correspond to a particular port number on a LAN hub or other network connection. You should keep a chart to match each extension or port number to each jack number.

The scheme in EIA/TIA-606 is a good basis for marking the jacks. The markings should be clearly visible with bold lettering about $3/8$ in high. The standard calls for marking at the front of the jack position, above or below the jack. However, you might also want to consider a neat marking on the top of the plate, so it can be read from above. (If you ever have to peer down behind a desk or a cabinet to identify an outlet, you will be thankful for the top markings.)

The outlet markings should identify the cable number, and, in a large facility, the wiring closet as well. For example, a marking for wiring closet number 3, cable 42, might be W3-42, or W3-J42 to differentiate it from a cable number. It is best if the outlet number corresponds to the cable number in the wiring closet. The cable itself can also be marked with a variety of vinyl labels, but the type of label that is part of a tie-wrap should be avoided on Category 5 cables, for the reasons previously mentioned. If all else fails, fine-point indelible markers will also effectively mark cable numbers on a cable jacket. Station cables are usually either in the wall or at the rear of termination blocks in the wiring closet, so the marking doesn't have to be pretty. Neatness counts, however, as you must be able to read the cable number later.

Outlet jacks are often color-coordinated with the room decor, which is fine as long as standard NEMA electrical colors are chosen. Nonstandard colors may cause a clash with other electrical outlet plates and may be difficult to obtain. If the electrical plates have already been selected, you should try to match those colors.

Modular outlet jacks offer a variety of colors for the modular inserts. In theory, the color markings allow one to easily differentiate between, say, voice and data jacks. However, there are really no universal standards for these colors. What may seem to be a logical color choice to you could be the opposite of what someone else might choose. You might simply want to leave all the colors the same neutral color as the faceplate and differentiate the jack functions with the plate markings. Alternatively, you may choose to leave the telephone connector neutral and use a differing color for the LAN jack.

Workmanship

The subject of workmanship is particularly important with regard to jack termination. The higher categories of wiring have strict guidelines for the amount of untwist that may be present at any point of termination. When a pair of wires is untwisted, even slightly, it can cause a greater disruption to the high-bandwidth signal transmission than almost any other noncontinuity defect. Furthermore, there are several points of unavoidable untwist in any cable link, and the effects of untwist are somewhat additive. You should carefully inspect at least a sampling of cable terminations to make sure they are done properly.

Figure 7.17 shows a sample of a cable link with one improperly untwisted connector termination. You can see that there is a very large cable anomaly at approximately 72 feet from the tester. This test link was made up with an intentional excess untwist of $\frac{3}{4}$ in at one of the

Figure 7.17
A sample of a cable link with one improperly untwisted connector termination. The resulting NEXT anomaly is shown in the scan graphic. (Courtesy Fluke)

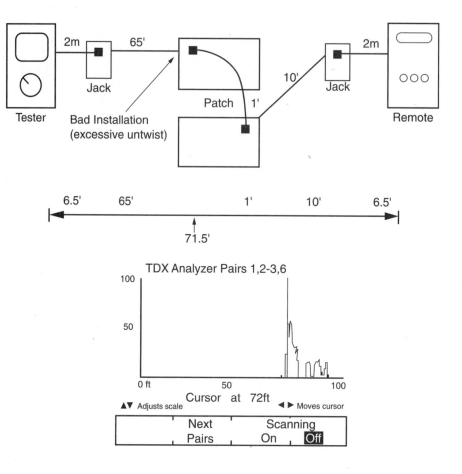

connectors of the 1 ft patch cord. When you compare the size of this anomaly to the other connectors (which are all properly made), you can clearly see the reason we emphasize good workmanship. Patch cords are easy to inspect visually, but your outlet jack wiring is hidden in the wall. Be sure that the person who installs your wiring is properly trained to do these connections well and test each horizontal cable run with a cable scanner.

Another workmanship issue that affects station outlets is the amount of excess cable that is left at the outlet. Standard practice is to leave no more than 18 in of extra cable length in wiring the outlet. This distance is measured from the wall opening to the jack before it is installed. Excess cable should be trimmed before the outlet is terminated. You should leave a reasonable length of extra cable to simplify the initial termination and any subsequent inspections or reterminations of a cable run.

Telecommunications Room Termination

Chapter 8 Highlights

- Punchdown blocks
- Other connecting systems
- Routing and dressing devices
- User equipment location
- Fiber optic termination
- Marking
- Workmanship

Another important area in the chain of wiring devices that makes up the LAN wiring connection is the cable termination in the telecommunications room. The telecommunications room is the point of termination of all of the cables that go to the workstation areas and the cables that interconnect to other telecommunications rooms. The telecommunications room will contain the punchdown blocks, patches, and other termination and wire-routing devices necessary for the interconnection of horizontal and backbone cabling. It may also contain various types of equipment, such as wiring hubs, routers, and even network servers. All the wiring termination methods and devices described in this chapter apply as well to similar wiring facilities, including the main cross-connect, intermediate cross-connect, equipment room, and entrance facility.

In the generalized model of universal, structured cabling, the telecommunications room also may contain telephone switching and distribution equipment. Much of what we cover in this chapter will be equally applicable to telephone wiring, but we will emphasize the LAN wiring aspects wherever possible. It is interesting that the telecommunications room is one place where the differing needs of telephone and LAN wiring are most apparent.

The very strict length and workmanship guidelines that are required for a successful LAN wiring system are very much relaxed if you are installing telephone wiring. For example, telephone wiring is often run through multiple sets of punchdown blocks in several wiring closets with little regard to total distance or care in routing. (Although the TIA standards do have strict guidelines, they are often excessively strict for telephone connections.) The much lower audio frequencies of most telephone installations allow these signals to run thousands of meters on very loosely twisted pairs. However, if you intend for your wiring system to be general purpose, so that you can run either voice or network signals, you must design your system to meet the more stringent operating rules embodied in the TIA or similar ISO standards. LAN wiring requires strict adherence to length, routing, rating, and workmanship rules in order to operate properly. The extra requirements for LAN operation are really the determining factors in the design of a modern structured wiring system.

In this chapter, we will cover all of the types of termination and routing devices, other than patch panels, that are used in the telecommunications room. Patch panels, jumpers, and cross-connects are covered in the next chapter. This chapter will also cover some of the mounting and location issues of the telecommunications room, including the options for locating hub equipment.

Punchdown Blocks

The fundamental component for wiring termination in the telecommunications room is the punchdown block. The punchdown block can take many forms, and has over the years evolved into a rather complex "system" component. The punchdown can even be incorporated into a patch panel, but we will talk about that later.

There are two main types of punchdown blocks in common use: the 66M block and the 110 block, both originated by AT&T. Both types are now offered by a number of companies and versions of these block terminations are incorporated into many products, including outlet connectors and patch panels. These two punchdown types dominate the installed and new markets, so we will cover them in some detail. We will also cover a couple of proprietary punchdown systems that are available from only one manufacturer or are not as widely used.

Type 66M Connecting Blocks

The original workhorse of telecommunications termination is the type 66 connecting block, shown in Figure 8.1. This style of termination block has been around for decades and its use in the telephone industry is pervasive.

There are several types of 66 blocks, but the most common is the 66M1-50. This block has 50 horizontal rows of wire termination contacts, with four bifurcated contact prongs in each row. (*Bifurcated* means the contact is split in two, so the wire can be held in place by the bifurcated fingers of the contact.) Each contact unit is called a *clip*, and the contact clips in this style of block each have two prongs, stamped from the same piece of metal. The four contact prongs in each row are paired 1-2, 3-4, with each pair of contacts electrically and mechanically connected. Figure 8.2 shows a cutaway view of one contact row. Some varieties of the 66M block have four common contacts in each row, while others have four totally independent contacts in each row, so be careful what you use.

Connectorized 66 blocks with 50-pin connectors are also available. as shown in Fig. 8.3. These blocks are usually manufactured with clips and may have a wire-wrap post protruding from the bottom of the clip. When the assembly is manufactured, a wire is wire-wrapped to each post on each clip with the other end connected to a 50-pin "telco" connector mounted on the side of the bracket. The connector may be

Figure 8.1
The original work-
horse of telecommu-
nications wiring
termination is the
type 66 connecting
block. (*Courtesy of
The Siemon
Company.*)

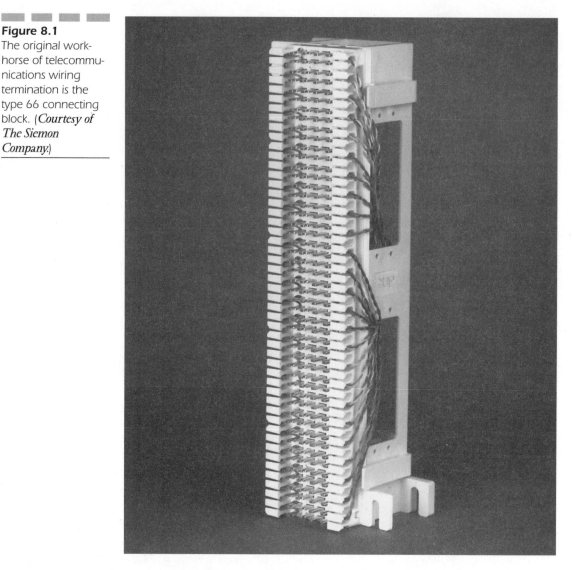

Figure 8.2
The four contacts of
a 66-block contact
row are paired, 1-2,
3-4, with each pair of
contacts electrically
and mechanically
connected.

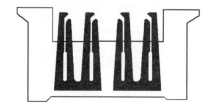

Figure 8.3
Connectorized 66
blocks with 50-pin
connectors are also
available. (*Courtesy of
The Siemon
Company.*)

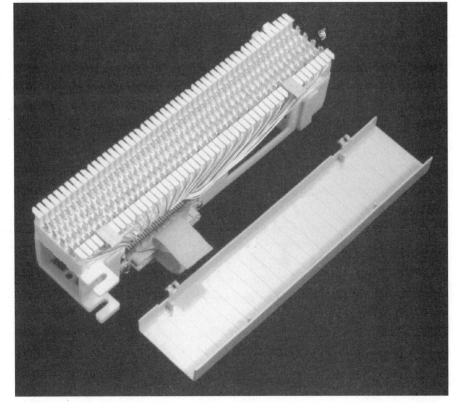

either male or female. The contact pins of the connector are typically wired to all 50 rows of contacts, with one pin on the telco connector corresponding to each row of the block. Connectorized 66 blocks are available with one or two telco connectors. Single-connector blocks are connected to only one column of paired contacts, while dual-connector blocks are connected to both, one on each side of the block. The connectors may be used to attach preassembled 25 pair jumper cables or to attach a modular fan-out (octopus) cable.

The fan-out cables, if used, come in several varieties, with 2, 3, or 4 pair modular legs (so they are often not "octo" cables at all). The 4 pair variety fans out to cable six legs, each terminated in an 8-pin modular plug. The legs may be ordered to length, but the wire should be stranded twisted pair wire and rated to the appropriate category in order to conform to LAN wiring guidelines. Because the 66M has 50 contact rows, and a 4 pair leg has eight wires to terminate, and eight divides into 50 rows only six times evenly), there are thus six legs from the fan-out when 4 pair cable legs are used. The last two rows of the 66M are not needed in most pairings, so only 48 of the 50 rows are used. Similarly, the 3 pair fan-out has eight legs [48 ÷ (2 × 3) = 8], and the 2 pair fan-out has 12 legs [48 ÷ (2 × 2) = 12]. The modular plugs of fan-out cables may be wired with any of several wiring patterns, including T568A and T568B. When ordering these fan-out cables, you should specify which pattern you want, and ensure that the corresponding outlet jack (at the other end of the station cable) is wired the same. See Chaps. 7 and 10 for more information.

Horizontal station cables (or backbone cables) are routed down the mounting backboard underneath the 66 block for termination. They are threaded out through the openings in the side of the mounting bracket where they are to be terminated. The outer jacket of the cable is removed to expose the wire pairs. The pairs are fanned out, sorted by color, and routed to the appropriate contact. Each wire passes through one of the narrow slots in the face of the block. The wires generally enter the slot that is just above the target contact row and are wrapped down into the jaws of the contact for termination. The contact has a hook on one prong to help position the wire until it is punched down. The side slots hold the wires in place-order, although they do not in any way support or strain-relieve the wires. (By the way, the process of fanning out, sorting, wrapping, and terminating wires on a block is called *stitching*. And one *punches down* the wire with the impact tool, equipped with an appropriate blade for 66 block termination.)

The 66 block has a TOP marking to indicate the way it should be positioned. If it is connectorized, the connector pins will be numbered upside down if the block is accidentally reversed. If the block is not connectorized, mounting it upside down just looks amateurish (and you will get comments). You should always punch down all the wires of a station cable, even if you do not intend to use all of them immediately. It is impractical to later use the extra pairs of a cable if they are not already punched down, because there is nowhere to terminate them. If for some reason you must install a cable that has more pairs than you generally use (such as a 4 pair cable in an installation that otherwise uses 3 pair cables), then simply cut the extra pairs flush at the point where the cable jacket was removed. You may be able to pull the jacket back slightly before you cut off the excess pairs. Then push the jacket back in place to make a neat installation.

The 66M block is designed to terminate unskinned *solid* copper conductors with plastic insulation. Conductors from AWG 20 to 26 can be accommodated. The use of stranded wire is not recommended under any circumstances. To terminate a wire, the wire is pressed down into the slot between the fingers of the contact with a punchdown impact tool, shown in Fig. 8.4. The tool causes the insulation to be displaced, leaving the copper wire and phosphorous-bronze clip in direct, gas-tight contact. The spring action of the tin-plated contact holds the wire in place. The tool also trims off the excess wire after termination. *Only one* conductor may be terminated in each contact slot. If a wire is removed from a contact slot, the slot may be reused, but only if the new wire is the same or larger size. Using a smaller wire, using a slot repeatedly, or punching two wires into a slot will cause the connection to fail, sooner or later.

As we have said, station wires are punched down from both sides of the block. Because the block has 48 usable row positions, that means that you can terminate 6 station cables per side of the block, or a total of 12 cables per block, assuming 4 pair cables. On a standard (nonconnectorized) 66M block, you should punch down the station wires on the outer contact of each dual pronged clip. (Some connectorized blocks wire to the outer contacts, instead of using a wire-wrap clip.) Cross-connect wire pairs are introduced to the row in the same way, through the slot just above each contact row, but punched down on the inner contact of each clip. The cross-connect wire overlaps the station wire that was punched into the outer contact, but the color code of each should still be visible. When punching, take care that the punchdown tool is turned the correct way so that the excess wire, and not the intended connection, gets trimmed off.

The tendency to fan the wires out, and the amount of untwist that exists where the wires of a pair pass through the side slots, creates a

Figure 8.4
To terminate a wire, the wire is pressed down into the slot between the fingers of the 66 block contact with a punch-down tool. The optional 110 blade is also shown. (*Courtesy of The Siemon Company.*)

problem for higher wiring categories, such as Category 5e. The contact clip also is a rather large piece of metal that can contribute to impedance mismatch and crosstalk at higher frequencies. For that reason, standard 66M blocks are not suitable for use above Category 3.

Some manufacturers have introduced low-crosstalk 66 blocks that are rated to Category 5e. The standard wiring technique of one-wire-per-slot can still cause an unacceptably high amount of crosstalk, even with these special blocks. If special 66 blocks are used for Category 5e, the wire pair should be inserted intact through the side slot between the two target contact rows, rather than split and run through two side slots. The wires are then wrapped up or down (whichever is appropriate) into the contact and terminated. This method maintains the amount of untwist well under the $\frac{1}{2}$-in maximum. Unfortunately, this technique also makes reading of the color stripes on the wires difficult after termination, so you should be very methodical when wrapping the wires to be sure you do it right the first time. A subsequent termination on the inner contact will make the color stripes of the wires on the outer contact of that row almost impossible to discern.

Many accessories, including adapters, bridging clips, and plug-on jacks, exist for the 66M block. Virtually all of these are inappropriate for use in permanently installed LAN wiring systems. The most stable connection system for LAN wiring is to use cross-connect wires between punchdown blocks and patches, or to terminate directly on the patch and skip the punchdown block altogether.

A version of the 66 block incorporates 8-pin modular jacks into the assembly. The jacks are mounted in groups of four or so at the side of the mounting bracket. This type of block may eliminate the need for a separate patch panel. However, it is more difficult to see and to access the jacks on the side of the block, and potential wire management methods are poor. In addition, you must be certain that the assembly is datagrade, so that some twist is maintained between the jacks and the 66 block's clips. This arrangement must not be used for Category 4 or 5 operation, unless it is certified to those specifications.

Several methods may be used to mount the 66M in a permanent location. Most commonly, the block may be directly mounted on a backboard. The typical backboard is a 4 × 8 sheet of $\frac{3}{4}$-in plywood, securely fastened to the wall and painted an appropriate color, such as battleship gray. The 66 block assembly is actually made up of two pieces, a front block that contains the contacts and molded side slots, and a stand-off bracket that mounts to the backboard. The bracket is normally mounted to the backboard before the 66M is attached. Two slotted, mounting holes are located on the upper left and lower right corners of the bracket. The slotted arrangement allows the mounting screws to be preinstalled on the backboard before the bracket is mounted. This two-piece structure, bracket and block, allows easy alignment and quick mounting, and facilitates the use of preassembled mounting frames. After the bracket is mounted, the 66 block is set on top and secured by four clips that are part of the bracket. Connectorized 66 blocks have the connector mounted to the bracket, with the front block already attached, and must be installed as a unit.

In addition to direct mounting on a backboard, the 66M may be mounted in a preassembled mounting frame. A typical distribution frame is shown in Figure 8.5. The frames are mounted as a unit onto wooden backboards, although some of the frames may be mounted on standard 19-in rails as well. These frames usually have preinstalled mounting brackets, into which the appropriate 66M block can be snapped. The frames may also have plastic stand-offs (sometimes called "mushrooms") and brackets for cable and cross-connect wire management. Frames are available in various sizes and can be used to implement full intermediate or main distribution frame (IDF or MDF) facilities.

Figure 8.5
A typical distribution frame for 66 block.
(*Courtesy of The Siemon Company.*)

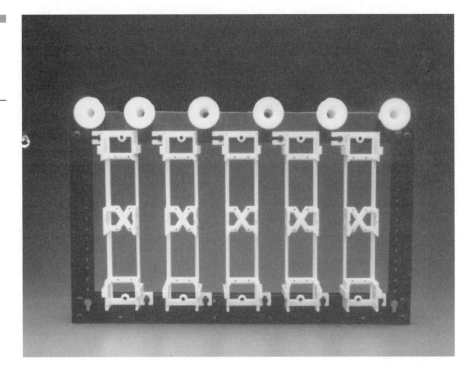

Various wiring color codes for the 66 block are shown in Fig. 8.6. The cables are terminated on the 66 block according to these standard color codes. The 4 pair code may be used to implement a TIA/EIA-568-B wiring system.

The overall color pattern is laid out in five "groups" of five pairs (10 wires) each. Each group of five pairs has one primary color that is the same for all

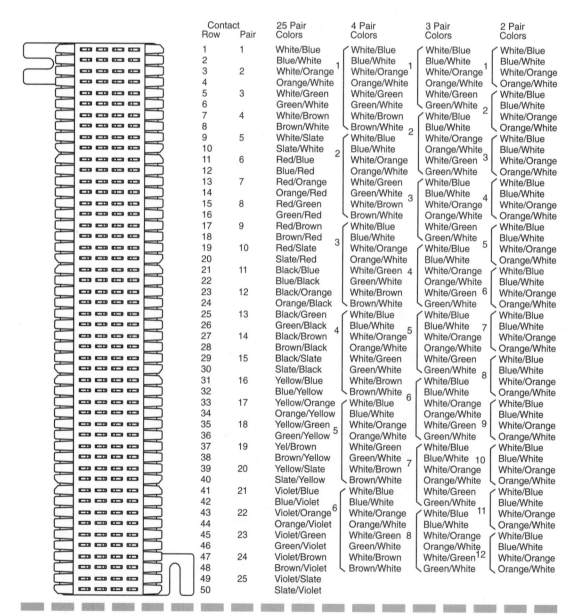

Contact Row	Pair	25 Pair Colors	4 Pair Colors	3 Pair Colors	2 Pair Colors
1	1	White/Blue	White/Blue	White/Blue	White/Blue
2		Blue/White	Blue/White	Blue/White	Blue/White
3	2	White/Orange	White/Orange	White/Orange	White/Orange
4		Orange/White	Orange/White	Orange/White	Orange/White
5	3	White/Green	White/Green	White/Green	White/Blue
6		Green/White	Green/White	Green/White	Blue/White
7	4	White/Brown	White/Brown	White/Blue	White/Orange
8		Brown/White	Brown/White	Blue/White	Orange/White
9	5	White/Slate	White/Blue	White/Orange	White/Blue
10		Slate/White	Blue/White	Orange/White	Blue/White
11	6	Red/Blue	White/Orange	White/Green	White/Orange
12		Blue/Red	Orange/White	Green/White	Orange/White
13	7	Red/Orange	White/Green	White/Blue	White/Blue
14		Orange/Red	Green/White	Blue/White	Blue/White
15	8	Red/Green	White/Brown	White/Orange	White/Orange
16		Green/Red	Brown/White	Orange/White	Orange/White
17	9	Red/Brown	White/Blue	White/Green	White/Blue
18		Brown/Red	Blue/White	Green/White	Blue/White
19	10	Red/Slate	White/Orange	White/Blue	White/Orange
20		Slate/Red	Orange/White	Blue/White	Orange/White
21	11	Black/Blue	White/Green	White/Orange	White/Blue
22		Blue/Black	Green/White	Orange/White	Blue/White
23	12	Black/Orange	White/Brown	White/Green	White/Orange
24		Orange/Black	Brown/White	Green/White	Orange/White
25	13	Black/Green	White/Blue	White/Blue	White/Blue
26		Green/Black	Blue/White	Blue/White	Blue/White
27	14	Black/Brown	White/Orange	White/Orange	White/Orange
28		Brown/Black	Orange/White	Orange/White	Orange/White
29	15	Black/Slate	White/Green	White/Green	White/Blue
30		Slate/Black	Green/White	Green/White	Blue/White
31	16	Yellow/Blue	White/Brown	White/Blue	White/Orange
32		Blue/Yellow	Brown/White	Blue/White	Orange/White
33	17	Yellow/Orange	White/Blue	White/Orange	White/Blue
34		Orange/Yellow	Blue/White	Orange/White	Blue/White
35	18	Yellow/Green	White/Orange	White/Green	White/Orange
36		Green/Yellow	Orange/White	Green/White	Orange/White
37	19	Yel/Brown	White/Green	White/Blue	White/Blue
38		Brown/Yellow	Green/White	Blue/White	Blue/White
39	20	Yellow/Slate	White/Brown	White/Orange	White/Orange
40		Slate/Yellow	Brown/White	Orange/White	Orange/White
41	21	Violet/Blue	White/Blue	White/Green	White/Blue
42		Blue/Violet	Blue/White	Green/White	Blue/White
43	22	Violet/Orange	White/Orange	White/Blue	White/Orange
44		Orange/Violet	Orange/White	Blue/White	Orange/White
45	23	Violet/Green	White/Green	White/Orange	White/Blue
46		Green/Violet	Green/White	Orange/White	Blue/White
47	24	Violet/Brown	White/Brown	White/Green	White/Orange
48		Brown/Violet	Brown/White	Green/White	Orange/White
49	25	Violet/Slate			
50		Slate/Violet			

Figure 8.6 Various wiring color codes for the 66 block. Patterns for the 110 block are the same.

the pairs. (Note that these are not "primary" colors, as taught in art class; rather, the term refers to the first color in a pair of wires.) The primary colors, in order, are white, red, black, yellow, and violet (sometimes called purple) and are commonly abbreviated W, R, BK, Y, V (or P), although other abbreviations are sometimes used for readability, such as WHT, RED, BLK, YEL, and VIO or PUR. The first wire group, for example, has a white wire in each pair, along with a second wire of a different color. The secondary wire colors are blue, orange, green, brown, and slate (sort of a silver/gray) and are abbreviated BL, O, G, BR, and S (or sometimes BLU, ORG, GRN, BRN, and SLT). Note that TIA/EIA-568-B uses the shortened abbreviations (the first letter of each color where there is no ambiguity and a second letter where needed). For example, for a 4 pair cable, the color abbreviations are W (the primary color) and BL, O, G, and BR (the secondary colors).

Each wire in a pair bears a helical or round stripe that is the same color as its mate. The pair is referred to by its primary and secondary colors. Thus the first pair of the first group is the W/BL pair (said "white-blue") and consists of a white wire with a blue stripe and a blue wire with a white stripe. The primary-colored wire is always punched down first (from top to bottom of the block). Thus the top row of the punchdown block will have a W/BL wire, the second row a BL/W wire, the third row a W/O, and so forth.

Four-pair wires are a special case because they only use four pairs from the first primary color group. Because of this, there is no confusion if the pairs are referred to by their secondary color only, such as "the blue pair" or "the green pair." Just make sure you punch down each pair's white wire first. TIA/EIA-568-B also allows an alternative color coding for patch cords. See Chap. 9.

Type 110 Connecting Blocks

Another style of cable termination block that is in wide use is the type 110 connecting block. The 110-style block is a relative newcomer, compared to the older 66 block. Even so, it has been in use for over two decades. The 110 system is designed for higher wiring density and a better separation of "input" and "output" cables than the older system. The 110 system is also important because the 4 pair connecting block component is used for the insulation-displacement termination in many patch panel and outlet jack systems by various manufacturers.

The 110 system consists of two basic components, the *110 wiring block* and the *110C connecting block*. The 110 wiring block, shown in Fig. 8.7, is a molded plastic mounting block with "horizontal index strips" that organize and secure 25 pairs (50 wires) each. Blocks are available that accommodate 100 and 300 pairs. The 100 pair block has four horizontal index strips, and the 300 pair block has 12. The 110A wiring block has a 3.25-in depth from the mounting plane and is used in normal applications. The empty space below the block makes feeder cable routing easy. A 110D block is available with a 1.4-in depth for low-profile special applications. The 300 pair wiring block illustrates the potential wiring density of the 110 system. It accommodates all 300 pairs in a 10.75-in square footprint. A 110T disconnect block is also available that provides the ability to disconnect a circuit for testing in either direction.

Space is available between the index strips to neatly organize incoming cables. In addition, the tip (secondary) colors are marked on the strips to help the installer place wires in the block. The wiring block does not make electrical contact with cable pairs, but merely secures them in place. Once the incoming cable wires are routed onto the wiring block, they are electrically terminated by the insertion of a 110C connecting block.

The 110C connecting block is a small, waferlike plastic housing containing metal contact clips at opposite edges, as shown in Fig. 8.8. The blocks are designed to snap onto the wiring block strips and connect to the wires that are held in place there. The back edge of the wafer con-

Figure 8.7
The 110 wiring block is a molded plastic mounting block with horizontal index strips. (*Courtesy of The Siemon Company.*)

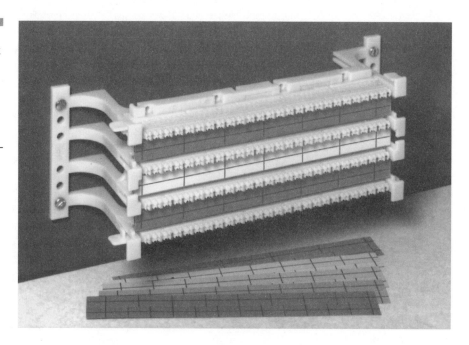

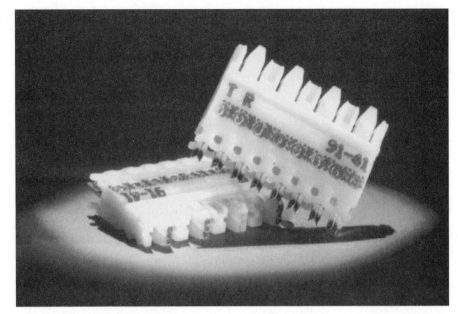

Figure 8.8
The 110C connecting block is a small, waferlike plastic housing containing metal contact clips at opposite edges. (*Courtesy of The Siemon Company.*)

tains quick clip IDC prongs that resemble the straight portion of the 66 block's clips. When the connecting block is pressed into place on the wiring block, the wires positioned on the wiring block are terminated, in mass. Each connecting block can terminate from three to five pairs (six to ten wires), depending upon the size of the block.

The front edge of the 110C connecting block is used to terminated cross-connect wires, or occasionally, other cables and adapters. It is color-coded to assist in the placement of wires, which are terminated on the top of the block one at a time. The IDC clips in the 110 block are designed to slice through the insulation as the wires are punched onto the top of the block. The action mechanically holds the conductors in place and makes electrical contact between the cross-connect wires and the underlying station cable wires that were previously terminated at the back of the block.

The routing of wires in and around the wiring block is one of the advantages of the 110 system. Station cables are routed to the wiring block from the rear of the wiring block. They are terminated on the back edge of the connecting blocks and are out of the way during the cross-connect operation. Cross-connects are made with unjacketed paired wire that passes in the troughs between horizontal index strips to reach the front edge of the appropriate 110 block. The positions on the wiring block may be easily identified by labels that attach to the block, unlike 66 blocks that have only a very small plastic surface for circuit marking. The labels fit into clear plastic label holders that snap onto the wiring block between the rows of the horizontal indexes strips. The labels are typically marked with thin vertical lines between station positions. Thus a label for 4 pair cable will have the lines every eight index positions. The labels do not interfere with the routing or tracking of cross-connect wires.

The 110 system is designed for AWG 22 to 26 plastic insulated wires. The 110 connectors are designed for *solid* wire only. Never use stranded-conductor wire with these connectors. The incoming station wires are inserted and trimmed using a 788-type impact tool, shown in Fig. 8.9. This tool can insert and trim up to five pairs (10 wires) at a time on the wiring block. It also seats the 110C connecting block onto a wiring block position. Cross-connect wires are terminated and trimmed one at a time with a punchdown tool that has a 110-type blade.

A 110 block system is used to terminate multipair station cables and allow the cross-connection to other punchdown locations. A typical LAN installation might have an appropriate number of wall-mounted or rack-mounted 110 blocks with cross-connect jumpers to a patch panel.

Figure 8.9
The station wires are inserted into the 110 wiring block and trimmed by using a 788-type impact tool. Cross-connect wires are punched down using a 110 blade (see Fig. 8.4). (*Courtesy of The Siemon Company.*)

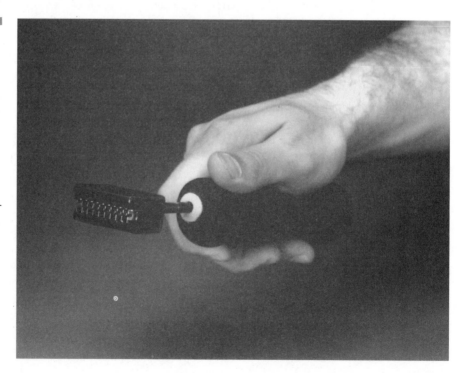

Distribution frame assemblies are available that have preterminated 25 pair multiconductor cables connected to the back positions of the connecting blocks. The multiconductor cables terminate in 50-pin Telco connectors that may provide mass connections to hub, telephone, or other equipment.

As we said before, each horizontal strip of the wiring block unit contains positions for 50 wires, or 25 pairs. This strip, in many ways, corresponds one side of the 66 block. Using the same math we did for 66 blocks, the 110 strip can accommodate six individual 4 pair station cables (or eight each 3 pair cables, or 12 each 2 pair cables). It can also terminate one 25 pair cable. As with the other type of block, 50 is an "odd" number, so the last two positions are not needed; only 48 positions are used. Contacts on adjacent 110 block strips may be connected together, one position at a time, by using the 110-type patch plug. The plug stretches vertically between two rows and connects the corresponding positions without having to use jumper wires. The plugs correspond to the bridging clips that are used to connect clips in adjacent columns on the 66 block.

In general, the 110 system is suitable for Category 5e use, as well as all lower categories. This is due the smaller geometry of the 110 clip and the ease of maintaining pair twist up to the very point of termination. Still, workmanship is an important consideration. Terminating wires on the 110 block is an individual operation, and the individual doing the termination must be properly trained and sufficiently methodical. The station wires at the back of the connecting block are more difficult to inspect than those on the front, so care must be taken to maintain good-quality terminations.

The 110 connecting block is also an important component in many outlet jacks and patch panels. It has the advantage of being extremely quick and easy to use, in addition to providing a high-quality connection. You can probably terminate a 110 block in half the time it takes with many custom IDC types, even though you must use a 110 punch-down tool. The comparison to conventional screw terminations is even better. The ready availability of 110 tools makes the block ideal for outlet and patch uses because it is likely to be in any installer's tool pouch. The quality of the connection is an issue for these jacks, just as it is for wiring closet terminations. The 110 block allows you to easily maintain the twist right up to the point of termination, ensuring a low-crosstalk and impedance-matched connection. The major disadvantage to the 110 block is that it positions the contacts in a line, and thus it might be a little wider than some of the custom IDC terminations. This restricts the side-to-side placement of jacks somewhat on the outlet plate. Some manufacturers of patch panels have dealt with this problem by simply placing the 110 strips in two conventional rows, with paired wires running to each jack.

Because of the design of the wiring blocks, the 110 system is not really suited for concurrent mounting of modular jacks, as is done with 66 blocks. However, models of patch panels with from 12 to 108 integral jacks are available. The jacks take up horizontal spaces, reducing the wiring density. It is difficult to offer a 50-pin telco jack mounting for the same reason. The 110 blocks tend to have cable-attached 50-pin jacks, rather than integral ones. Any 110 block that is thus wired to a 50-pin jack could use any of the fan-out cables that were mentioned in conjunction with the 66 blocks, but you would lose much of the simplicity of this latter method. A better route would be to just use a patch panel with 110-type connections and either terminate directly to the patch panel or wire cross-connects cover to the 110 wiring block.

The 110 system also offers a number of clip-on adapters that connect adapter jacks, cables, and other devices to the front of a 110C connecting block. Few, if any, of these are really practical for the orderly, structured wiring approach we have advocated for LAN wiring. In most cases, these adapters cannot be used for higher categories of operations, such as Category 5e/6. However, they are handy for testing, as you can easily adapt cable scanners and other testers to the block wiring.

The design of 110 wiring blocks makes direct wall mounting, without a backboard, somewhat easier. This is due to the multiple rows of index strips that are an integral part of the wiring block. A 300 pair block could be mounted all at once, whereas the 66 block equivalent would require mounting at least six separate blocks. That would require a lot of wall anchors to mount to a hollow wall. However, we are not at all advocating abandoning the backboard. It is a part of the recommended standards for telecommunications rooms, and is a very convenient way to provide a structurally sound mounting for many wiring components.

Relay rack and equipment mounting options are also available for 110 system wiring blocks. These methods are very good for large installations with the need for significant wire management.

The wiring color codes were covered in detail in the earlier section on 66 blocks and are identical for the 110 block system. The length of the horizontal index strips, on which the connecting blocks mount, are the same 50 wiring positions as the 66 block. However, the 110 system has the added advantage of placing color coding on some of the 110 components, to make placing the wires much easier. When you get to a very large installation, this can be quite an advantage, as it can also be used to visually inspect the terminations for correct routing.

Other Connecting Systems

We will mention two other connecting systems that are beginning to be used in more and more installations. One is the NORDX/CDT (formerly Nortel Northern Telecom) BIX system and the other is the Krone system. Both offer equivalent functionality to the 110 system and significant advantages over the older 66 system.

BIX Connecting Blocks

The NORDX/CDT BIX system is very similar in concept to the 110 system with a dual-sided 50-clip connector wafer that mounts horizontally in a mounting frame. Unlike the 110 system, the station or feeder cables are punched down directly into the back edge contacts of the connector. Some examples of BIX hardware are shown in Fig. 8.10. An array of BIB connectors, installed in a 50, 250, or 300 pair mount, forms a "module." Modules may be mounted directly on walls, backboards, or in racks with an optional rack-mount kit. Modules may also be combined into a mounting frame to form an intermediate distribution frame (IDF) or main distributing frame (MDF).

Cross-connects are made from the front of the BIX connector. Connectors are available premarked at either 4 or 5 pair intervals. The 4 pair version is used in most LAN wiring, while the 5 pair version is useful for termination of cables of 25 or more pairs.

All connections are made with a BIX connecting tool, which is similar in function to the tools used for 66 and 110 blocks. Some standard impact tools can be equipped with BIX blades. Cables enter from the sides of the mount, either above or below the connector. The mounts

Figure 8.10
A BIX connector.
(*Courtesy of NORDX/CDT.*)

include a provision for a marking strip (or designation strip), which mounts between a pair of BIX connectors. The connectors are used in pairs, with a marking strip in between. A common 300 pair mount, for example, contains a maximum of 12 connectors and six designation strips. Connector labels are available for the 4 pair connectors in colors to match the usage group designation in EIA/TIA 569. (see Table 15.1 in Chap. 15.)

Bridging clips are available that connect vertically between two adjacent connector positions. A special wiring fixture can also be used to assist in terminating cables. The fixture is snapped into a connector position and is moved to the next connector position to be terminated, in turn.

Modular jack assemblies that mount in the BIX mounts are also available in a variety of jack configurations. The assemblies occupy two connector slots in the mount. Category 3, 5e, and 6 versions are offered.

The BIX-type connecting blocks are also used in modular jack outlet and on patch panels for wire termination. They are common options from several manufacturers, although all of the blocks come from NORDX/CDT. Distribution modules, jacks, and patches using BIX connecting hardware may be certified for Category 5e, although some accessories and assemblies are certified to lower categories. The usual caveats apply to maintaining pair twist up to the point of termination, as with the other systems.

Krone Connecting Blocks

The Krone (pronounced krohn-a) connector system, shown in Fig. 8.11, features 8, 10, and 25 pair basic connector modules that can be mounted in several configurations. The 8 and 10 pair connectors can be installed in separate module mounts that each have a capacity of 20 connectors. These mounts can therefore terminate a total of either 160 or 200 pairs. The 8 pair connector is typically used to terminate two 4 pair station cables. The module mounts can be indivdiually wall-mounted or can be installed in triples in a 19-inch wall or rack-mount frame.

For 25 pair applications, a 25 pair connector and mount is available. The 25 pair connector module is a feed-through design with a front and back piece, similar to the 110 system. As with the 8 and 10 pair connectors, the 25 pair connector is installed horizontally into its mounting bracket. A special assembly of two of the 25 pair connector modules is available with a mounting bracket that is the same size and footprint as the standard 66 block. These assemblies terminate 50 pairs, as do 66

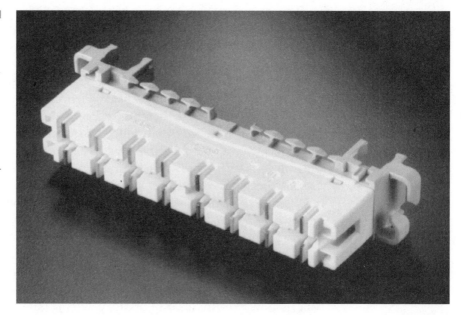

Figure 8.11
The Krone connector system features 8, 10, and 25 pair basic connector modules that can be mounted in several configurations. An 8 pair module is shown. (*Courtesy of Krone, Inc.*)

blocks, and are ideal for replacement of the older blocks for higher performance applications. They also mount in distribution frames intended for 66 blocks. The availability of 66-type mounts means that versions of the block can be offered with integral 50-pin telco connectors, just as with the 66 blocks.

Station and cross-connect wires are terminated on the bottom and on the top, respectively, of a Krone connector. The connector offers a unique "disconnect" contact as an option. Connectors with the disconnect feature can be temporarily opened for isolated testing to either leg of the circuit.

Krone connectors may also be found in outlet jacks and patch panels for wire termination. Many manufacturers offer the Krone wire termination system as an option. The connector hardware may be certified to Category 5e/6, but again, the usual guidelines for wire pair termination must be observed.

Routing and Dressing Devices

Cabling, cross-connects, and patching in wiring closets should result in an installation that is neat and orderly. Unfortunately, all these wires and cords have a natural tendency to be very disorderly, it seems. To avoid the mess and stress it brings, you should use the proper wire-management devices to put the wire in its place. Proper planning, layout, and dressing-in of the wire and patch cords can result in a very nice installation.

The process of properly routing and dressing your cable can keep your wiring closet much more "user friendly." If you are the user, you will really appreciate the neatness every time you need access to your cable system. If you are the installer, your customer will be much more pleased with your installation. The system of routing and dressing-in (putting in place) twisted pair wiring is relatively simple. The principle is that all the wires and cables should be run along wiring channels or trays, secured in place with cable ties or other devices, make relatively square corners, and be out of sight as much as possible when securing the wire is not practical.

You can divide the wire management problems into two separate issues. One issue is how you should deal with relatively permanent wiring, such as horizontal station cables and cross-connects. The other issue is how you manage temporary wiring, such as patch cords (and equipment cords).

Station cable and cross-connects consist of solid core wires and cables. These wires are relatively easy to bend into position, wrap around standoffs or brackets, and secure with a few cable ties. The solid wire tends to bend into place and stay with a minimum of restraint.

The stranded-wire cable used in patch cords is not as well behaved. It is used specifically because it is so flexible, relative to solid-wire cable. These patch cords are truly a mess to deal with, leading some to try to avoid their use entirely and use semipermanent cross-connect wire instead. In LAN wiring, flexible equipment cords are often used between a patch panel and the LAN hub. Many hub ports must be connected with an 8-pin modular cord, unless they are equipped with mass-terminated connectors, such as the 50-pin telco connector. Thus, the use of stranded-wire cords may be unavoidable.

How you handle cable management will be a function of the size of your telecommunications room and how it is laid out. If you have a small, wall-mounted wire termination system with a few wall-mounted hubs, you may be able to deal with the cables by securing the station wires with tie wraps and standoffs, running cross-connects (if any) around standoff posts or through cable rings, and letting the patch or equipment cords droop (neatly, we hope). On the other hand, if you have a large facility, or have equipment and wiring devices that are mounted in floor racks, you should use a system of cable trays, panels, raceways, and brackets that routes all types of cable and keeps them very neat. There is nothing more troublesome than having to pick your way through a curtain of patch cords to find a patch jack. There are much better ways to install your cable and cords, and we will cover some of them here.

We should caution you first that the use of tie wraps (cable ties) that are excessively tightened can cause performance problems on Category 5e/6 wiring and should be avoided. However, relatively loose ties that do not distort the cable jacket should be all right. The effect is cumulative, so 20 tie wraps are worse than five. You can use a new type of wrap made with hook-and-loop mesh (Velcro-type) in lieu of traditional nylon tie wraps. This new wrap is wider, so it does not pinch the cable as badly. It is also easily removable so that new cables can be added without adding tie-wrap clutter.

Standoffs and Distribution Rings

The most basic accessories for wire management are plastic standoffs and plastic or metal distribution rings. Some examples of standoffs are shown in Fig. 8.12. The plastic standoffs (sometimes called *wire spools* or *mushrooms)* are designed to hold cables or cross-connect wires

Figure 8.12
A basic accessory for wire management is the plastic standoff. (*Courtesy of The Siemon Company.*)

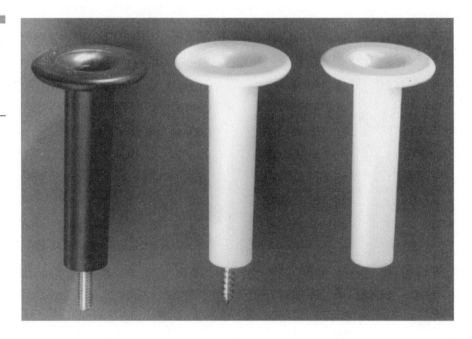

underneath the outer lip (which gives it the mushroom look). Wires are wrapped over a standoff, or down the side of a line of standoffs and routed to their destination. The wire is usually bent around the post very slightly to give the wire a "set" and hold it in place. Tie wraps may also be used to hold wires in place, subject to precautions applicable to Category 5e/6 use. The standoffs are hollow and may be secured to the wiring board with a captive wood screw or machine screw. Standoffs are also supplied without a screw. These devices are very widely used in cross-connect fields, especially with the 66-style connecting block, which has little native cable management.

Metal distribution rings are often used on wooden backboards to secure station cables and cross-connects as they are run across the board. Cables may be quickly laid in, sorted, wrapped, and secured to the rings. The rings are available in several sizes, although the 2-, 4-, and 6-in widths are the most common. The rings mount with two screws at either side. Some installers split the rings in two with a tubing cutter to provide an open half-ring for cross-connect wiring. This method allows more wires to be controlled than with the plastic standoffs and avoids the problem of threading the jumper wires through each closed ring as the wires are run from point to point.

The half-ring idea probably inspired a version of split plastic ring, also called a wire hanger or bracket, shown in Fig. 8.13. This bracket has a flat, solid back with side loops that almost meet at the top to form a ring. The small opening at the top allows a cable to be placed directly into the ring without threading. These brackets work well with stranded-wire-cable or cross-connect wire, but the solid, plenum-style wire tends to work out of the bracket if not secured with a wrap.

Figure 8.13
Split plastic rings are useful for managing wiring with frequent changes or additions. (*Courtesy of The Siemon Company.*)

Wire Management Panels

Wire management panels are available that mount horizontally across a relay rack or cabinet to offer orderly routing for patch cords and equipment cords. These panels prevent cable droop that would obscure equipment front panels and connections. The panels may consist of a series of split-ring loops similar to the wire hanger described above, or they may be a semienclosed slotted raceway with a removable cover. The styles with the cover offer the ultimate in out-of-sight cable management. They are great for patch cords, which present the greatest challenge to neatness. The covers are quite easy to remove, so you can get to the cords easily for new connections and moves. Covered cable raceways are also available for the vertical runs, although the vertical pathways are not as much of a problem.

With the increased concern over tie wraps and sharp bends when using Category 5e or 6 cabling, the gentle side of wire management really gets a boost. Now that many in the industry have heard of actual cable failures that were cured when tight tie wraps were cut, you can bet that installers will pay more attention to cable management issues.

User Equipment Location

The proper location of user equipment, such as LAN hubs, is a key part of creating a successful telecommunications room. While the telecommunications room does not necessarily contain any equipment or hubs at all, at least in the grand scheme of structured wiring, your LAN telecommunications room almost certainly will. These hubs and other equipment must be connected to the horizontal wiring for your workstations in order for your network to function.

The needs of LAN hubs and other active network components are a little different from the considerations for horizontal wiring terminations, cross-connects, and patches. Wiring and termination components simply need a fixed mounting location, wire management devices, and accessibility. The hubs and other equipment need power, ventilation, and connectivity to their respective station links. The wiring components tend to be fixed in place for the life of the installation. The equipment components, however, may need room to grow and to be reconfigured as network needs change and as technology changes.

The best location for the equipment is one that places the hubs and the wiring components as close together as practical, while maintaining a logical separation of the two functions. Remember, too, that some expandability of horizontal wiring connections should be allowed, and that correspondingly more room should be allowed for future changes in equipment technology.

In a small installation of less than 100 station terminations, you have a choice of wall or rack mounting for both the equipment and the cable terminations. There are advantages (and disadvantages) to both. Wall mounting will give you the most free and open space in the center of a small wiring closet. However, it limits access and cable routing options for the wiring components, and wall mounting of hubs and other equipment may be difficult. Rack mounting may be used both for wiring components and for hubs, but you will have added expense for the rack, cable trays, and other accessories.

In larger installations, serious consideration should be given to rack-mounting all components. This is particularly appropriate with the wide range of wire management devices for cable racks. You may face an interesting dilemma trying to decide in which racks to put the hubs. Should they be in the same racks, or should they be in adjacent racks? While there may be a certain logic to separating the wiring termination blocks and patches from the hubs, remember that you must generally run lots of patch cords between the two. With separate racks, all those patch cords would have to pass up through one rack and across and down into the other rack to make connections. Not only is such a long run probably unnecessary but it may stretch you to the limit for the channel link. Remember that you are allowed only 10 m (33 ft) for all equipment cords in a channel, including both the user cords in the work area and the patch/equipment cords in the telecommunications room.[1] If you required a 5-m (16-ft) equipment cord to reach to the hub rack, you would only have a 5-m (16-ft) allowance in the work area. It is not unusual for larger work area offices to need 6- to 8-m (18- to 25-ft) user cords, so you would be better off using as short a telecommunications room equipment cord as possible.

The easy way to keep patch and equipment cords short is to locate the horizontal wire terminations and the hubs in the same rack. The terminations might be on connecting blocks, with cross-connects to patch panels, or directly to patch panels. You can use wire management panels

---

[1]TIA/EIA 568-B actually now limits the total to 5 m (16 ft) for user cords in the work area, and 5 m (16 ft) for all equipment cords, patch cords, and cross-connect jumpers in the telecommunications room.

to route the patch wires to the side rails and down to the hub. Most modern wire termination blocks feature a very high connection density, so they should not take up much rack space. Also, patch panels with 96 or more jacks consume very little vertical rack space. So it should be possible to place both equipment and terminations in the same rack, with plenty of room for future expansion of both. A popular alternative mounting scheme is to place the wiring terminations in wall mountings and place the LAN hubs in relay racks. This hybrid approach is serviceable, as long as the equipment cord runs are not too long from the wall to the racked hubs. In a variation to this approach, you could place the connecting block terminations on the wall, as before, and run cross-connect jumpers to patch panels in relay racks. Then, relatively short equipment cords could be connected between the patch panels and hubs in the same rack. Remember that the cross-connect jumpers count as part of the 5-m (16-ft) limit.

"To patch or not to patch, that is the question." You can get into a considerable debate with some wiring managers of large facilities about this very question. There is definitely a downside to the use of patch panels in a large facility. The problems with patch panels are that they are an added expense, provide an additional point of failure, may provide questionable long-term connections, and add lots of spaghetti-mess (from the myriad of patch cords that must be used). Their modular connectors may also add unnecessary crosstalk. If connections to the hubs can be made with cross-connect wires, you may be able to avoid the use of patch panels altogether. Simply terminate your station cables on the connecting block of your choice and run a cross-connect jumper over to a mass-termination connector for the hub. The larger card-cage hubs offer port connections on 50-pin Telco connectors, among other options. The hub's 50-pin connectors can be extended to connecting blocks, such as the 110 system blocks, and the cross-connects can be made directly to the station punchdowns, without using patch panels.

The patchless approach has at least three disadvantages. First, troubleshooting is more difficult, as you cannot simply unplug a modular equipment cord and insert a tester. You must either disconnect the cross-connect wire and use a test jack adapter that fits onto the connecting block, or you must have used one of the types of connecting blocks with a built-in disconnect capability and still use a test adapter. Second, it is much more difficult, on average, to locate the connections for a particular station cable, since the circuit markings are imbedded in the high-density connecting block. Third, you may double the potential workmanship problems. For example, if you had directly terminated on a patch panel, that would be the only field termination that might have excessive untwist or other problems. The patch cords would hopefully be factory terminated, carefully inspected, and certified. With cross-connection, you add four points for potential workmanship problems: the connecting block, the two end connections of the cross-connect jumper, and the multicircuit termination leading to the hub.

Which is right for you? That may depend upon the size of your facility and the training of LAN maintenance personnel. The patchless approach should probably be avoided in telecommunications rooms having under 100 to 200 connections, unless you have trained people with the ability to do complex wire tracing and the workmanship skills to make high-quality connections.

Fiber Optic Termination

Fiber optic cable are treated a little differently in the telecommunications room than conventional metallic cables. However, the concepts are the same. Basically, in the telecommunications room, cables from the horizontal distribution to work areas are all brought together in some fixed, orderly arrangement for interconnection to network hubs and other equipment. Fiber optic cables may be used for the horizontal cabling, in place of traditional metallic wire. Fiber optic cables are also widely used for the backbone cabling that runs between telecommunications rooms. Differences in the telecommunications room terminations for fiber optic cables exist because the nature of the signals (light) and the fact that the transmission medium (glass fiber) requires special handling. These differences limit the number of ways the cables may be terminated, severely limit routing and handling considerations, and cause fundamental differences in the structure of patch panels.

In addition, fiber optic connections have what might be called "polarity." Because a single fiber sends its signal in only one direction, a fiber optic link always requires two fiber connections, one to transmit the signal in each direction. To operate properly, the transmitted signal from one end must connect to the receiver port at the other end, and vice versa. This is called a crossover connection.

In contrast, metallic cable link standards use straight-through wiring for most connections. This is accomplished by defining two types of equipment interfaces, such as the DTE and DCE interfaces of RS-232 that are meant to connect "terminal" and "communication" devices. The same type of pattern exists in 10/100/1000BaseT (adapter versus hub) and AUI (repeater versus transceiver). However, the fiber optic equipment interfaces have traditionally been identical, providing signal identification only by the label beside each of the two fiber optic connectors. That means that the user/installer must provide fiber cabling that accomplishes the crossover.

AB/BA Fiber Polarity Orientation

Fiber optic cabling standards, including 568SC and FDDI standards, specify a dual fiber connector that helps maintain polarity. The TIA/EIA-568-B method uses a polarized AB/BA orientation of the dual connector to incorporate polarity into the fiber connector. The method proposes to make the user connections foolproof, so that a user never has to worry about polarity. If the fiber cables are properly installed, the proper polarities at the user connections are assured. Unfortunately, that places all of the burden of maintaining proper polarity on the installer.

The polarity with the SC connection[2] is achieved by using a nonsymmetrical dual-connector body and labeling each position either A or B, as shown in Fig. 8.14. The actual SC connector fiber termination was discussed in Chap. 7. The fibers of a given cable are not literally called "transmit" or "receive," but are numbered consecutively, paired, and alternately placed into either the A or B position at each cable end in such a

Figure 8.14
The polarity with the SC connection is achieved by using a nonsymmetrical dual-connector body and labeling each position either A or B. The two orientations are called AB and BA, reading either left to right or down.

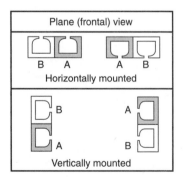

[2]TIA/EIA-568-B still uses the SC connector for illustration purposes, although SFF connector may be used.

way that the transmit connector on the equipment at one end is connected to the receive connector on the equipment at the other end. This
is a variation of the old metallic cable proposal that would have made all
equipment interfaces the same and all cables crossed over (transmit and
receive reversed end-to-end). It is perhaps ironic that the metallic version
was abandoned only to be resurrected in fiber.

If you are accustomed to placing wire in connectors according to pin
numbers and color codes, the fiber optic polarity method at first may
seem confusing. Don't worry, the confusion will probably not go away, it
will just become more tolerable. Figure 8.15 shows the interconnection
of two duplex cables at an adapter interface. The term *adapter* is defined
as a fiber optic coupling between two fiber connectors of the same type,
that is, two SC connectors, in this case. (The term *hybrid adapter* is used to
describe a coupling designed to mate two unlike-like fiber connectors,
e.g., an ST and an SC.) Because of the narrow fiber spacing, SFF connections require an adapter cable instead of a hybrid adapter. Note that the
adapter actually does a polarity reversal by having opposite orientations
between its two sides. The A connector position is always placed into the
A adapter position, but that passes through to a B connector position,
through the panel to the opposite side of the adapter.

The end-to-end connections of all fiber cables are also reversed. This
includes the horizontal and backbone fiber cables as well as all patch cords
and user equipment cords. The pattern of reversals should be standardized
in an installation, but it is up to the installer to maintain this discipline.
One way to accomplish this is to designate a correspondence between the

Figure 8.15
The interconnection
of two duplex cables
at an "adapter"
interface.

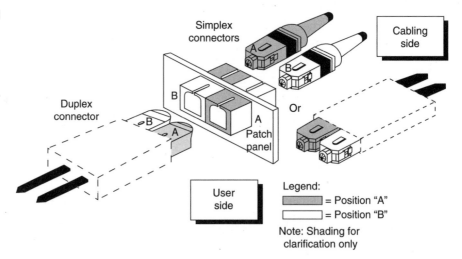

fiber buffer color and a fiber number. Then terminate all odd numbered fibers as A fibers and even numbered fibers as B fibers in the telecommunications room. Do the opposite at the station outlet. For example, in a two-fiber cable, you could designate fiber 1 as an A connection in the telecommunications room and as a B connection at the outlet. (Of course, the adapters at each end reverse the A and B, but that takes care of itself.) Be very consistent in your fiber polarity designations at every fiber termination. The scheme for backbone interconnections should be determined before any work begins and should be consistently applied. Remember that all fiber cable connections must reverse end-to-end for this scheme to work.

A couple of typical fiber optic connection patterns are shown in Fig. 8.16. Note that the scheme must have an odd number of reversals to work, but that this is automatically accomplished by the fact that the coupling devices at the end of each run do reversals, too. Because cables always have the same type of male connectors, each patch cord, horizontal cable, or backbone cable always provides one reversal. The adapter couplings at either end provide two additional reversals, restoring the signal to the original polarity. All that you (or your installer) must remember is where to clip each SC connectorized fiber into its duplex housing to provide the proper AB/BA orientation. SFF connectors must provide a means to maintain fiber AB polarity, and cable polarity is not easily reversed, so the installation must be checked carefully.

If you do fiber to the desktop, you will want to install a fiber patch panel in your telecommunications room. This panel will provide a point of termination for each station drop and will also provide labeling and identification for the horizontal run. It will also provide a handy point for testing and troubleshooting of fiber runs. An all-fiber installation will usually involve locating network hubs with fiber optic interfaces in the telecommunications room. You will connect a duplex fiber optic patch cord between the station location on the patch panel and the hub port. You may have to use an adapter cable or hybrid adapter if your equipment does not have the new SC connectors. All of the standard cable management trays and panels may be used with fiber patch cords, in general, although you should pay particular attention that the cords are not bent sharply or otherwise stressed. The tiny glass fibers in these cables are subject to stress fractures when excessively bent, and may cause signal loss that prevents a good connection. A sharp bend may actually break the fiber entirely. Suspected cords should be tested and scrapped if bad.

You will want to carefully plan the location of your fiber optic terminations in the telecommunications room. Because of the nature of

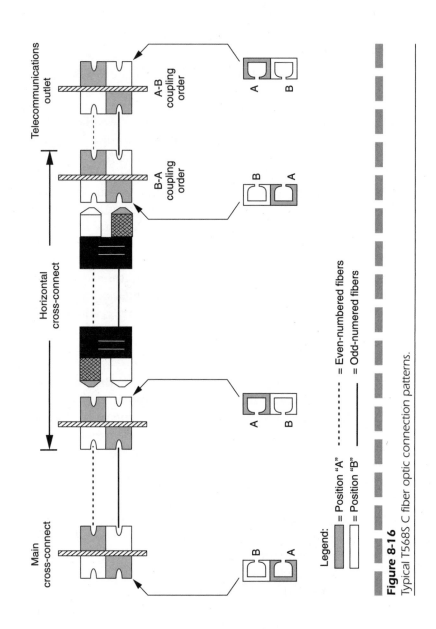

Figure 8-16

Typical T568S C fiber optic connection patterns.

fiber optic cable, it is difficult to reterminate when you move the termination location. Fiber optic cable is often terminated in a special fixture in the telecommunications room. Fiber optic termination fixtures, as shown in Fig. 8.17 often have space for extra fiber to be wound. However, the extra fiber is really intended to allow some slack in case a damaged connection needs to be replaced, and to assist in the termination process by allowing greater access to the fiber end. Usually, the amount of extra cable is only 1 m or so, and may have already had the outer jacket removed to expose the fibers, so it would not be useful in a relocation.

The use of a termination fixture, sometimes called a break-out box or fan-out enclosure, makes freestanding racks the most conventional location for large numbers of fiber optic terminations. This certainly is the case if you are using fiber for the horizontal connections. The fiber is typically routed across cable ladders or closed-bottom trays to each rack, and then down along the inside or back of a rail to the termination location. Fiber optic cables may be carried in a brightly colored, corrugated plastic "innerduct" for additional protection. The use of innerduct is mandatory if

Figure 8.17
Fiber optic termination fixtures often have space for extra fiber to be wound. This multimedia example also has copper wire jacks. (*Courtesy of MOD-TAP.*)

you use unjacketed fibers or highly flexible fiber cables, such as "zip cord" cable. The innerduct has an additional advantage because of its fairly rigid walls. It can be tie-wrapped in place with little chance of damage to the interior fiber cables. Also, it bends easily, but keeps a fairly large bend radius because of its construction. Fiber cables can often be placed in existing interduct much as you would in a rigid conduit.

Caution is the watchword when you are working with fiber optic cable, whether terminated or not. Terminated fibers may be connected to equipment that emits hazardous levels of laser light, especially if you are using single mode fiber. Appropriate eye protection should be worn and appropriate precautions taken when you are dealing with laser emitters that transmit above Hazard Level 1. ANSI Z136.2 contains more information on these precautions. Even nonlaser light may be a hazard if held close to the eye. A good rule is to *never look into a fiber.* The fiber wavelengths are in the invisible infrared spectrum anyhow, so looking only subjects you to a hazard, as you cannot see the light. The hazard exists even with unterminated fiber ends, so treat them as you would terminated fiber. Although there are visible light testers that are sometimes used during troubleshooting, you would be safer when using these if you simply point the connector or fiber at your hand and look for a dot of light, rather than trying to look directly at the fiber. Bad habits are hard to break, so don't start one.

The tiny glass slivers that result from cutting, cleaving, and sometimes polishing fibers are also an eye hazard. Always wear safety glasses when terminating fiber. You also don't want to inhale any of the fiber dust, but you should be able to avoid it when field terminating fiber cables under normal circumstances.

See Chap. 12 for more information on fiber optic installation practices.

Marking

As we discussed for station outlets, proper identification of telecommunications room terminations is crucial to a high-quality, serviceable wiring installation. This includes a consistent method of cable and termination numbering.

Each and every cable, termination fixture, patch panel, and cross-connect should be clearly identified. EIA/TIA 606 provides a method of marking just about everything in the telecommunications wiring world, with wiring terminations being no exception. The guidelines call for each item to be marked with a unique identification that is clearly visible. That generally means you should use $^3/_8$-in-high lettering and designate each cable termination and patch position in some orderly fashion, such as with the telecommunications room number, cable number, position, etc.

Some connecting blocks make marking difficult, while others seem to have marking panels as part of the design. Contrasting examples are 66 blocks and 110 blocks. Some older patch panels leave little room for markings, other than the small factory-marked jack number. In a very small installation, this does not cause too much difficulty, but larger installations require more detailed, explanatory markings.

You also may choose to use color to help designate wire termination areas. The official colors for grouping different types of cable terminations are given in Chap. 15. The colors are defined in EIA/TIA 569-A. *Commercial Building Standard for Telecommunications Pathways and Spaces.*

Workmanship

One of the keys to a successful LAN wiring installation is good workmanship. You can use all the finest components, the most expensive mountings, and the highest ratings, and still have a poor installation because of improper installation techniques. The telecommunications room is as much in need of good installation practice as any other area. In some ways, there are more places in a typical telecommunications room to do things that might adversely affect your cable link performance than elsewhere. For example, you may make one station cable termination, two cross-connects, and one patch termination all in the same cable link. Add to that the patch cords and equipment cords, and there is a lot of room for error.

The first workmanship guideline is to properly route the cables. You should minimize the sharp bends, avoid tight tie-wraps, avoid proximity to magnetic fields, and limit exposure to mechanical damage. You should ensure that the pairs are kept twisted as close to the point of electrical termination as possible. See Chap. 7 for an example of the effect of slightly untwisted wire.

The standards for preserving wire pair twists are defined in TIA/EIA-568-B. For Category 5e and Category 6 terminations, the maximum amount of untwist is 13 mm (0.5 in). For Category 3, it is 75 mm (3.0 in). The twist must be preserved within these limits at each point of termination to connecting hardware, including connecting blocks. The practice is not optional; it is mandatory. Even Category 3 links can suffer from the increased crosstalk and impedance mismatch problems of excessive untwisted wire.

The cable jacket should only be removed to the length necessary to terminate the pairs. This means that you do not strip back the jacket 4 or 5 in but only about 25 mm (1 in) more than is needed for the termination of a single pair.

When punching down wires in a 66-type block, there is a possibility that another adjacent wire may accidentally get cut, along with the wire you intend to terminate and trim. The punchdown tools usually have one bright yellow side with the word "CUT." If you wrap the wire in from the top of the contact, the cutting edge of the blade should trim excess wire on the bottom. To do this, the CUT side is down. If you see the word CUT, turn the tool around so you don't cut the wire in the wrong place. If you are using the minimum-untwist technique for Category 5e we described earlier (see Type 66 connecting blocks at the beginning of this chapter, you must use the tool in both orientations, as one wire of a pair will enter its contact from the bottom and the other from the top. This takes a little extra thinking and dexterity to master, but can be done quickly once you get used to it.

Do not leave the cut-off excess wire in between the contact rows of the block. You may accidentally force it into a contact slot or cause a short when you terminate another wire. Neglected trimmings will also make wire verification and tracing more difficult.

The requirements of TIA/EIA-568-B dictate a minimum bending radius of 4 times the cable diameter for certain categories of operation. The time-honored technique of pulling the station cable very tight and bending it back to the block before terminating the wires will have to be abandoned. You should leave the cable somewhat loose if you are trying to maintain this minimum bend radius. Many newer types of connecting blocks have wire management pathways, clips, ties, and other devices to prevent sharp bends in the cables. If your termination blocks do not have these accessories, you will need to use good habits when you terminate cables, in order to avoid problems. Remember that these guidelines also apply to cross-connect cable and patch cords. Pair twist must be maintained and the bend radius guidelines observed to maintain maximum rated cable link performance. A summary of workmanship and installation rules is shown in Table 8.1.

TABLE 8.1

Typical Workmanship and Installation Rules in the Standard

Document	Reference	Rule	Measure	Applicable to:
TIA/EIA-568-B	10.2.3	Terminate w/connecting hardware of same of higher category		Horizontal cables
	10.2.3˙	Minimum amount of untwist at termination, for 100-ohm cable:	13 mm (0.5 in)	Cat 5
		Minimum amount of untwist at termination, for 100-ohm cable:	75 mm (3.0 in)	Cat 3
	10.2.2	Maximum pulling tension	110 N (25-lbf)	4-pair UTP
	10.2.1	Minimum bend radius	> 4× the cable diameter > 8× the cable diameter	4-pair UTP 4-pair Sctp†

˙Mandatory requirement.
†ScTP = screened twisted pair.

Patch Panels, Cross-Connects, and Patch Cords

Chapter 9 Highlights

- Patch panels versus cross-connects
- Cross-connects
- Patch panels
- Patch cords
- Screened patch cords
- Fiber optic patches and cords

Patch panels, patch cords, and cross-connect jumpers form a vital part of LAN wiring. In this chapter, we will describe the wiring, cables, and devices used for cross-connects, patch cords, and circuit patching in cabling systems designed for LAN use. In the last chapter, we covered the actual termination of the station cables into punchdown blocks in the telecommunications room. We did not yet cover the cross-connect wiring or jumper cables. In addition, although we also included the wiring method of terminating cables directly into patch panels, we have waited until now to describe the patches themselves.

Patch Panels versus Cross-Connects

There is an ongoing debate about the alternatives of patch panels and cross-connects. Figure 9.1 shows the two alternatives to cable termination and the interconnection to equipment in the telecom room.

Figure 9.1
Two alternatives to cable termination and interconnection in the wiring closet. (*a*) Traditional punchdown and cross-connect. (*b*) Direct patch-panel termination. The asterisk (*) indicates optional components; patch cord may be connected directly to the common equipment.

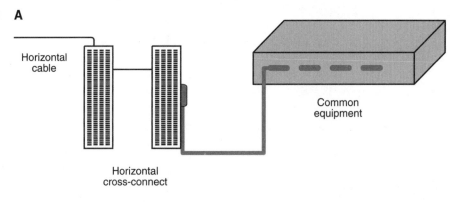

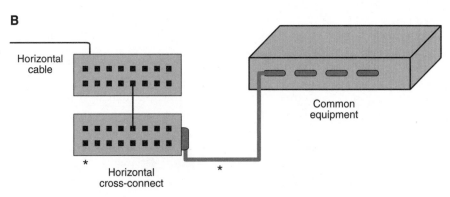

The first alternative shown in the figure copies the cross-connect methodology used for traditional telephone equipment room wiring. In this method, horizontal (station) cable is terminated on a connecting block, such as a 110 or 66 block. Then cross-connect wires are run from each station position on the horizontal connecting blocks to a second field of connecting blocks, which is a point of termination for multipair cables from the common equipment such as a telephone switch or PBX. This common equipment connection in a telephone system corresponds to the network hub connection in a local area network.

The common equipment connections and the station cable connections are rather like the input/output of the cross-connect field. Both types of cables are hard wired to the connecting blocks and then cross-connected with unjacketed wire pairs called *jumpers*. To make a connection across the field, a cross-connect jumper (consisting of one to four twisted pairs) is punched down on the appropriate contacts of a common-equipment connecting block, routed across the frame, and the other end punched down on the contacts for the station cable (the horizontal cable). Changes require you to rip out a punched-down cross-connect jumper and install a jumper to the new location.

Connections in the punchdowns are theoretically gastight, so corrosion and signal deterioration over the long term are rare. This method works fine for LAN wiring, although you must pay attention to the issues of pair twist, minimum bend radius, and other cable-care considerations when wiring LAN cross-connects, just as you did with the installation of the station cable. This first method is pure cross-connect wiring, without the use of any patch panels.

With the second alternative, the horizontal cables are directly terminated into modular patch panels (unless you use a hybrid method of termination, described later). Patch cords are connected between the patch panel jacks and the equipment hubs, and connections may be reconfigured at will. This method is very convenient for smaller installations that use hubs (or other equipment) with discrete modular jacks. You must use some sort of modular connection to these hubs, and the patch allows the use of individual modular cords, and avoids the use of multipair fan-out cables and connectorized punchdown blocks. It is also handy for companies that frequently move connections or add stations to the network.

Testing a patch panel installation is very easy. You simply disconnect a patch cord and connect your tester. The patch jacks are also more spread out than the cable termination areas on cross-connects, and this may make it easier to provide proper circuit marking. (If you have ever tried to mark on the little tabs of a 66 block, you can appreciate this a lot.) Newer designs for patch panel terminations use IDC connections that are similar to those on traditional connecting blocks or outlet jacks. These designs help preserve the pair twist on terminated cables and use internal jack wiring that is designed to meet the new categories of performance.

Direct patch panel termination of station cables, however, limits the flexibility of a terminated cable, because the station cables pairs are available only on the corresponding patch jack. To gain access to unused cable pairs, an external adapter or splitter must be used. Because of this limitation and other factors, there are differing opinions on whether you should use patch panels or cross-connect fields.

Obviously, there are valid arguments on each side of this question. On one hand, when you terminate your horizontal cables directly onto a patch panel, you avoid the muss and fuss of at least three terminations (the horizontal punchdown and the two cross-connect terminations). Using the direct patch approach also saves space in the wiring closet, saves installation labor, and provides for quick reconfiguration of your network connections. On the other hand, you still have all the patch cables to deal with, and each of your station cables is forever terminated in a single 8-pin jack. You have no easy way to grab a pair or two to use for another purpose, such as a modem line or a second network connection. In addition, the patch cords are normally available only in standard length increments (unless you make your own), and thus rarely fit a connection exactly.

On some points, the two opposing cable termination techniques each have disadvantages that compensate for their advantages. For example, the exclusive use of cross-connect fields eliminates the crosstalk and contact resistance problems of patch panel jacks, but it compensates by introducing two more terminations (each end of the cross-connect wire) where the finest workmanship may still introduce untwist to the link (thus causing more crosstalk, etc.). Also, the ease of moving connections on the patch belies the fact that an average mature installation may actually need only 2–5% of its connections moved in a year.

Although we have mentioned two termination techniques—direct patch termination and cross-connect field termination—a third hybrid method uses a combination of both. In this method, the horizontal cable is terminated in a connecting block (such as a 110 block). Then a cross-connect is made between the connecting block and a patch panel. The patch panel may allow for directly connecting the cross-connect wire, or it may be connectorized, requiring multipair cables that terminate in a second connecting block. The patch cords are connected from the patch panel to the hub equipment. This hybrid method offers wiring flexibility between the station cable and the patch: A cable may easily be reconfigured, shared, or moved to another patch position. Equipment connections may be made quickly or reconfigured at will, because of the use of patch cords. However, the patch jacks and cords are still there, with all their other potential disadvantages.

It's almost impossible to take sides in this argument, as there are valid reasons behind both viewpoints. However, you should know that there is a definite choice to be made and have an understanding of the issues surrounding that choice.

The next few sections will describe cross-connect wiring (the connecting blocks were covered in Chap. 8), multipair jumper cables, patch panels, and patch cords.

Cross-Connects

If your station cables are terminated on punchdown blocks, you must connect each cable to its ultimate destination by using a cross-connect jumper. Cross-connect jumpers consist of unjacketed, twisted pairs that are loosely spiraled around each other. The pairs are usually heat treated to make wires of each pair adhere slightly together. Cross-connect wire, as with station cable, is made from solid copper wire with plastic insulation. Unlike station cable, no covering jacket is used. The overall construction of each pair is similar to pairs used in jacketed cables, although the insulation may be a little stiffer to provide additional resistance to abrasion and cuts.

Cross-connect wire (it is usually not called cable) is available in 1, 2, 3, and 4 pair configurations. The pairs are usually color coded in the same manner as regular jacketed cables. The complete color code for these cables is shown in Chap. 3, but the short version is blue, orange, green, and brown pairs, each paired with a white wire. If you have only one pair, the blue pair (blue/white marking) is usually used. However, single pair cable is also available in a blue/yellow combination, as that is often used for Telco central-office lines, as well as other uses. If you are doing LAN wiring, you will want to use 2 pair of 4 pair cross-connect wire that is color coded the same as the station cable. To implement the full TIA/EIA-568-B, a 4 pair wiring system will require 4 pair cross-connect wire, unless you make two runs of 2 pairs for each cross-connect. That standard suggests that cross-connect wire consist of pairs with one wire having white insulation and another having a distinct color such as red or blue. The normal color code scheme complies with that recommendation.

Wire Size and Type

Cross-connect wire is available in various wire gauges including 22 AWG to 26 AWG, but the most commonly used gauge is 24 AWG, as with jacketed station cables. The wire guidelines in TIA/EIA-568-B state that standard-compliant UTP horizontal cable should use 24 gauge wire, but that 22 gauge wire that meets the transmission requirements of the standard may also be used. It is vague about the wire size of cross-connect wire, requiring only that the wire meet the same transmission characteristics as regular horizontal UTP cable.

Cross-connect wire is normally used in wiring closets so it does not need to be plenum rated. The normal insulation for cross-connect wire is PVC. Some local codes may require riser or plenum rating for these locations.

Wire Category

The performance categories for cross-connect wire are the same as for horizontal cable. The proper category of wire should always be used when making your cross-connects. This means that a Category 5e installation will require Category 5e cross-connect wire and that lower category installations will require wire that meets at least the specified link category. You may always use a higher category rating, as with any cable or component. For example, you may use Category 5e or 6 wire in a Category 3 installation.[1]

Because there is no jacket to show the wire ratings, the category of installed cross-connect wire is difficult to determine, unless it is Category 5e or higher wire. Category 5e cross-connect wire has an extreme amount of twist to each pair, so it is easily recognized, because there is no plastic jacket to hide the twists. Category 5e and Category 6 are difficult to distinguish, as are Category 3 wire and Category 2 (or uncategorized wire), since the appearances may be similar. You may need to keep

---

[1]Category 6 cable is simply a very high performance version of Category 5e and may be freely substituted without the 5e/6 connector compatibility issues.

track of all the cross-connect wire used in your facility. The category, if any, should be clearly marked on the wire spool. If you are in an environment where you have total control over all the wire and cable, you may want to use Category 5e or 6, the highest category of cross-connect wire available, for all your cross-connects. Keep in mind, however, that the tight twists also make this wire a little harder to terminate, particularly in 66 blocks.

There are subtle differences between category ratings from different manufacturers. As is the case with jacketed pairs, consistency in the geometry of the pair has a big influence on the transmission characteristics of cross-connect wire. This means that pairs manufactured to very tight specifications can produce a superior jumper that will have a very low return loss, indicating very consistent impedance. This carefully controlled manufacturing also achieves better performance in some of the other transmission parameters. You can indeed buy better wire and cable, but you will probably pay more for the privilege.

Routing and Marking

Cross-connect wires are routed between two termination points using the standoffs, brackets, and other wire management devices that were described in Chap. 8. Because of the lack of a jacket, cross-connect wiring may require a little extra care in handling to maintain performance characteristics. Some of the standard practices that are used in running telephone-grade cross-connects are not at all appropriate for the higher performance categories of LAN wiring. For example, it is common practice to sharply bend jumpers when they are wrapped around wire management brackets. These sharp bends may be less than the minimum bend radius recommended in cabling standards such as TIA/EIA-568-B. This standard calls for a bend radius to be not less than four times the diameter of the cable. Cross-connect wire has no jacket, so its overall diameter is not as easily determined, but it is probably less than jacketed cable. It is certainly easier to bend, as the jacket imparts additional stiffness to regular cables.

Connections on wooden backboards are fairly easy to route. If you need additional routing paths, you simply add the appropriate brackets or standoffs to the board. Wire is routed around or through the devices from start to end of a jumper. Distant blocks or patches just require longer cross-connect jumpers. If you use connecting blocks or patch panels that are in separate relay racks, as may be the case with large installations, you will probably need to route the wire up the side rail of the rack, across a wiring tray to the other rack, and down the other side rail. Wire management brackets, hangers, raceways, and trays are essential for proper protection and containment of the wire.

Cross-connects must be marked just as with all other wiring components, pathways, and cables. Obviously, the cross-connect wire itself is difficult to mark, so you will have to ensure that the points of termination of the cross-connect are marked clearly to indicate the identification of the cross-connect. The particular type of connecting block that you use will make a big difference in your ability to provide marking for each connection. Ideally, you would mark the termination in such a way that identifies both the point of termination and the associated connection at the other end of the cross-connect. In some cases, it may not be practical to mark anything other than the termination ID on the block. Some connecting blocks have additional labels, clips, covers, and brackets that can contain additional identification. However, some of these items have to be removed to terminate wires or to test connections, and thus they are easily lost.

You should keep comprehensive paper (or database) records of your interconnections to compensate for the shortcomings of connector block and jumper marking. The standards do allow placing a listing of circuit identifications adjacent to a point of termination, if the design of the devices makes direct marking impractical. Keep your records up to date, including all adds and moves, so that you can easily find the needed circuit links if you need to do troubleshooting or changes.

Workmanship

In addition to the routing and cable handling issues mentioned in the previous section, several other installation practices need to be observed in using cross-connect wiring for LAN cabling. Preserving pair twist is probably the most important single requirement to maintain the transmission characteristics of a link. Pair twist integrity becomes more critical with increasing link performance category. The very low speed links, such as LocalTalk and Arcnet, will operate on very slightly twisted pairs. The modest speeds of 10BaseT and 4 Mbps Token-Ring are not very particular about maintaining cable twist, as long as the amount of untwisted wire is minimal and the link does not push the distance limit. However, at the higher wire speeds of 16 Mbps Token-Ring, 100VG, 100/1000BaseT, and ATM, the preservation of pair twist becomes increasingly more critical.

The effect of adding connecting blocks necessarily adds some amount of untwisted conductor paths to any cable run. However, you can certainly minimize the link degradation by maintaining the natural twist of wire pairs as close to the actual point of termination as possible. The maximum amount of untwist allowed by TIA/EIA-568-B is 75 mm (3.0 in) for Category 3 terminations and only 13 mm (0.5 in) for Category 5e terminations. Category 6 untwist should be the same as that for Category 5e. That should certainly be easy to accomplish. If you use even less untwisted wire at any termination, you will simply have a higher-quality connection. The effects of untwist are somewhat cumulative, so neatness counts at every termination point in a link. You will find that some types of connecting blocks are more difficult to terminate with minimum twist than others. For this reason, many authorities do not recommend the use of traditional 66 blocks, even for Category 3 operation. However, some manufacturers do offer 66 blocks rated at Category 5e that should operate at the prescribed level of performance if the proper termination procedures are used. Chapter 8 describes a technique for minimizing the amount of untwisted wire when using a 66 block.

Many multipair cables are manufactured with a carefully controlled variation in twist pitch and pair spiraling in order to meet tough performance parameters such as NEXT (near-end crosstalk) and SRL (structural return loss). These parameters are maintained, at least in part, by the cable jacket that holds the pairs in their relative places. Cross-connect wire has no jacket, which means that you may have more difficulty keeping the pairs in approximately the same geometry, but you should try to do so if possible. It may be beneficial to use cross-connect wire that already has the number of pairs spiraled together that you need for the jumper connection, rather than making the connection with two or more individual jumper pairs. Using jumpers with the proper number of pairs will also make wire tracing and connection easier.

Remember to remove any unwanted jumpers completely. Nothing is more frustrating than having to fight through abandoned jumpers to trace down a connection or to make a new cross-connect. The cross-connect field should always be neat and orderly. The jumpers should be placed in wire management brackets or other devices and loosely bundled. (Hook and loop straps are good for this.) Never use tight tie wraps to bind jumpers together. The tie wraps distort the pair geometry and can produce the same performance degradation as with regular jacketed cables.

Punchdown (connecting) blocks are intended to have only one wire connection per contact. That means you should never punch down a second crossconnect wire in a slot where another wire is already punched. The integrity of the insulation displacement method used in these contacts depends upon the tight positioning of the contact jaws. The jaws are opened somewhat during the termination of the first wire, and a second wire will often not make good electrical or mechanical connection with the contact. The traditional 66 block is the most prone to this type of wiring error, because incoming and outgoing wires of a connection are lapped over each other as they are introduced to the block for termination. The cross-connect wires should be placed on the second (inner) set of contacts on this type of block. The 110 and similar types of connecting blocks are less subject to this type of error, because they physically separate incoming and outgoing wires on opposite sides of the contact block.

Patch Panels

Patch panels are used to provide flexible connections between horizontal station cables and the equipment ports in the wiring closet. Patch panels have modular jacks that are connected so that they correspond to work area outlet jacks or to equipment ports. An example of patch panels is shown in Fig. 9.2. The ports on LAN equipment, such as hubs, often use 8-pin modular (RJ-style) jacks, so it is easy to connect a port to a patch panel by using a modular equipment cord (or patch cord). This provides sort of a natural LAN connecting environment, where all connections that the user makes involve familiar modular jacks and cords.

Patch panels are easy to install and are a little less confusing to the user than punchdowns, because each jack on the horizontal wiring patch corresponds to an identical jack at the workstation outlet. Patches also provide easy access, quick connections, and clear circuit marking (at least until they become covered with patch cords). Another aspect often overlooked is that adds, moves, and tests are all accomplished without using any tools. If appropriate wire management devices are used, the patched network can really be kept quite neat.

Specifications

A patch panel intended for LAN wiring use has an array of modular jacks that are mounted in a metal panel. The panel is generally wide enough to mount directly in a 19-in equipment rack, while its height

Figure 9.2
Typical patch panels.
(*Courtesy of The
Siemon Company.*)

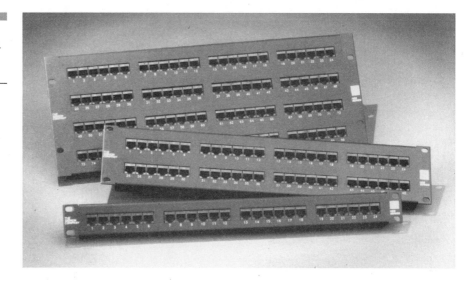

Figure 9.2
Typical patch panels.
(*Courtesy of The
Siemon Company.*)

varies with the number of jacks on the panel. Panels may be of any size, with multiples of 12 to 24 jacks being common. Patch jacks often correspond to punchdown block connections, which dictate the multiple-of-12 convention. For example, a group of twelve 4 pair circuits can be accommodated by both sides of a 66 block (50 pairs) whereas a group of twenty-four 4 pair circuits can be accommodated by a 100 pair 110 wiring block. Additionally, the number of 24 is convenient to use, because a row of 24 modular jacks fits nicely across a 19-in panel width.

The connectors on a LAN patch panel are standard 8-pin modular jacks, wired either to T568A or B. The connections for station cables at the rear of the panel are usually color coded by pair number, which corresponds to the wiring pattern at the jack. It is possible to use other wiring patterns for special applications, but that isn't covered here. The connectors are spaced according to the manufacturer's preference, and you may find individually spaced connectors or connectors that are placed in groups of four or six adjacent jacks. The connectors are usually numbered according to their position on the panel. Optional circuit marking is described later in this section.

If you still have questions on whether to use the T568A or the T568B wiring pattern, see the discussion in Chap. 7 under "Standard Jack Pinouts." For station cable wiring, it basically doesn't matter which you use, as long as the outlet jacks use the same pattern. Patch panels that use color-coded wire terminations are prewired to support either the T568A or the T568B pattern, so make certain which you order. You could use the opposite pattern for your needs, but you would have to reverse the orange and green pair (Pairs 2 and 3) and be forever confused when it came time to add new cables.

Standard patch panels will generally use some type of insulation-displacement connection (IDC) at the rear of the panel to terminate station cables (or other cables, such as those from nonmodular hubs and other equipment). The two basic design variations for patch panels either use groups of connectors that have a common termination block, or use individual connectors that have their own terminations at the rear of each connector. Another type of panel uses multicircuit connectors and is described in the next section.

The terminations for patch panels (Fig. 9.3) are identical in function (and often in appearance) to the terminations used in workstation outlet jacks or in connecting blocks. Variations in panels are available that use individual IDC connections, 110C blocks, 66 blocks, BIX blocks, and Krone blocks. Each method has the advantages and disadvantages that were described in Chaps. 7 and 8, and you should choose according to your preferences. It should be pointed out that the use of standard 66 blocks is not generally recommended for LAN wiring, particularly

Figure 9.3
IDC patch-panel termination.

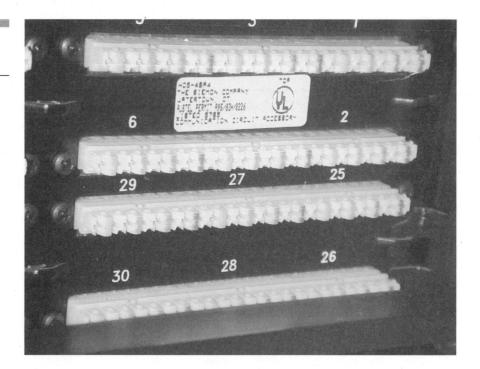

above Category 3. The use of newer Category 5e-rated 66 blocks is still somewhat controversial, but if the manufacturer has UL-verified performance, you can probably use the blocks in this application.

Patch panels are connecting hardware, as far as the standards are concerned, and must be performance rated according to the category of operation you intend. Older-style panels paid little or no attention to preserving the pairing and twist of wires between the termination points and the modular jacks. You could get a nice glitch on a tester when you looked at an older patch panel in a LAN link. Fortunately, most modern panels are rated for Category 3, 5e, or 6 operation. The TIA/EIA-568-B standard, as well as others, requires that all connecting hardware be marked as to its category of performance. Either the markings may be the words "Category n" or "Cat n," where n is the category number 3, 5e, or 6, or they may be a large capital letter "C" with the category number enclosed. If you do not see the marking, you should consider the panel unrated, and not suitable for LAN wiring use. Some manufacturers are providing a small category rating note along with the panel, certifying the category of performance. The only problem with this method is that it may be impossible to determine the rating of the panel after installation. You could provide the postinstallation assurance by affixing the notice to the rear of the panel, if a clear area is available. Still, clear factory marking of the category is the best bet.

The workmanship issues discussed in the previous sections on cross-connects and punchdown blocks certainly apply to patch panels. You must ensure that the cables to the back of the patch panel are properly routed and terminated. The termination procedures are the same as for the corresponding type of connection at a workstation outlet or punchdown block. The termination techniques and procedures to minimize untwist, kinks, and bends are described in Chaps. 7 and 8.

Wire Management

Many of the potential problems that you will have with patch panels involve the routing and dressing (putting in place) of terminated cables and connected patch cords. The wide selection of wire management devices can help get the myriad of wires under control and out of the way (Fig. 9.4). Some patch panels come with ingenious methods of cable routing and strain relief for the station cable terminations at the rear of the panel. The use of external wire management trays, such as the slotted

Figure 9.4
Wire management
on patch panels.

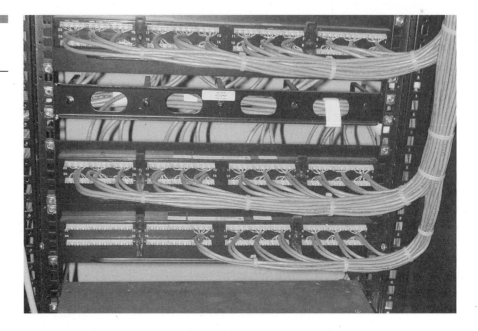

raceways from several vendors, can eliminate the need for individual wire support and binding. Similar slotted raceways are available for management of the patch cords on the front of the panel. If the panel is not too many rows high, the patch cables can be routed vertically from the jack to the nearest cable raceway and tucked out of sight. From two to four rows of patch jacks can be accommodated easily, representing 24 to 48 patch positions on an average panel. At the side rails, vertical raceways can be provided to route the cords to patch or equipment positions on other levels of a rack. Cable trays and ladders (open bottom trays with cross pieces between the side runners) can be used to route the cords from rack to rack across the top of the racks. Computer rooms with raised floor panels can route the cables beneath the floor, but the above-rack method is much easier to maintain and can be made to look as neat.

Connectorized Patch Panels

A type of patch panel is available with preconnectorized jack connections, rather than terminations for individual cables. This type of patch is used extensively with the connectorized punchdown blocks, as described earlier, and it may be useful for the multiport equipment connections of larger hubs. The connectorized patch panel has one or more multicircuit connectors mounted at the rear of the panel. The modular jacks are each connected to the appropriate pins on the connector. Six 4 pair, 8-pin modular jacks can be connected to a single 50-pin "telco" connector (also called an Amphenol connector), if all pins of the jack are connected. This means that you would have to use four 50-pin connectors for a 24-position patch panel. Some patch panels connect only two pairs to each jack position and support 12 jacks per 50-pin connector. These panels are wired to support only one style of network connection, such as 10/100BaseT (which uses Pins 1, 2, 3, and 6) or Token-Ring (which uses Pins 3, 4, 5, and 6). Obviously, the styles may not mix, so you may have to replace the panel if you change your network hardware.

The mapping of connections from the connector to the jack is critical, as each pair on the jack generally has a preassigned circuit function, such as Transmit +/−. If used to provide connectivity to a multiport hub card, the connectorized patch panel takes the place of the modular equipment jacks. In a conventional modular jack hub, the hub connector pins are each assigned a specific network function, and any 4 pair equipment cord wired straight-through will automatically put the proper signals on the pins of the horizontal patch panel (and thus the cable). If a connectorized patch panel (Fig. 9.5) is used, the signals are mapped

Figure 9.5
Connectorized patch panel.

from a multicircuit connector to each modular jack on the panel. Instead of an equipment cord from the hub to the horizontal patch, a patch cord is connected from each hub-port jack on the connectorized patch to the corresponding station jack on the horizontal patch.

If the mapping from the 50-pin connector to the jack is wrong, the signals will be reversed or totally miswired and the connection will not work. The standard mapping for connections between the 50-pin connector and each jack is dependent upon pair numbering. It is customary to connect the pair color codes in the order given in Chap. 3 for the wire pairs. In other words, pair 1 of a 25 pair (50-wire) cable connects to Pins 1 and 26 on the connector, pair 2 connects to pins 2 and 27, etc. However, the connections do not go to a 25 pair cable but lead instead to individual jacks, either four or eight wires at a time. So a 4 pair jack panel connection would have pairs 1 through 4 connected to the first jack, pairs 5 through 8 to the second jack, and so on, in turn, up to the sixth jack (all that can be served by one 50-pin connector).

Then at each modular jack, the pairs are connected in a pattern consistent with either T568A or T568B. This distinction is important, as the two patterns reverse the connections for pairs 2 and 3. Remember that T568B is equivalent to Lucent (AT&T) 258A and is probably what you will get if you don't specify. You should always specify the patch jacks and connectors to be mapped to T568B or T568A, whichever matches your hub cord. You are still free to use a different pattern for the horizontal cable and patch, since it is separate from the hub-to-patch connection. Panels that use all eight pins of each jack are clearly identified as to pattern mapping, but panels that use only four pins use a skeleton of one of the TIA patterns. As we said, hub connections often use only two pairs per port (four wires) to minimize the number of ports per connector.

Here is an example of the problem of connection mapping. Let's say you intend to install a 10/100BaseT network using hubs with multicircuit connectors instead of single port jacks. Such hubs typically use a card rack with plug-in cards. You will run cables from your hub cards to a patch panel to fan out the individual ports to modular jacks. Naturally, you will want the 8-pin modular jacks on the patch to have the same pinouts as a standard 8-pin modular jack on a hub with individual connectors. A standard 10/100BaseT modular hub jack defines pin 1 as transmit +, pin 2 as transmit −, pin 3 as receive +, and pin 4 as receive −. A multicircuit hub connector could assign eight pins to each port, but only four pins are needed for 10/100BaseT, so assigning four pins per port achieves twice the density on the connector. For our example, let's say that your hub card connector assigns four pins per port, representing two pairs of the multiconductor cable. This gives you a total of twelve 10/100BaseT ports per 50-pin connector. The cable connects between the hub card and the patch.

Now, let's look at the way the ports are mapped from the 50-pin connector to each modular jack on the patch panel. Standard 10/100BaseT uses four pins of the 8-pin jack. These four pins correspond to two of the regular 4 pair groupings of the jack, pairs 2 and 3. Remember that T568A and B simply reverse the order of these two pairs. The "first" pair for each of the ports on the 50-pin connector usually corresponds to "transmit +/−". Now, here is the mapping part: the transmit +/− pair must connect to pins 1 and 2 on the 8-pin jack, while the receive +/− goes to Pins 3 and 6, for proper 10/100BaseT connectivity. If the patch uses the T568B pattern to connect the patch jack, the transmit and receive pairs will appear on the proper pins. But if the patch is wired for the T568A pattern, the two pairs will be in reverse order and the connection will not work (unless the pair order on the 50-pin connector has also been reversed—a rare circumstance because the TIA recognizes only the standard color code pattern for that connector, not a pair order).

This is one place where the TIA's preferred T568A pattern will get you into trouble. Note, however, that all of your modular patch cords, horizontal cable patch panels, and workstation outlet jacks can still use the A pattern. The rule is that both ends of a connection must be the same, whether that connection is a horizontal cable, a patch cord, or (yes) a patch panel fan-out of a multicircuit hub port. Always ensure that you match the LAN connections on the hub card's connector with the connector-to-jack mapping on the patch panel so that the correct LAN signals will be on the pins of the patch panel jacks.

Connectorized patch panels may be rated by category as with any connecting hardware and should be properly marked. The actual internal wiring between the connector and the jacks must be appropriate for the category of operation, such as Category 5e/6. The manufacturer will normally provide testing and verification by an independent laboratory, such as UL, to attest to their claimed category of operation. As with regular patch panels, connectorized panels that are not marked or otherwise certified, should not be used in any application that requires Category 5e or 6 performance.

Location and Marking

Patch panels are usually designed to be mounted in 19-in relay racks. They also mount on simple, freestanding mounting rails, in equipment cabinets, and on wall mounts. You should carefully plan the location of your patch panels, as you should all your connectivity hardware. If you choose to use patch panels, rather than direct cross-connection, you will have also to plan for routing and dressing of the patch cords from origin to destination jacks. If you route the cords using wire management brackets and accessories, you will avoid the "cable curtain" that sometimes infects patch panel installations.

As we said before, patch panels may be used for both the horizontal cable and the equipment (hub) port connections. If you use patches for both, you will have to provide a telecommunications room layout that brings the two types of panels conveniently together for reasonable patch cord lengths. If your installation involves the frequent reconfiguration of networks and workstations, you should definitely plan to have the two panels as close together as possible. In large installations, each type of patch location may actually be composed of several panels. You may want to place them on multibay relay racks (side by side) so that patch cords can go directly across between racks to make connections, rather than climb the rails in a route across the top of the racks. Wire management accessories for adjacent racks are available to make this job easier (Fig. 9.6).

If you terminate the horizontal cables in punchdown blocks and then cross-connect to patch panels, you should plan the layout in such a way that you minimize both the length of cross-connect wire and the

Figure 9.6
Patch panel marking.

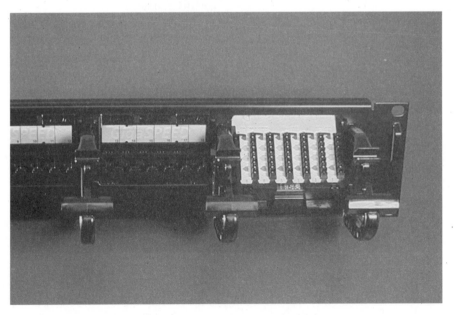

length of equipment cords (or patch cords) to the network hubs. You can easily mount punchdown blocks and patch panels on wiring closet backboards (wall mount), but the hubs will probably mount in an equipment rack, unless the installation is very small. This may increase the length of the cords from the horizontal patches to the equipment connections. Proper planning will minimize this distance.

You should carefully plan your telecommunications room installation, including the location of the patch panels and hubs, to keep the total cross-connect and cord lengths within the TIA/EIA-568-B guidelines (Fig. 9.7). You will recall that the maximum length of horizontal cabling is 90 m (295 ft). An additional length of up to 10 m (33 ft) for each link is allowed for horizontal cross-connect, jumpers, patch cords, patch panels, and user/equipment cords. That makes the recommended total length for all of the wiring in a horizontal link (Channel) 100 m (328 ft). This is the distance from the workstation, all the way to the final hub or equipment connection in the wiring closet, including the user cords on each end. (The actual connectors at each end are excluded in the Channel measurements.)

The standard also recommends maximum lengths cable in the work area and wiring closet which make up the 10-m (33-ft) allowance. For example, the total length of the horizontal cross-connect facilities should be no more than 5 m (16 ft). This distance includes the cross-connects, the patch cords or cross-connect jumpers, and the equipment cords in the wiring closet. The standard also recommends that the user cords in the work area be no more than 5 m (16 ft). Combining these two lengths yields 10 m (33 ft), although the rounding off of conversions by the standard makes the math look rather strange.

Work area cords of only 5 m (16 ft) are rather short for many office environments. As these cord and jumper lengths are only recommendations, you could adjust them to fit your needs, as long as you keep the

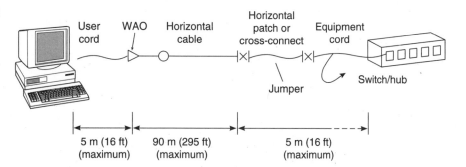

Figure 9.7
User and equipment cord distances.

total length under 10 m (33 ft). However, the only way to guarantee that you can make this tradeoff is to carefully plan the wiring closet installation so that you minimize the patch cord and equipment cord lengths. That will give you more length for user cords in the work area.

What difference does it make if you exceed these lengths a little bit? If the total Channel length is less than 100 m (328 ft), you will probably have no problems; however, if you run a longer link, it might not work properly. Network equipment manufacturers design their network adapters and hubs to operate over the standard link distances and characteristics. If you exceed those distances, or in other ways provide a substandard link, the networking components may not operate on that link. Also, keep in mind that some of the performance parameters of patch cords and cross-connect wire are worse than those of regular horizontal cable, so you may not be able to just borrow some of the allowed horizontal cable length and add it to the patch cord allowance.

Open office work areas are now allowed to exceed the user-cord limits by decreasing the horizontal run to a consolidation point. See Chap. 11 for details.

In addition, a link that is too long will fail a cable scan. Just how long is too long? Cable length measurements are a part of the tests done by a cable scanner. These test procedures are supplied in a supplement TIA/EIA-568-B (originally TSB-67). TIA/EIA-568-B has length test limits for both the Basic Link and the Channel. If you exceed the test limits, the tester is required to fail the link. The test requirements for cable links are fairly detailed and are covered in Chap. 16.

If you test the part of the horizontal link called the Basic Links,[2] it excludes the user cords, jumpers, and patch cords. When these cords are added in, the total link is defined as a Channel. You might later add cords that make the Channel exceed the allowable length, even though the Basic Link test was passed. If you exceed these lengths, your network equipment may not operate properly and it will be nobody's fault but your own.

Proper marking of the patch positions is also an important part of an installation. Generally, each wiring closet, cable termination, rack, patch, and hub is required to have a unique identification, according to EIA/TIA 606. In the case of patch panels, you will have to provide your own circuit or jack position identification separate from the panel's numbering, unless you have a very small installation and can use the jack numbers as your station numbers. The detailed requirements for

---

[2]The Basic Link and the newer Permanent Link are specified for similar measurements, as detailed in Chap. 2, Fig. 2.8, and in Chaps. 5 and 16.

circuit identification have occurred fairly recently, and it is also difficult to provide labeling for closely spaced jacks, so many patch panel manufacturers just provide a stick-on label holder that can be positioned above or below a row of jacks.

A typical patch position might need to be marked with a wiring closet number, patch number, cable number, and jack position number such as "W4-P10-C321-J15." However, the space allowed for each jack might require this to be shortened to "C321-J15," with the patch number marked on the panel itself and the closet number obvious from where the panel was located. The rules for marking identification numbers are part of an emerging standard and will get more detailed as time goes on. Panel manufacturers will begin to incorporate more of the marking guidelines in their product's features. For the time being, we may have to make do with adhesive label strips and other marking methods.

Patch Cords

A patch cord consists of a length of flexible cable, terminated at either end by an 8-pin modular plug. The patch cord, in reality, is identical to the user cables in the work area and the wiring closet. As a matter of fact, all of these cords go by a number of names that reflect their purpose, rather than their construction. For example, a cord from a patch to another patch is called a *patch cord,* but an identical cord from a patch to a hub is technically an *equipment cord.* The cord from the workstation to the outlet connector is also an equipment cord. Equipment cords are sometimes called *user cords,* because they are presumably connected by the user, not the cable installer, to the user's equipment. Oh, yes, all of these cords are sometimes called *cables,* instead.

The one distinguishing feature of the patch cord is that it is usually shorter than the user cords. We will describe the wire and connectors for these types of cables in the next chapter. In this section, we will concentrate on the characteristics that make a good patch cord, with the understanding that these same characteristics apply to all user cables.

UTP Patch Cords

Now that we have all that terminology out of the way, what makes a good UTP patch cord? As we mentioned, a patch cord should be flexible—

which means it should use stranded wire with a flexible plastic jacket. Patch cords usually consist of four pairs of 24-AWG, 100-ohm, stranded copper, thermoplastic-insulated wires with an overall thermoplastic jacket. As with horizontal cable, 22 gauge wire is allowed but rarely used. The plastic insulation is often PVC or a compound with similar characteristics. The cords are used in a work area, not a plenum space, and usually do not require the stiffer plastic insulations that are plenum-rated.

Patch cord cable may have any wire colors, but the colors of the standard 4 pair color code are normally used. TIA/EIA-568-B also offers an alternative color code, consisting of eight unique solid colors. Actually, the color code is only important if you make up your own patch cords from bulk cable. The colors will help you get the wires in the correct pin positions before you crimp on the modular plug.

Patch cords have their own set of performance characteristics that are slightly different from those for the horizontal cabling. Most of the transmission requirements are the same, but an additional 20% increase[3] in attenuation is allowed for the stranded construction and "design differences." This is one of the reasons that the lengths of the patch and user cords are limited by the standards. The attenuation limits vary for the three categories of performance and are prorated from the values per 100 m (328 ft). The allowable attenuation values for patch cords are shown in Table 9.1

If you purchase preassembled patch cords, you should be sure that the cords are certified by the manufacturer to comply with TIA/EIA-568-B for the appropriate category of performance. Certification testing by an outside agency, such as UL, is a good indication of quality and assurance. The testing of patch cords presents somewhat of a dilemma, both for the manufacturer and for you. The standard contains detailed performance requirements for the cable and the connectors that are used to make up patch cords, but there are no specifications for assembled cords. In addition, certain of the tests, such as NEXT, are known to be unreliable for links less than 15 m, because of a phenomenon called resonance. Many testers will not test a cable that is less than 6 m, except for wire map. Because of these factors, the only assurance of patch cord performance you will have (until the cord becomes part of a larger link) is the use of quality components and cable, and good workmanship. Workmanship is important because the cable pairs must be untwisted before the modular plug can be attached. If the amount of untwist is not minimized, the plug will contribute even more degradation to the link than is inherent

---

[3]ISO/IEC 11801 allows a 50 percent attenuation increase on stranded cords, which results in some parametric differences in the acceptable links.

TABLE 9.1

Patch Cord Maximum Attenuation Values

Frequency	Cat. 3	Cat. 4	Cat. 5	Cat. 5e	ScTP/26 Cat. 5
0.772	2.7	2.3	2.2	2.2	2.7
1	3.1	2.6	2.4	2.4	3.0
4	6.7	5.2	4.9	4.9	6.2
10	11.7	8.3	7.8	7.8	9.8
20	n/a	12.0	11.1	11.1	14.0
31.25	n/a	n/a	14.1	14.1	17.6
100	n/a	n/a	26.4	26.4	33.0

NOTE: Attenuation values above are given in dB per 100 m (328 ft) at 20 degrees Celsius. The measured attenuation of a particular cable will depend upon its actual length, but the value (using the worst-case pair) should be less than the prorated value at 100 m. The values in the chart are calculated from a formula at specific frequencies. The allowable attenuation of a stranded wire cable is 1.2 times the value of the formula in Paragraph 6.4.1 in TIA/EIA-568-B.2, prorated to the actual length of the cable under test.

SOURCE: TIA/EIA 568-A and -B.2.

in its design. By the way, the unavoidable amount of untwist in the modular connector design is one of the reasons that these connectors are not looked upon favorably for high-frequency performance. Annex B of TIA/EIA-568-B.2 shows a detailed procedure for assembling modular cable ends. Although this procedure is for the preparation of connectors that are to be used in the testing of connecting hardware, it represents an ideal method to maximize the performance of patch cord plugs.

Screened Twisted Pair (ScTP) Patch Cords

Screened Twisted Pair (ScTP) Patch Cords are a new addition to the TIA/EIA-568-B standard, and are described in the normative Annex E, which means the annex sets mandatory performance and installation requirements when ScTP is used in lieu of UTP. As frequencies rise on network cabling, the use of ScTP is increasing to minimize interference emissions and to decrease susceptibility to outside interference sources.

TIA-recognized ScTP consists of 4 twisted pairs with an overall shield. The cable impedance is specified at 100 ohms, and either AWG 24 or AWG 26 plastic-insulated, stranded conductors may be used. The standard allows a 50% increase in attenuation for cords made with AWG 26 stranded wire, which requires a commensurate decrease in the allowable user and equipment cord lengths. Some sample attenuation values are given in Table 9.1.

The total length of the work area cord plus the total of the cross-connect/jumper/patch cords must not exceed 8 m (26 ft) for AWG 26 cords. The total of the cross-connect/jumper/patch cords must be 5 m (16 ft) or less, while the user cords are limited to 3 m (10 ft). If AWG 24 ScTP cords are used, the same lengths of UTP (10, 5, and 5 m) are still allowed. This makes a case for using the slightly bulkier AWG 24 cords in many applications.

You should remember that a ScTP installation requires 100% use of shielded components. This includes the cable, connectors, patch panels, and even equipment connectors and NIC interface jacks. In addition, you now have a ground (earth) potential conductor running throughout your cabling system, requiring special care for grounding and bonding of all cabling components and equipment.

Fiber Optic Patches and Cords

Fiber optic links may also use patch panels and patch cords in the wiring closet. The very nature of fiber optic interconnection makes almost every fiber connection an optical patch, since the light path must be preserved. Optical cables from workstations or from other wiring closets typically terminate in termination boxes that allow excess fiber to be protected from accidental damage. The termination boxes may serve as the patch point for a small number of connections, but larger installations will have a separate patching location that serves all of the incoming and outgoing fiber cables.

Optical patch panels that are designed for TIA/EIA-568-B compatibility may use the very same passive duplex SC adapter that may be used at the work area outlet. Alternatively, any of the SFF connectors described in Chaps. 7 and 12 may be used. The fibers of optical cables are terminated directly into fiber connectors instead of passing through some intermediate device, such as the connecting block used with copper wiring. The fiber terminations and connector types are described in detail in Chaps. 7 and 8. We will concentrate on the physical arrangement of the patch panels and the characteristics of fiber patch cords in this section.

A fiber optic patch panel consists of an array of duplex SC adapters, hybrid adapters, or SFF jacks. If the entire installation, including the fiber optic hubs, repeaters, or network adapters, uses the same type of fiber connector, then the array can be made up of compatible adapters or jacks. However, it may be necessary to convert fiber connections between other connector types. There are hundreds of types of LAN equipment with fiber optic interfaces that use one of the other connector types. For the foreseeable future, there will be a need to provide a conversion between the various connectors. The TIA/EIA-568-B standard contains suggestions for migrating to the new connector systems. The reason the standard has chosen to allow the SFF connectors is to encourage innovation and allow quick connection/disconnection and easily combined polarized pairs.

To convert between fiber connector types, you need to use a hybrid adapter or a conversion cable. A hybrid adapter is a passive coupler that joins two different connector types, while a conversion cable simply has one connector type on one end and the other type on the opposite end. For example, to convert an ST equipment interface to SC, you need a fiber cable with one of the connectors on each end. The duplex (two-fiber) cord described later is convenient to use, because you will need two fibers for each link connection. Some of the other types of fiber connectors usually have no built-in polarization, so you will have to pay attention to which connector goes to transmit and which to receive. The standard, at present, gives no guidance as to which A or B position of each orientation (AB or BA) should contain the transmit signal and which the receive. The designation of one fiber as transmit and one as receive is a little difficult anyway, because the transmit from the one end of a cable is the receive from the other. You will have to create your own standard practices for which connector position at the patch panel is transmit or receive.

Fiber optic patch cords are usually made from flexible duplex fiber cables, often referred to as "zip cords" because of their resemblance to the household electrical cords by the same name. Fiber zip cords differ in that the conductors are replaced by fibers, insulation does not have to meet any electrical specifications, and color is usually bright. Because the fibers are always patched in pairs, the dual-fiber construction is handy in that it easily splits out for two individual connectors.

Most SFF connectors are duplex by nature, and A/B orientation is part of the design. A duplex clip can be used to attach the two SC connectors together. Fiber connector pairs that are thus clipped together have an orientation just as the adapters do. This is so that the A and B connections will be placed into the proper position in the mating adapter. There are two orientations for the connectors of a fiber patch cord, the AB and BA orientation, which were described in Chap. 8. It does not really matter which orientation you use for each end of the patch cord, as long as the two ends of the cord have opposite orientations, and the fiber polarity is flipped between the two ends.

Adapters and connectors may be marked with a distinctive color, such as red or white, to assist in the installation and later connection of mating plugs. The TIA standard does not specify or define these "shadings" at the present time, but you can use them in your facility to help differentiate the A and B positions of the mating connections. Remember that neither the A nor B designation specifies the direction of the optical signal but merely acts similar to pin numbers in a conventional connector. Furthermore, since every fiber cable and adapter coupling reverses the A/B order, you could not tell which fiber carried which signal, even if you knew how the signals started out, unless you traced every fiber through all the polarity reversals. As a user, the only thing that you would know is that the reversals had been arranged in such a way that whatever signal started out on the A position at one end of the link will be on the B position at the other end. This is done to create the transmit/receive reversal that must be done for a fiber link to function.

Remember that some fiber optic links may use laser illumination that is potentially hazardous. Most LAN connections that you encounter will use optical power levels far below that which would actually cause a problem, but you should still use reasonable precautions when working around fiber. As we cover in Chaps. 8 and 12, you should never look directly into an open fiber connector (or an unterminated fiber that might be part of an active circuit). The light wavelengths used for these signals are in the infrared part of the spectrum and are quite invisible to the eye. Some fiber transmitters that are intended for long distances may actually have levels of light that would be hazardous, so just don't get into a bad habit.

10

User Cords and Connectors

Chapter 10 Highlights

- Modular cable-end connectors
- RJ-45: What's in a name?
- Connector pin-outs
- User cords
- STP-A Data Connectors

This chapter will describe the various types of connectors that are used to terminate cable ends. We have previously covered station cable termination, so this section will concentrate on the modular plugs, coax cable-end connectors, and Data Connectors that are used on freestanding cables such as patch cords and user cords (sometimes called user cables). We will also look at the user cords that are used in the work areas and the equipment rooms for connection to network hardware.

This is a good time to point out the distinction between "cables" and "cords" as used here and in many of the standards. Obviously, both cables and cords used in telecommunications employ jacketed, multi-paired cables, but it is customary to refer to the flexible, stranded-wire, plug-terminated cables that connect to workstations, hubs, and patches as "cords." This is the convention followed by most of the standards. The usage rule is probably more consistent when referring to patch cords, which are inevitably referred to as "cords." The terms "user cord" and "user cable" are used interchangeably, as are the terms "equipment cord" and "equipment cable." For consistency in this book, we have tried to use the word "cable" to refer to any solid-conductor cables that are terminated in a jack or punchdown, and use the word "cord" to refer to any stranded-conductor cables that are terminated in modular plugs. With regard to coax cables, they are normally referred to as "cables," regardless of their location. By the same token (pun intended), IBM Data Connector cables are normally referred to as "cables." Both of these last cable types are identical in composition and termination whether they are used in the wall or by the equipment user.

Cable-End Connectors

We will cover three general types of cable connectors in this section: modular connectors, coax connectors, and IBM Data Connectors. Fiber optic connectors are described in Chaps. 7, 8, and 16 and will not be covered here. Of the three other types of connectors, the modular connector is obviously the most favored for modern LAN wiring, because of the increasing use of unshielded twisted pair cable. Coax has long been used for traditional Ethernet and Arcnet LANs, but is gradually being phased out in most locations. The Data Connector is an important connector in shielded twisted pair wiring systems and it is now included in the TIA/EIA-568-B standard.

RJ-45: What's in a Name?

The 8-pin modular connector is often improperly referred to as an "RJ-45." The terminology is in such wide use that even manufacturers who know better frequently use the RJ-45 name in their marketing materials. Some will call it RJ-45 "style" or "type." In truth, the RJ-45 designation is for a particular interface most often used for programmable analog modem connections to leased telephone lines. This true RJ-45 just happens to use the 8-pin (8-position, 8-contact), 8P8C, modular jack and plug that are found in many other LAN and telephone connections that have nothing whatever to do with the original intent of the RJ-45 connection. Many other common interfaces use the same type of modular connector, including RJ-48, 10/100/1000BaseT, Token-Ring/UTP, UNI-PMD, T1, ISDN, and on and on.

This common usage error would matter very little if you were never going to use high-speed data networks. However, the TIA/EIA-568-B version of the 8-pin modular connector interface has very specific performance requirements that may not be met by the generic nonrated connector that simply goes by the RJ-45 name. It would have been convenient if the TIA had given this connector a new name commensurate with its specifications. Perhaps it could have been called the "T568-8/Cat5e" connector, or how about just "T568-RJ" to borrow from both the past and the present.

When will people stop calling it an "RJ-45"? Don't hold your breath! Don't grate your teeth either! Just get used to thinking that a very special type of connector is meant when a "RJ-45" is used in your LAN wiring system. And, oh yes, never ever use the "RJ" word on the Net!

Modular Connectors

The modular connector used in LAN systems is the same connector that is used in modular telephone wiring systems. This connector is available in several possible sizes and pin configurations, ranging from four to eight positions, with from two to eight contact pins. The popular styles are often referred to by a USOC (Universal Service Order Code) "RJ" number, even though the connector may not actually be used in the designated application called for by the USOC code. For example, the common 6-pin modular telephone plug is often called an RJ-11, and the 8-pin modular plug is often called an RJ-45. The 8-pin modular plug is used in the TIA/EIA-568-B wiring standard for both telephone and data. The 8-pin modular plug is also the connector that is used for 10BaseT, 100BaseT, 1000BaseT, 100VG-AnyLAN, Token-Ring/UTP, and many other LAN applications.

This 8-pin modular plug is probably the most subject to name abuse, because it resembles the specialized RJ-45 connector. However the RJ-45 wiring pattern (which includes an interface programming resistor) is so radically different from that of T568A and B that it really should not be called by that name at all. In this book, we have tried to always use the "8-pin modular," or simply "modular," description, as it leads to less confusion. In truth, the plugs are identical until terminated. Then, they should probably be called by a name that reflects their wiring pattern. For example, you could certainly connect only four pins of an 8-pin modular plug in a 10/100BaseT pattern. It would not truly be either T568A or B, because both require eight connected pins. It would not be an RJ-45, because the wiring pattern would be wrong and the programming resistor would be missing. Why not call it a 10/100BaseT connector, as that would be its purpose and its unique wiring pattern? If all eight wires are connected in an approved pattern, you could call it a "TIA" connector, or even "T568A or B." Well, until the habit is changed, we will probably still find people who call the connectors "RJ-45's" however wrong the term might be.

The 8-pin modular plug that is used in standard LAN wiring is specified by IEC 603-7, "Detailed specification for connectors, 8-way, including fixed and free connectors with common mating features." This is the plug that is specified in TIA/EIA-568-B and related documents, and in ISO/IEC IS-11801.

Modular connectors are primarily intended for terminating cables with stranded conductors. In fact, the original connector was designed to terminate a flat cable containing from two to eight stranded conductors. It was also designed primarily for the audio frequencies of telephone lines, although the connector is officially rated for use up to 3 MHz. Unfortunately, the industry not only uses the connector at frequencies far above that, but also needs to place twisted pair conductors, encased in a round cable jacket, into the modular plug. To allow the use of the modular connector at LAN frequencies from 10 to 100 MHz, the TIA has simply specified performance criteria (primarily attenuation and NEXT) that the connector must meet. As long as these criteria are met, the connector can be used in applications up to Category 5e.

The Category 6 plugs and jacks, while physically compatible with their Category 5 cousins, are specially configured to reduce NEXT (near-end crosstalk) when mated. This electrical compensation can actually cause a link failure when Category 6 connectors are mated to Category 5 connectors.

Modular connectors are available for solid wire, although some authorities discourage using solid wire for cords, even with these special connectors. The difference between the two connector designs is shown in Fig. 10.1. As you can see, the modular pin is a flat contact with a

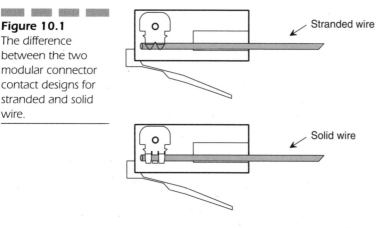

Figure 10.1
The difference
between the two
modular connector
contact designs for
stranded and solid
wire.

pointed end that pierces through the insulation of the wire and makes electrical contact with the stranded wires. The contact may have one or more points. If this contact is used on a solid wire, the conductor may slip off to the side of the points and make intermittent contact or no contact at all. For this reason, the solid wire contact has three pointed fingers at the bottom that are offset slightly, so that the wire is centered, the insulation displaced, and the wire held in constant contact with the opposing fingers. (You can also see why it is bad practice to use the wrong type of connector for the type of wire you are using.)

Shielded modular plugs have been developed that use a metal sleeve around the plug to provide shielding of the connection. These plugs require jacks that are compatible with the shielded plugs for proper functioning of the shield. Shielded cables can also have their shield drain connected to one of the unused pins on the regular 8-pin plug, but the standard connection pattern of four balanced pairs is lost when this configuration is used. Shielded connectors should be used with ScTP cable and shielded equipment jacks.

The standard pin numbering and wiring patterns for the 8-pin modular connector were shown in Fig. 7.10. The T568A and T568B patterns featured in TIA/EIA-568-B are also shown at the top in Fig. 10.2, along with other patterns for specific applications. The two-wire 10BaseT pattern is also applicable to 100BaseTX (but not 100BaseT4 1000BaseT, or VG, which require all four pairs). Notice that the T568A and B patterns actually support several other applications, including virtually all of the LAN connections for Ethernet, Arcnet, Token-Ring, and the PMD patterns of ATM. The key to whether an application can be supported is the pairing of the wires that connect to the plug. For example, the TIA

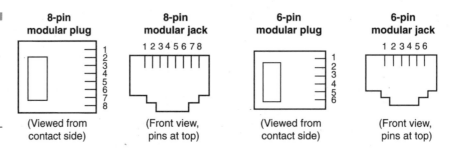

Figure 10.2
The standard pin numbering and wiring patterns for the 8-pin and 6-pin modular connectors. (See also Fig. 7.11.)

Standard	Plug	Pairs	Pins
T568A	8	Pair 1	4,5
		Pair 2	3,6
		Pair 3	1,2
		Pair 4	7,8
T568B	8	Pair 1	4,5
		Pair 2	1,2
		Pair 3	3,6
		Pair 4	7,8
10BaseT, 100BaseTX	8	Pair 1	1,2
		Pair 2	3,6
100BaseT4, 1000BaseT(4), 1000BaseTX	8	Pair 1	Same as T568A or B
		Pair 2	
		Pair 3	
		Pair 4	
Token-Ring	8	Pair 1	4,5
		Pair 2	3,6
Arcnet/UTP	6	Pair 1	4,5
Analog 2-wire voice, ISDN-u	6	Pair 1	4,5
DDS	8	Pair 1	1,2
		Pair 2	7,8
T1	8	Pair 1	4,5
		Pair 2	1,2

patterns both put pairs on pin combinations 1-2, 3-6, 4-5, and 7-8, although in differing pair order. Some patterns, however, pair the wires 1-8, 2-7, 3-6, and 4-5, which places the outer four pins on different pairings. If your connection required pairings on a different pattern than was wired in the cable, the result would be a split-pair condition, which causes severe performance impairment above 1 MHz or so. That is why choosing the correct pattern for your equipment is so important. Fortunately, virtually all LAN equipment can be supported on either of the TIA patterns.

Figure 10.3 shows the recommended dimensions for an optimum modular plug termination. The dimensions are taken from Annex B of TIA/EIA-568-B. Notice that the crossover of conductor 6 occurs inside the plug, forward of the strain-relief tab. This is a little different from the standard practice, which places the crossover just outside the tab. Also notice the length of untwisted wire that exists inside the plug body. It is this untwist that hampers the performance of the plug in very high frequency environments.

The step-by-step procedure for terminating a cable in a modular plug is as follows. Refer to Fig. 10.3 and Fig. 10.4. Begin by removing the cable jacket a minimum of 20 mm (0.8 in) from the end of the conductors. Place the pairs in the order of the pins they will be terminated on (1-2, 3-6, 4-5, and 7-8). The color of the first two pairs depends upon whether you use T568A or T568B. Flatten the cable jacket at the end, to place the pairs side by side. Untwist the pairs back to the jacket, placing the wires of each pair in the proper order, according to the color code, as shown in Fig. 10.4. Arrange the wires so that they are parallel and lay flat. Cross the wire for pin 6 over the pin 4 and 5 wires such that the crossover is no more than 4 mm (0.16 in) from the edge of the jacket. Trim the conductors to approximately 14 mm (0.55 in) from the edge of the jacket. Place the plug over the conductors so that they extend all the way to the bottom of the termination slot and the jacket extends at least 6 mm (0.24 in)

Figure 10.3
The recommended dimensions for an optimum modular plug termination. The dimensions are taken from Annex B of TIA/EIA-568-B.

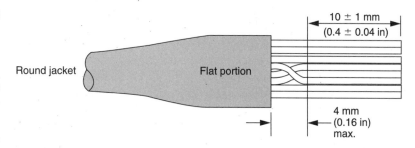

Figure 10.4
Untwist the pairs back to the jacket, placing the wires of each pair in the proper order, according to the color code. Notice that conductor 4 crosses over conductors 5 and 6 inside the strain relief and terminates at position 6.

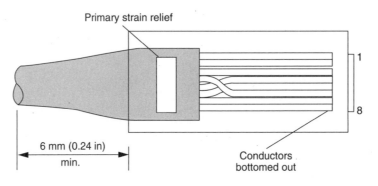

Primary strain relief

1

8

6 mm (0.24 in) min.

Conductors bottomed out

Figure 10.5
Typical crimp tool for modular connectors.

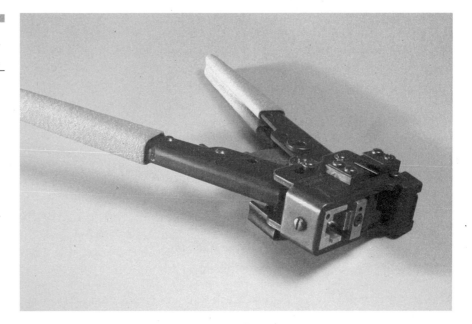

inside the plug body. Crimp the plug using a modular crimp tool. After both ends of the cable have been terminated, test the connections for continuity and proper conductor placement.

A crimp tool with a machined crimping die should be used for the best connections. A typical crimp tool is shown in Fig. 10.5. A proper crimp tool will cost as much as $100 to $200, but will be well worth the expense. The tool should have a ratchet action that will not allow the jaws to open until the plug has been fully crimped. Inexpensive tool designs that were adequate for the smaller 6-pin modular telephone connectors will typically fail to properly crimp the inside pins (pins 3, 4, 5, and 6) when the design is extended to 8-pin widths. This may or may not cause immediate failure of the connection, but you will have problems over time, even if the connection initially tests good. You can inspect a plug to see if it is properly crimped by looking at the plug end. A properly crimped plug will have all eight contacts fully engaged and even on the tops, which will be very slightly below the plastic channels between the contact pins. An improperly crimped plug will show a distinct rounding upward of the center contacts. Side-action crimp tools may favor the contacts on one side of the plug. The plug should also be inspected from the front and top to ensure that the contacts have not drifted to the side, as can happen with inexpensive modular plugs. As simple as these plugs are, there are still significant differences in quality.

It would not be proper in a discussion of LAN cable terminations to ignore the problems that result from the choice of the classic 8-pin modular plug and its mating jack as the standard UTP connector. These components are responsible for the majority of limitations with this generation of high-speed networking. The connector, as we said before, was designed for frequencies well below 3 MHz, and it has very poor characteristics at frequencies approaching 100 MHz. The NEXT performance of the plug and jack is so bad at 100 MHz that you will see few manufacturers quote performance margins that exceed the TIA performance standards by more than 1 or 2 dB, if they quote a margin at all. Test equipment manufacturers even create specialized low-NEXT adapters or digital processing techniques to compensate for the performance deficiencies of the modular connector in their own test equipment.

Nevertheless, with quality materials, proper assembly procedures, careful workmanship, and proper testing, the modular connector can support the highest performance metallic cable networks in wide use today.

Coax Connectors

Coax connectors have been used for LAN connections for a very long time. The two most common LAN coax connectors are the BNC and the Type N connectors. The Type N was used on the original "thick Ethernet" or thicknet cable that first introduced Ethernet to the world. The connector is designed for the large cable diameter of coax cables such as that used for thicknet. The Type N is shown in Fig. 10.6. Type N connectors are expensive and bulky, but they are commonly available

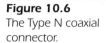

Figure 10.6
The Type N coaxial
connector.

and are very reliable. The connectors are designed for use at many times the frequencies of Ethernet LANs, so their use in networks is no problem. The use of thicknet cable and the corresponding Type N connector has decreased greatly with the advent of thinnet cabling and its BNC connectors.

BNC coax connectors are smaller, less expensive connectors compared to the Type N. BNC connectors are used with "thin Ethernet" or thinnet cable that was introduced during the 1980s as a cheaper alternative to thicknet. The BNC connector is also used for Arcnet coax connections. The typical measurements to prepare a coax cable for BNC termination are shown in Fig. 10.7. BNC connectors are available in two- and three-piece constructions that are designed for termination using screw-on, crimp-on, or solder assembly.

The crimp-on style is the most popular for field termination of LAN cables. The assembly process is as follows. The crimp-on sleeve (of a three-piece connector) is placed over the end of the coax. The coax cable is stripped in the manner shown in the figure, exposing the center conductor and the braided shield. The center pin is crimped onto the center conductor. Next, the connector body is slid over the coax so that the back barrel of the connector slides under the braid and the center pin is properly positioned in the insulator in the center of the connector. The sleeve is slid down over the exposed braid until it contacts the connector body. Finally, the sleeve is crimped onto the connector using an

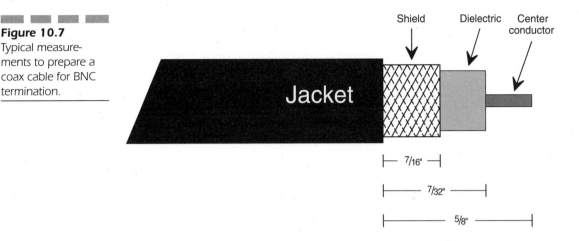

Shield   Dielectric   Center
conductor

Figure 10.7
Typical measure-
ments to prepare a
coax cable for BNC
termination.

Jacket

7/16"

7/32"

5/8"

appropriate crimp tool. The assembly is tested for electrical connectivity
and mechanical soundness.

The proper tools are essential for anyone doing coax connector termination. A rotary coax stripper makes preparation of the cable ends a
snap and eliminates the need for measurement. The tool has two blades
that cut through the insulation to the precise depth to expose the center
conductor and the braid. Crimping the center conductor and the outer
sleeve also require the proper crimp tool. As with crimp tools for modular connectors, the best tools are machined-die tools with a ratchet-type
action that ensures the tool is completely closed, and the sleeve or pin
fully crimped, before the handles release.

A properly crimped connector can easily withstand a moderate
amount of stress without pulling loose. If you use plenum-rated cable,
be aware that the dimensions of this type of cable are much smaller
than standard nonplenum coax. In addition, plenum coax dimensions
may vary slightly with different manufacturers as the dielectric insulation and jacket composition may differ. It is essential that you use coax
connectors that are designed for the style of coax you are using. Plenum-
style connectors have a smaller barrel that matches the smaller diameter
of the dielectric layer of plenum coax.

STP-A Data Connectors

The hermaphroditic IBM Data Connector is the recommended connector for 150-ohm shielded twisted pair (STP) cabling systems. It has long

been standardized in IBM, ANSI, and EIA documents and was made a part of TIA-568-A.[1] The connector design, shown in Fig. 10.8, is defined in EIC 807-8, *Rectangular Connectors for Frequencies Below 3 MHz,* with additional transmission requirements in ANSI/IEEE 802.5 (ISO 8802.5) and TIA/EIA 568-A.

Although the title of the IEC document implies a 3 MHz frequency limit, the connector is actually designed for much higher frequency operation. The TIA/EIA 568-A characterizes the connector parameters to 300 MHz, three times that of the humble 8-pin modular connector. The connector incorporates an electromagnetic shield to complement the shielded cable structure. This connector could theoretically support data rates beyond the current 100 MHz offerings.

The Data Connector is configured for four signal wires and a shield. The connector design uses IDC contacts for the cable wires. The contact pins are color coded red, green, orange, and black to match the color coding of the STP cable. The wires terminate in the pins with a plastic clip and the cable body snaps together, making the installation almost tool-less. A rotary cable stripper is handy to remove the outer jacket and trim the shield.

The step-by-step assembly instructions are as follows. Place the threaded connector sleeve over the cable. Then remove $1\frac{1}{8}$ in of the outer

Figure 10.8
The Data Connector design is defined in IEC 807-8 *Rectangular Connectors.*

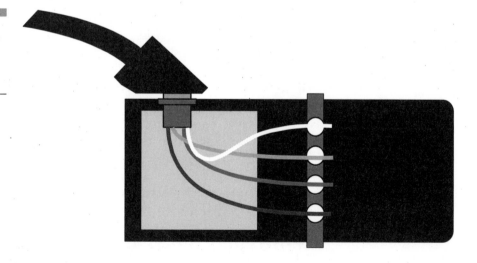

---

[1]Recognition of 150-ohm STP-A and its associated connectors was removed in TIA/EIA 568-B, other than test methods in Sec. 11 and a brief reference in Annex D of Part B.1.

jacket. Trim the braided shield and the foil pair shields to within approximately $\frac{1}{8}$ in of the edge of the jacket. Place the cable into the connector body and bend the braid back against the connector shield. Route the wires to the proper color-coded termination posts, push the wires into the slotted jaws of the posts, and snap the retaining clip over the posts with the recommended tool or with a pair of pliers. Place the side cover onto the connector body and snap it into place. Test the electrical integrity of the connector.

Two versions of the Data Connector are available that have slightly different mechanisms to lock the connectors together. They are referred to as style A (the original IBM version manufactured by AMP) and style B. If you use this connector in legacy applications, you should decide which style works best for you and stick with it throughout your installation.

User Cords

User cords are the final connection step in a LAN wiring system. They are the cables that connect between the installed universal wiring system and the network equipment. They exist on both ends of a link, at the work area and in the wiring closet. In most modern LAN installations, the telecommunications room will contain wiring hubs that consolidate all the station cables into a contiguous network. Telecommunications rooms that contain network equipment in the form of hubs, repeaters, and servers are also "equipment rooms," which add a few requirements over and above those for a simple telecommunications room, but we will just call them telecommunications rooms to keep things simple.

User cords include workstation cords and equipment cords. Technically, user cords do not include patch cords, which are part of the horizontal cross-connect. Sometimes all modular cords are called patch cords, regardless of their function, but we will make a distinction between true patch cords and user cords. Patch cords, which were covered in the last chapter, are essentially the same as user cords, except for length and color.

User cords are a required part of the Channel, the total end-to-end LAN connection, but they are not part of the Basic Link or Permanent Link, both of which exclude user cords. These terms are defined in TIA/EIA-568-B.1, Sec. 11, "Cabling Transmission Performance and Test Requirements," and they are very useful in describing the component parts of a cable link. The theory is that a cable plant could be installed and tested before any of the user network hardware is installed. To do the testing, we need to have performance specifications for the portion of the horizontal cabling that excludes the user cables. That portion is called the Basic Link or Permanent Link. However, Channel performance levels are also defined, so we could also test with the user cables in place. In fact, that type of test is the ultimate assurance that a link meets the requirements for a particular category of operation. More information on Channel, Basic Link, and Permanent Link is covered in Chaps. 2 and 16.

Construction

User cords are flexible, 4 pair cables, terminated at either end in an 8-pin (IEC 603-7) modular plug. The cords may be provided in various lengths, as needed for the application. The cords should meet the transmission and construction requirements for patch cords in TIA/EIA-568-A, as user cords are not explicitly defined elsewhere in that standard. The color code for the conductors and the wiring pattern connections for the modular plug are the standard configuration as shown in the first part of this chapter, as well as in Chaps. 3 and 5. Either the T568A or B pattern may be used, as long as the same pattern is used at both ends of the cord. The standard also allows an alternative color pattern consisting of eight solid colors that can be used for user cords.

User cords are made from AWG 24, stranded, thermoplastic insulated conductors, arranged into four pairs, with an outer thermoplastic jacket (Fig. 10.9). Conductors of 22 gauge wire may be used, if the cord meets the same performance requirements, but you must be certain that the connector plugs are designed for the larger wire size. The various jacketing materials are described in Chap. 3. Stranded wire is used because of its flexibility. From a technical standpoint, solid wire could be used, but it lacks flexibility and is more difficult to connect with modular plugs. In most locations, nonplenum insulation, such as PVC, is acceptable. PVC is also much more flexible than most plenum-rated plastics and it is more easily tinted.

The outer jacket color should be appropriate for the office environment. Some of the cable manufacturers offer Category 5 cable in a bright color, to distinguish it from lower category wire. There is no need to use this bright color in the work areas. The cable should be marked to indicate the category if there is any doubt. Compatibility with work area electrical cord colors is desirable for user cords. Many colors are available but customary colors of beige (or ivory), gray, and black are commonly available for user cords.

Figure 10.9
User cord.

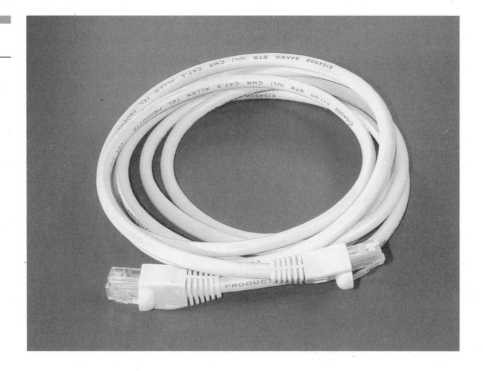

Screened (ScTP) user cord construction and performance are described in Chap. 9 under "Screened (ScTP) Patch Cords."

Category

User cords have the same performance category designations as the other portions of the horizontal cabling system. However, since the 8-pin modular cords are very similar to nonrated telephone cords, you should be certain that the cords you use are certified to at least Category 3 or 5e levels, as appropriate. Cables are often called "data grade" by suppliers. This is an older classification system that simply indicates that the cables offer a little higher performance than "telephone grade" cables. Unfortunately, you need to know exactly how much better the performance will be, so insist on the proper certification of Category 3, 5e, or 6, as needed.

Of course, if you are actually connecting telephone hardware, which is certainly permitted by the "universal telecommunications cabling," then telephone grade user cables are adequate. However, they do not actually meet the requirements of the TIA and other standards and may eventually get mixed in with the high-performance categorized cables.

User cords should be marked with the rated category of operation (Fig. 10.10). The proper marking is "Category *n*" (*n* = 3, 5e, or 6), "Cat *n*," or a letter "C" with the category number enclosed. Categorized cable is normally marked on the jacket, while the connectors may or may not have their own markings. Preassembled cable should be made from properly certified components, and should bear an additional marking indicating that the assembly meets workmanship standards or has been tested to the appropriate category.

Length and Routing

According to the guidelines you must limit the length of your user cords in both the work area and in the wiring closet. The work area cords should be less than 5 m (16 ft), if possible. The cords in the wiring closet require a little simple math, as the length guidelines for the user cords are not separately specified. The total length of equipment cords, patch cords, and cross-connect jumpers in the telecommunications room

Figure 10.10
Category markings.

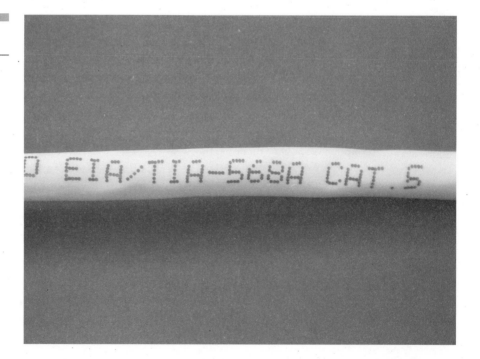

should be less than 5 m (16 ft). The length of the user/equipment cords will have to be taken from that total allowance. The standards allow 10 m (33 ft) patch cords and jumpers. The two end-area lengths do not quite add up to this total because of rounding.

You might think that 5 m of user cable for the workstation might be a little short in some locations, and you would be right. This number is a recommended, not mandatory, requirement. The 10-m limit is mandatory, however, so if you need longer cables in the work area, you will have to subtract from the allowance in the telecommunications room.

In reality, the total length of each horizontal link is the real operating limit of a LAN circuit. This limit is 100 m (328 ft) for the entire Channel, including all of the cables, cords, and jumpers that run all the way from the back of the workstation to the network hub. The 100 m (328 ft) includes 10 m (33 ft) for all the cords and jumpers, with the understanding that the stranded wires used in these cords have a 20% greater attenuation. Although the standard requires that all horizontal cables be solid wire, you could safely exceed the 10-m (33 ft) limit for user cords and jumpers by a little bit if you were well below the 90-m (295 ft) limit for the horizontal cable.

User cords in the work area should be routed with the same guidelines as for the horizontal cable runs. You should avoid sharp bends, kinks, and abrasions of the cable, and you should route it away from potential interference sources such as fluorescent lights and motors. Modular furniture usually requires that user cords be placed in raceways that are built in to the furniture. Remember that the guidelines suggest a separation of 2 to 5 in from power conductors, such as the AC outlets (or mains) in the furniture channels. Furniture manufacturers are beginning to recognize the need for totally separate telecommunications raceways, so you may be able to specify furniture that meets this requirement. The placement of Category 5e, and 6 cable is much more sensitive than lower categories. If you are still operating a 10BaseT or Token-Ring 16 network, you really do not need Category 5 performance, and you may want to relax the rules somewhat. Routing of user cables in the telecommunications room will not be covered here, as it is identical to patch cord routing, which was covered in Chaps. 8 and 9.

Workmanship and Quality

If you build your own user cords, or if you manufacture them for others, the quality of construction is a very important factor in the performance of the cord. You should first ensure that all connectors and cable used in the assembly of the cord are certified by the manufacturer for the level of performance you desire for the completed cord. Then you should use proper techniques to assemble the cord for maximum performance. It is known to be difficult to test short lengths of cable, particularly with regard to NEXT, so your best performance guarantee is good workmanship. Basic continuity and wire map, of course, can be completely tested without regard to length.

The section on patch cords in Chap. 9 and the earlier section on modular connectors in this chapter describe some techniques to ensure a high-quality termination that will perform up to expectations.

Open-Office Wiring

Chapter 11 Highlights

- Modular offices
- TSB-75 guidelines
- Consolidation point
- Multiuser telecommunications outlets
- Fiber optic cabling
- Cable separation

This chapter deals with the special wiring environment of open offices. Open offices include the popular modular-furniture rooms, as well as classrooms, training facilities, and shops. These office arrangements are used in many medium to large enterprises that have a need for compact workspace for their office workers. Office cubicles form a familiar type of modular furniture arrangement.

The very flexible nature of these open offices means that the fixed wiring that is used in other office environments is not practical. In addition, the modular furniture (Fig. 11.1) is frequently placed into the rooms at the last step, long after in-place wiring should have been completed. There are a number of other issues that complicate the LAN-scape and create problems for the cabling installer. We will go over two of the most popular approaches to open office wiring, along with the guidelines for solving these issues according to the standards.

Modular Offices—Modern Problems

In the original concept of structured cabling embodied in EIA/TIA-568 and TIA/EIA-568-A, horizontal wiring is run directly between the telecommunications closet and the work area telecommunications outlet/connector.[1] The outlet is presumed to be permanently mounted in or on an office wall, and is expected to be within a relatively short distance from the actual workstation device (such as a computer).

This presents several difficult issues with regard to modern modular offices. These office environments use modular furniture designs and modular walls to form cubicles with built-in work desks. They effectively replace both fixed-wall offices and totally open desk "pools." The arrangements are popular, because they make much more effective use of office area by eliminating closed-in offices, and they afford a measure of privacy and personalization that is unavailable in the open room with a "desk array."

Modular furniture systems typically place power and communications cables in narrow pathways within the panels of the modules (Fig. 11.2). The modules must be fully assembled before the cable can be inserted, which makes the process of providing home-run cables to the

---

[1]This is commonly referred to as the work area outlet or the workstation outlet.

Figure 11.1
Modular furniture.

telecommunications room very cumbersome. This means that portions of the data and telephone cabling process must be postponed until the modular furniture has been assembled. At this stage in an office build-out, the ceiling tiles are already in, the walls are finished, the carpet is laid, and there is modular furniture all over the place. Placing ladders to install standard horizontal runs is difficult, even if the cable has been prepulled to the expected area of the power/communications riser (commonly called the power pole) of the module. If prepulled, enough wire must be left coiled in the ceiling to run down later through the pole and through the wire pathways to an outlet/connector. This leads to wasted wire, in the best case, and insufficient wire in the worst case.

In addition to the wire-pulling problems, the installer has the delight-ful task of crawling down on the floor and under the modular desk, with flashlight and tools, to terminate the horizontal cable at an outlet. After all this is done, the installer can finally use a cable scanner to see if the link passes. Then, any problems that have occurred must be solved at great time and expense.

The next significant cabling problem that occurs with the use of modular offices, is that they can, theoretically, be easily rearranged. The furniture modules are very easy to move about. The modules

Figure 11.2
Cable channel in
modular furniture.

ED: Shouldn't
the Footnote be
number "2" See
page 262.

simply unbolt the same way they originally bolted together, and then are moved as needed to reconfigure the office to meet new personnel requirements. In most of these modular furniture systems, the electrical connections either snap together as the modules are joined, or they are easily cabled and extended by a system of plugs and receptacles. Unfortunately, the communications wiring, including our LAN wiring, must be painstakingly removed and rerouted into the new furniture layout.

In many instances, the relatively delicate fiber and copper cables are the wrong length in the new arrangement and cannot be reused. The only option in such a home-run environment is to pull an entirely new cable to each relocated workstation. At an earlier time, this was the only approved method to run and rerun standards-compliant Category 5 and higher links, as the standard allowed no splices or intermediate couplings.[2]

However, a much better alternative that was found was to run the horizontal cables to a series of distribution points within an open area. Typically, these points of termination are along the walls in the same

---

[2]A transition point is allowed; however, this is narrowly defined as a connection between the round horizontal cable and flat undercarpet cable.

area where electrical power will connect to a set of modular furniture. In some instances, the distribution point might be in a ceiling area with the communications wires distributed to the furniture modules nearby. Then, after the furniture is installed, cables can be extended from the distribution point to the workstation outlets within the modules. This became almost a universal practice in open-office installations.

This method was not only much more practical, but also a good deal less expensive, laborwise, than dealing with home runs. Despite the standards to the contrary, these links normally worked fine, and passed the applicable link tests. A few links failed, however, and it became obvious that some standard practices needed to be created to ensure that the needed performance parameters were preserved for the appropriate category of operation.

TSB-75 Guidelines

TSB-75, *Additional Horizontal Cabling Practices for Open Offices 1996,* adds the rules for handling modular furniture. A TSB (Telecommunications Service Bulletin) provides additional information and practices to supplement an approved standard, in this case TIA/EIA-568-A[1]. TSA-75 puts down on paper the information that installers need to properly provide standards-compliant cabling in an open-office environment. In investigating the open-office practices that were in use, it was determined that essentially two variations were popular and successful.

The *consolidation point* method initially terminates the horizontal cable at the distribution point on an appropriate punchdown block. When it is time to extend the link to the office module, another length of horizontal cable is connected to the punchdown and run on to the workstation outlet. The *multiuser outlet* method actually places the workstation outlet at the distribution point (in multiples to match the served modules) and runs an extended-length user cord to the work area.

We will go into each of these methods in some detail in the sections that follow. The provision of this standards bulletin gives the system designer and installer the ability to provide approved intermediate interconnection points for modular offices. Guidelines for cable length allowances that will meet the performance requirements are given in the bulletin, and we will summarize these here.

---

[1]The practices in TSB-75 have now been incorporated in the newly revised TIA/EIA-568-B standard.

Consolidation Point

A consolidation point is a point within the horizontal cable run where two cables are interconnected (Fig. 11.3). In reality, it is a reusable connector, such as a punchdown block, that effectively joins the two cable segments without being called a "splice." In fact, cable splices are still forbidden, presumably because they do not allow reusability and may make poor connections when viewed from the perspective of crosstalk and return loss.

The connecting hardware for the consolidation point is required to function for at least 200 mating cycles. This requirement is already the case with all the normal connecting hardware under the standard, but is a more significant limit for the connecting blocks used with punchdowns. You will need to be sure that the hardware meets this limit. This number of cycles is quite sufficient in allowing for the periodic movement and reattachment of the portion of the horizontal cabling that connects to the work area outlet.

All cables and connecting hardware used in the consolidation point should meet all the performance measures of TIA/EIA-568-B, including the transmission parameters for the applicable category of operation. In addition, the actual location of the consolidation point should be placed *more* than 15 m (49 ft) from the telecommunications room, because of the "short cable" problem and the added crosstalk effect. If you need to connect to a consolidation point that is closer to the TR termination than 15 m, the recommended method is to coil extra cable until that minimum distance is met.

All connecting hardware for the consolidation point must meet or exceed the requirements for the category of operation that you need. In other words, a Category 5e installation must use a punchdown that is

Figure 11.3
The consolidation point method.

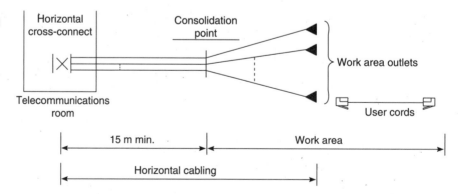

rated at Category 5e or higher. Additionally, installation methods must meet the same category of operation, just as with any connection within the link.

The placement of the consolidation point within the open office is another important consideration. Clearly, the points of consolidation of all these cables must be easy to access and in a permanent location. It would do little good to provide a handy distribution point for the modular furniture outlets and face the possibility of that location being moved. As you can see from the diagram, this method does redefine the work area to encompass the consolidation point. This allows one to say that the horizontal run to the work area still can remain splice-free.

All other cabling distance guidelines remain the same. The horizontal limit is still 90 m, and the user cord limit is still 5 m (16 ft). The consolidation point is simply placed somewhere within that 90-m run. The main advantage to this cabling method is that it does allow for an intermediate connection point for the last part of the horizontal run to the work area outlet. In the event that the furniture is moved or reconfigured, only the portion of the horizontal run from the workstation outlet to the consolidation point has to be replaced, and most of that cable would have to be rerun within the modular furniture's pathways, anyhow.

The requirements for marking the cables in the consolidation point are the same as in any other cabling location. The details on the marking of these facilities are covered in ANSI/TIA/EIA-606 and elsewhere in this book. Keep in mind that you may have an additional opportunity to lose track of cable connections because of the consolidation point. The best practice is to be very orderly in placing the modular-furniture ends of the horizontal cables onto the consolidation hardware. Mark both cable ends as you go and try to keep the numbering consistent. For example, rather than label an entire cable run with the number C310, you could differentiate between the two legs on either side of the consolidation point by designating one as C310a and the other as C310b.

Multiuser Telecommunications Outlet Assembly

The consolidation point architecture is one of the most common methods of providing for open-office cabling. However, it is not the only alternative. Another logical method is to simply extend the length of the user cord far beyond the normal 5-m limit. It turns out that lengthening the user cord can be successfully accomplished on shorter horizontal cable runs. This accounts for the early success of some open-office wiring methods, prior to the existence of these standard practices.

The most logical arrangement for an open office with extended user cords is to bring all the cords together at a wall or other permanent structure upon which the appropriate number of workstation outlets can be mounted. This description essentially defines a multiuser outlet arrangement that TSB-75 calls a *multiuser telecommunications outlet assembly*. Fortunately, the industry has nicknamed this a MUTOA, or just a multiuser outlet. The multiuser outlet method is illustrated in Fig. 11.4.

The basic premise of the multiuser outlet method is that the operating parameters of the Channel must be maintained for proper operation of equipment that is connected according to TIA/EIA-568-B. So, as long as the installed-link criteria are met, there can be allowable variations in the absolute lengths of the horizontal cabling and the user cords.

It is well established, and is included in the component standard, that the stranded-conductor cable used for user and patch cords has a lower velocity of propagation and higher losses than the solid-conductor cable of the horizontal runs. This TSB utilizes that effect in determining how much to decrease the horizontal cable allowance as the user cord is lengthened. This derating factor turns out to be a very good approximation of how much the link performance is degraded by unusually lengthy user cords.

These long user cords can be threaded through the modular furniture raceways in much the same way that the unterminated horizontal cable was in the consolidation method. The cable raceways in modular furniture normally have removable panels along their length, so the

Figure 11.4
MUTOA method.

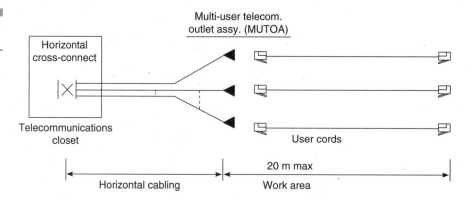

only difficulty is to run the connectorized cable ends through the openings in the structural columns placed for this purpose. Care should be exercised in this process, as there is ample opportunity to bend, kink, or cut the cable during the installation into modular furniture. Some of these removable panels are constructed of metal, so the installer must also be cautious of sharp edges. Also, the electrical power to the module should be turned off for safety reasons during the installation of cable in the module.

Note that there is no workstation outlet in the normal area of the workstation. The work area outlet is in fact part of the multiuser outlet assembly that serves all the nearby modules. The extended user cord is run through the modular furniture and plugged directly into the user workstation.

Excess User Cord Length

The other issue that must be dealt with is the excess cable that will be left over. Unlike horizontal cable, which must be trimmed to within 18 in of excess cable at the workstation outlet, the user cords are normally fixed-length, manufactured cords. Category 5/5e or 6 field termination of stranded cable into modular plugs may be outside the ability of the average cable technician, because of the narrow tolerance in placement of the conductors in the plug. If you will trim to length, follow the methods in Chap. 10 for terminating the cable.

If you use manufactured cables, the excess must either be stored in the work area or in the MUTOA location. One possible storage method is to use one of the cable management devices to store the excess cord length. These can be slotted raceways, a series of cable management rings or standoffs, or one of the snap-in-place patch cable management devices. In general, any device that could be used in the telecommunications room to manage patch cables can be used here.

Horizontal Distances for the MUTOA Method

The formula for this calculation is given in the standard, but the maximum horizontal runs for several lengths of work area cable are shown in Table 11.1. As you can see, the length of work area cord is the normal 5-m length at the maximum horizontal distance of 90 m. This is a total of 95 m, and when added to the cross-connect/equipment cable length

TABLE 11.1

Horizontal Distances versus Work Area Cable Lengths for Multiuser Telecommunications Outlet Assembly

Work area cable	Horizontal cable
5 m (16 ft)	90 m (295 ft)
11 m (36 ft)	80 m (262 ft)
20 m (66 ft)*	70 m (230 ft)

*This is the maximum length for work area cords.

of 5 m, yields the magic total of 100 m, on which most of our LAN topologies are based. However, at the maximum user cord length of 20 m, only 70 m of horizontal cable is allowed (rather than the expected 75 m), which is to compensate for the poorer transmission parameters of stranded-wire cables.

The total length allowed for the Channel in open-office wiring is always at or below the length in normal horizontal links. In testing a link that uses the consolidation point method, the normal tests for Basic Link and Channel are still completely valid. However, if you use the multiuser outlet method, the differences in the allowable horizontal run and in the user cord lengths are significant. You should do a preliminary check of the run of the MUTOA in accordance with the Basic Link test parameters, but you should allow an extended "pass" margin to compensate for the slightly poorer performance of the extended user cord, which is excluded from the Basic Link.

In the end, you really must check the entire Channel link, as that is the ultimate connection path on which the high-speed network must operate. If equipment and patch cords are not yet installed in the TC, you might consider testing the Channel using the actual extended user cord and the 5-m maximum equipment/patch cord.

Fiber Optic Cabling in Open Offices

No discussion of the shortcomings of twisted-pair cabling would be complete without mentioning fiber optic cabling. Most of the problems with copper cabling occur at the points of termination and interconnection. This is the place where the impedance mismatches, untwist, NEXT, and FEXT occur. The only similar problem for fiber is the very small additional attenuation and mismatch that occurs at a termination.

For that reason, the only structured cabling limitation placed on fiber optics for open-office use is that the total length of cables and patch cords may not exceed the 100-m (328-ft) limit that is otherwise stated in the standard. With a typical connector loss of about 0.2 dB, you can add a lot of terminations, patches, and splices with little or no effect at only 100 m.

In addition, you can use the TSB-72 guidelines for length if you are connecting open-office workstations using a centralized fiber network scheme. Consult Chap. 12 for more information on fiber cabling.

Fiber Optic MUTOA Cabling

Fiber optic cabling in the open office is always considered to be the multiuser outlet method. Fiber cabling does not use the punchdown method of copper wire, and thus cannot use the consolidation point method. However, there are still several important considerations for using fiber with open offices.

Fiber optic cable must be very carefully managed, to prevent any possible damage to the fiber during and after installation. At the MUTOA, the fiber cable will have to be separated out to individual fibers and each terminated in a fiber connector. At this stage, the fibers are tiny, relatively unprotected jacketed strands that are highly susceptible to damage. An enclosure, such as a fiber termination tray, is the best way to deal with this problem.

The best fiber termination enclosures are the same ones that are used in the telecommunications room. Smaller versions are available for use with multiuser distribution points. In very limited modular furniture terminations, it may be possible to use some of the innovative multiple connector outlet boxes that are designed to terminate from four to eight fiber pairs. These outlet boxes are designed to contain the extra lengths of unjacketed fiber strands that are required at the point of termination. They are often available in surface-mount versions that are easy to place above the ceiling grid on the walls surrounding the open-office space.

The availability of small form factor (SFF) connectors for fiber makes this method even more practical. The SFF connectors place two fiber strands in an RJ-size modular jack that snaps into the outlet box in the same way as the modular jacks designed for copper.

Because fiber is terminated and interconnected in a fundamentally different manner from copper, all open-office wiring systems for fiber are considered to be the MUTOA method. This is true whether you are using the standard horizontal run from a telecommunications closet, or the optional centralized cabling method covered in Chap 12.

Fiber cable strands are generally terminated in a male connector, which is interconnected by aligning the fibers in an adapter receptacle. The terminations and connections at the hub/switch in the telecommunications room, at the patch panel, at the MUTOA, and at the work area outlet are identical. The insertion and return loss that result from each termination are comparable, and the only parameters that must be watched are the total path loss and the optical bandwidth. As a result, the point of interconnection of fiber runs can be just as easily placed in the open-office area as in the telecommunications closet.

Separation Anxiety

One of the side issues that arises in open-office cabling is the separation of communications cables from power cables. The reason for this is that there is some concern regarding the potential interference from power cables, and there are naturally safety concerns about having dangerous voltages in close proximity to low-voltage communications wiring.

In other sections of this book, we have discussed the issue of separation from power cables. Many authorities have attempted to create guidelines on this issue. The recommendations range from a 2-in separation in electrical codes all the way to a 3-ft guideline from a building standards organization. This leaves a lot to the imagination. What are the issues that should be considered?

The interference issue is fairly straightforward. Electrical power lines are not limited to carrying only 50 or 60 Hz electrical energy. They can in fact carry interfering signals at hundreds of megahertz. In one famous case, a 243 MHz interfering signal, generated incredibly by a malfunctioning air-conditioning relay, was carried hundreds of feet along power lines from the offending source. Commercial and military aircraft use 243.0 MHz as an emergency beacon and communications frequency, thus this unusual power-line interference was very serious. Eventually, the periodic nature of the interference caused the researcher to suspect that it might be an HVAC system that was the culprit.

We routinely use LAN equipment that generates and is susceptible to interference to/from power lines. The closer we allow the wires to be, and the longer we allow them to run in proximity, the greater the potential for interference. Thus, the basic guideline here is to limit both the run length and the separation of these types of cables. In a divided power/communications cable tray, it is very easy to run the two types of cables beside each other for hundreds of feet. This increases the chance for interference to couple between the cables, even if the separation is moderate.

A good rule of thumb is to place power and communications cables in separate runs, to avoid long parallel runs. The cables can generally cross at right angles or briefly run together with little or no coupling. Fortunately, the tight twisting and controlled geometry of modern LAN wiring minimizes the coupling to and from outside conductors.

The safety issue is a little more ominous. Electrical codes, such as the NEC, forbid placing power conductors and communications cables in the same raceway. Power conductors are considered Class 1 circuits, and are expected to have hazardous voltages and the ability to deliver dangerous levels of current. Communications cables, on the other hand, are Class 2 circuits, as defined by NEC 1999, Article 800 Communications Circuits.

The NEC also specifies a 2-in (50.8-mm) separation from electric light, power, and other Class 1 circuits, except when they are in a raceway or metallic sheath, or where they are separated permanently. For example, a LAN cable and a power conductor can be contained in the same power pole, or modular furniture pathway, if they are run in totally separate raceway sections.

The main idea to get from this discussion of power/data separation is that you need to keep power and data cables widely separated, except in that last run through the modular furniture fixtures, where the cables must at least be placed in separate raceways.

Modular furniture presents challenges for LAN wiring, but it is an office technology that is here to stay. As more manufacturers respond to the special needs of data and communications cabling, the process of running the needed cabling in this type of furniture will become routine.

12

Fiber Optic Techniques

Chapter 12 Highlights

- Fiber optic basics
- Fiber optic cable configurations
- Fiber optic connectors
- SFF connectors
- Structured/centralized fiber
- Equipment considerations
- Installation practices
- Safety considerations

Fiber optic cabling has occupied an increasingly important place in the networking arena over the past two decades. Optical fiber was initially used for intercity telecommunications links to replace antiquated and overloaded microwave links at a lower cost than satellite links. In this application, T-carrier links, and later SONET links, provided extensions of the telecommunications services over long distances. The intercity fiber featured advantages of long range, wide bandwidth, interference immunity, and relatively low installation cost, compared to the bandwidth provided.

In fact, it was this lower-cost, relatively high bandwidth fiber optic technology that fueled the modern wide area communications revolution. In the wide area, the availability of right-of-way for fiber cable installation was the key to building a fiber network. One of the earliest implementers utilized partnerships with railway companies to place fiber along rail lines from coast to coast. International communications also benefited from this technology, as literally thousands of fibers of fiber optic cable were placed along the ocean beds between continents.

The development of this high-volume intercity fiber network required significant advancements in the technology of both optical fiber and optical transceivers. As the technology moved forward, devices and materials became available for use in the bandwidth-hungry local area network domain as well.

Local area networks (LANs) have quickly adopted faster data transmission technologies, during the same period of time. These data topologies, such as Fast and Gigabit Ethernet, require high transmission bandwidths and really put twisted-pair copper to the test. Fiber optic cable, in contrast, has a vastly superior bandwidth capability, particularly at the shorter distances prevalent in LANs.

In this chapter, we will explore the many facets of fiber optic technology as they relate to LAN applications. Naturally, some of these topics are covered in other parts of this book. We will repeat certain key points here, and reference other chapters in the book for supplementation in other instances. The focus will be to provide a concise explanation of the utility of fiber optic technology in the LAN environment.

Fiber Optic Basics

Before we delve into the LAN applications, let's review the basics of fiber optic technology.

Fiber Optic Transmission

The phenomenon of the transmission of light through fiber optic media comes from the optical principles of refraction and reflection. As you know, a beam of light bends when it passes from one medium into another. The most familiar example of this is the air-water transition. Fill a pan with water and place a coin at the midpoint of the pan's bottom. Now, stand back from the pan and place a straw (or pencil) into the water, along your line of vision to the coin. The straw appears to bend upward from the straight-line path at the surface of the water. Place the straw into the water from the right or left side, and you will see the same upward bend. The straw has obviously not bent, but the reflected light from the part of the straw below the water's surface has bent at the air-water interface, making the straw appear to bend. This bending of light as it passes from material to material is called *refraction*. Materials that transmit light have a property called the *index of refraction*[1] that is proportional to the amount of bending that would occur between a given material and a vacuum reference. Both air and water have respective indexes that indicate this behavior.

While the air-water refraction is trivial, the same principle is used in forming lenses, such as those used for vision correction, and can be used to collimate light, as well. As illustrated in Fig. 12.1, the path of a ray of light bends at the interface between two transmissive materials. The amount of the bend is determined by the index of refraction. However, beyond a certain incident angle, called the *critical angle*, the ray of light is reflected from the interface, rather than refracted into the medium. That is exactly the reason that the coin in the water seems to disappear when you try to view it from a low angle to the water's surface. By the way, the angle at which the light is bent depends on the wavelength (color) of the light, which is how prisms (and rainbows) work. This phenomenon actually causes lots of problems for multimode fiber, as we will see later.

For years we concentrated on the incident angles that were necessary for the effective transmission and refraction of light from one medium to another, *through* the interface. This was useful for the design of eyeglasses, binoculars, telescopes, and microscopes. However, if the transmission medium is formed into a solid strand of glass or plastic, we can use the reflective and refractive principles to guide the rays of light within the strand, along its length.

---

[1]The index of refraction is defined as the ratio of the speed of light in a vacuum to the speed of light in the medium under consideration, thus $n = c/v_m$.

Figure 12.1
Reflection and refraction of light at an interface.

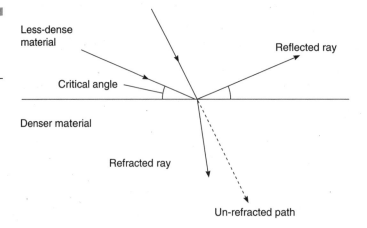

Carefully constructed fibers of glass or plastic can efficiently transfer light for long distances along their length, with very little loss of light outside the fiber. You are probably familiar with the decorative lamps made with hundreds of short plastic fibers. The fiber strands glow very slightly along their length when a light source illuminates the strand ends placed at the base of the lamp, but the fiber ends shine brightly. Similar fibers are used to transmit light along the network pathway of local area networks. We simply modulate (vary) the light beam in order to carry data.

Optical fibers can be either glass or plastic, but for most of this discussion, we will refer to the fiber optic material as glass, since silica compounds are used in most current LAN implementations.

These silica fibers are sometimes called *light pipes,* because light seems to be conducted through the flexible fiber, just like water through a tube. Optical fiber is not actually a pipe for light, as we have said. Light passes through the glass, not through a pipe formed from the glass. The transmission properties of the glass are used to direct the light through the center portion of the fiber, as shown in Fig. 12.2. Depending on the type of fiber core, the incident light beam reflects or refracts its way down the length of the fiber, staying roughly in the center portion of the core, much as water in a pipe (thus the analogy).

An appropriate light source is used to send a communications signal through an optical fiber. Here is how data transmission over a fiber works. A data-modulated electrical signal, such as a LAN packet, is first applied to the light source, which converts it into light. The beam of light from the source is then directed at one end of the fiber, which

Figure 12.2
Light transmission
through an optical
fiber.

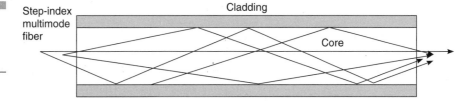

Step-index
multimode
fiber

Cladding

Core

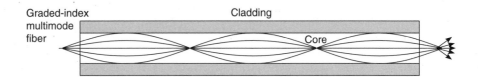

Graded-index
multimode
fiber

Cladding

Core

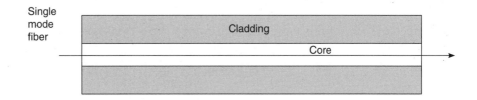

Single
mode
fiber

Cladding

Core

transmits it through the fiber all the way to the other end. At the far end of the fiber, a light-sensitive device is used to detect the light from the fiber and convert it back to the electrical signal.

The device that actually creates the light is called the *source* and the device that receives the light is called the *detector.* Light-emitting diodes (LEDs) and laser diodes are two examples of semiconductor light sources that are used in fiber optics. Photodiodes and phototransistors are examples of fiber optic detectors.

The entire portion of circuitry that converts the electrical signal to light is called simply the *transmitter,* and it includes the light source. Likewise, the circuitry that detects the incident light and converts it back to an electrical signal is called the *receiver,* and includes the light detector. In many fiber optic applications, the transmitters and receivers are actually prepackaged devices that are simply mounted on the circuit board of the optical equipment.

Most fiber optic interfaces are technically *simplex* connections—that is, either for transmitting or for receiving, but not both. Two interface connections, one for transmit and one for receive, are required to make a complete connection. This two-connector arrangement forms a *duplex* connection.[2] The optical connections to an equipment interface are thus paired, and are usually marked Tx and Rx (or T and R) to indicate transmit and receive, respectively. Taken together, the transmit and receive optics are called an *optical transceiver.*

Some optical connector pairs are arranged with an exact spacing so that a pair of optical plugs can attach simultaneously. Fibers are terminated into two separate plugs, and then snapped into a duplex adapter. Other connector types, such as the new small, modular fiber connectors, terminate two fibers in the same plug or jack. Paired fiber cords, called duplex cords or zip cords, are normally used for fiber optic jumpers. The cords should be terminated with the appropriate type of fiber optic connector for the application.

For transmission over the fiber, the light from the optical source may be collimated into a narrow beam to more efficiently couple its energy into the fiber. In the case of the laser sources, the light beam is already quite collimated and must simply be aimed accurately at the fiber core. As a matter of fact, the coupling of the optical source and the interconnection of terminated fibers are critical parts of fiber optic transmission technology. If the termination is properly done, the connector hardware will normally ensure that the fiber-to-fiber or fiber-to-source coupling is maximally effective, resulting in an absolute minimum coupling loss.

Fiber Types

Optical fiber is available in both glass and plastic compositions. Glass is the most prevalent in LAN uses, because of its inherently greater bandwidth. Several uses for plastic fiber exist, and this is an area that has had more development over the past several years.[3] However, at this point in time, the higher bandwidth and longer distance requirements of modern networking still demand the use of silica-based fiber in most applications.

---

[2] The reason for using this two-fiber duplexing scheme is that it is simple and cheap, with respect to transceiver optics. However, there are other methods for duplexing onto a single fiber. One method uses special optics to split the transmit and receive beams of light at each end of the cable. This method, technically called wave-division multiplexing (WDM), allows different wavelengths (colors) of light to be used for transmit and receive. A coarse version of this uses 850 nm for transmit and 1300 nm for receive functions at one end of the cable, and, of course, reverses the functions at the opposite end. Beam splitters and optical filters separate the two light wavelengths and allow both to share the single fiber.

[3] Graduated-index plastic optical fiber (GIPOF), for example, can be used to transmit moderately fast LAN signals over shorter distances.

The fiber's *core* actually transmits the light beam, while the *cladding* provides strength, flexibility, and a controlled stop in refractive index. All are necessary for proper fiber optic operation. Do not mistake the cladding for the plastic *buffer* and the *primary coating* that coat the fiber strand in a tight-buffer cable construction. The buffer and the primary coating are removed by a stripping tool in preparation for the termination of a strand. The cladding, on the other hand, is an integral part of the fiber strand and cannot be removed.

Optical fiber is also available in two modal compositions, single mode and multimode, described by the light modes that each will propagate (see Fig. 12.2). Single mode fiber has a very narrow core, so that only a narrow beam of light can propagate down the fiber, essentially in a single mode of light. Multimode fiber has a much larger core diameter, and multiple modes of light propagate either by reflection from the core walls or by a combination of reflection and refraction in the core. Table 12.1 lists optical fiber types approved for LAN wiring.

The single mode fiber used in LAN technology has a core diameter of only 4 to 8 μm and is surrounded by a cladding that is generally 125 μm in diameter. Multimode fiber used for LANs has a relatively large core diameter that is either 50 μm or 62.5 μm in diameter, again surrounded by a cladding of 125 μm in diameter. These dimensions hold for most LAN and campus fiber currently in use, although a variety of other types are available for other applications.

Multimode fiber is further subdivided into two primary types, step-index and graded-index fiber. These terms refer to the gradation of the refractive index, as measured radially from the center of the core. As illustrated in Fig. 12.2, the rays of light bounce off the "walls" formed by the radical step in refractive index between the core and the cladding of

TABLE 12.1 Optical Fiber Types Approved for LAN Wiring

Modal type	Refractive type	Core diameter, μm	Cladding diameter, μm	Optical windows, nm
Single mode	N/A*	4–8	125	1310, 1550
Multimode	Graded index	50	125	850, 1300
Multimode	Graded index	62.5	125	850, 1300

*Single mode optical fiber is technically a step-index refractive type, although the term is primarily used to distinguish multimode construction.

the step-index multimode fiber. In contrast, in graduated-index multimode fiber, the rays primarily bend because of the gradual change in refractive index from the center of the core to the edge of the cladding.

Fiber Optic Cable Configurations

Fiber optic cables are available in a wide variety of physical constructions. Fiber cable can be anything from simple two-fiber zip cord used for jumpers to 144-fiber cable (often layered in rows of 12) for intercity transmission. Cable for outdoor use (called *outside plant*) may have integral steel cables or even armor to resist rodents. However, most of the fiber used in LAN wiring is ordinary indoor-rated multifiber cable of one or two configurations. This discussion supplements the information in Chap. 3, "Wire and Cable Technology for LANs."

Tight-Buffer Cable   One of the two main fiber cable constructions is called tight buffer. As you can see in Fig. 12.3, which illustrates a single-strand cable, the buffer layer is between the fiber's primary coating and the outer strength and jacket layers. In a tight-buffer cable, the buffer is a plastic coating that separates the bare fiber from these outer layers, and protects the fiber as the outer layers are removed. A multifiber tight-buffer cable has two or more of the buffered fibers, which may be additionally covered by an individual jacket and one or more spacer/strength members.

During installation (called *termination*), the outer jacket and the fine strands of the strength layer are removed with a sharp blade or a cable-stripping tool, leaving what appears to be a small-diameter flexible strand. This strand is the bare fiber (consisting of the fiber core,

Figure 12.3
Tight-buffer fiber optic cord construction.

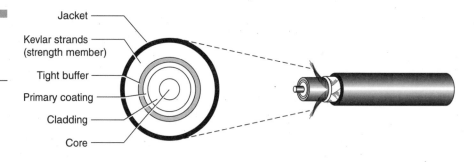

Jacket

Kevlar strands
(strength member)

Tight buffer

Primary coating

Cladding

Core

Cross-section

Side view

cladding, and primary coating) plus the tight buffer. A precision tool, which is basically like a wire stripper, is used to exactly cut the buffer sleeve without nicking the bare fiber. The plastic buffer is then removed by gently pulling the stripping tool toward the cut end of the fiber. The plastic primary coating may also be removed during this process. The fiber is then cleaned with an appropriate solvent, such as alcohol. After some additional steps, the remaining bare fiber is trimmed and terminated into an appropriate fiber connector.

Tight-buffer cables are often used where a small number of fibers are to be pulled between two points without the use of protective raceways. In these applications, the cable usually includes a tough outer jacket. The dual (duplex or zip cord) type of cable is also used for patch cords, user cords, and jumpers. At the interconnection location for a multistrand tight-buffer cable, you must be careful to provide proper strain relief for the main cord and the individually connectorized fibers. In many cases, the jacketed cable is very stiff, while the bare fibers are extremely flexible, yet quite fragile.

Loose-Buffer Cable   In loose-buffer cable, one or more bare fibers (fiber with the primary coating) are placed loosely in plastic tubes (Fig. 12.4). Several tubes may be placed together with strength members and fillers inside a plastic outer jacket. Because of the construction of the cable, it is often referred to as loose-tube cable. From one to twelve fibers are generally placed in each tube. Hybrid cable constructions may place both multimode and single mode fibers in separate tubes.

At the end of the cable, the outer jacket and strength layers are removed and any filler or central structural members are cut and discarded. The tubes are then carefully cut at an appropriate length for terminating, and the individual fibers are prepared and terminated in the

Figure 12.4
Loose-buffer fiber
optic cord
construction.

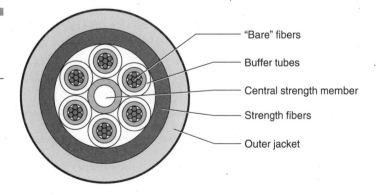

"Bare" fibers

Buffer tubes

Central strength member

Strength fibers

Outer jacket

appropriate connectors. Strain relief, fiber routing, and excess fiber storage are provided in the fiber termination tray.

Loose-buffer fiber optic cables are most often used in multiple-fiber runs between distribution points, such as telecommunications rooms. This type of cable is appropriate for structured-cabling backbone cables, for campus backbone runs, and for runs to the multiuser telecommunications outlet assembly (MUTOA) of open-office environments. Centralized cabling systems may use this type of cable effectively to run to a point of interconnection, such as a telecommunications room on a building floor. Individual cables may then be run to each user outlet.

The major advantage of loose-buffer cable construction is that it takes less time to terminate the same number of fibers than with tight-buffer cables. An application with a relatively small number of fibers will not see much difference, but an installation with large numbers of fiber terminations will go much faster with loose-buffer construction.

In addition to the inside plant cables, several types of loose-buffer cables are offered for outside plant. A nonmetallic direct-burial cable (sometimes called *dielectric cable*) is one type. This cable is somewhat more impervious to lightning damage, but be assured that if the cable is run alongside metallic cables, as is frequently the case in aerial runs (between telephone poles), a direct lightning strike will blow a hole in this type of cable just the same. An interesting technique wraps a steel wire around the length of the nonmetallic cable to add strength to the cable runs between poles. Outside plant cable may also be duplexed with a steel cable (called a *messenger cable*) for aerial installations. For direct burial applications, a metal-armored cable is used. The armor is often a wound layer of copper, steel, or lead (Pb) tape that is placed between two outer jacket layers.

Fiber Optic Connectors

Fiber optic technology is notable for its use of an incredible number of incompatible connectors. In defense of the fiber industry, traditional fiber optic applications have been more esoteric than copper applications,[4] and the interactions of the fiber, the connectors, and the

---

[4]Ethernet cabling has encompassed three incompatible connector types: Type N, BNC, and 8P8C modular (RJ type) for the medium-dependent interface. A fourth type that is semi-compatible with the 8P8C is under consideration for Cat 6 and Cat 7 applications, and a noncompatible Cat 7 variation has also been chosen. Token-Ring has used three: Data Connector, 9-pin D-shell, and 8P8C modular. Some readers may be familiar only with the newer 8P8C modular connectors used with 10/100BaseTX. Copper applications have rapidly obsolesced the older connector types as the technology has progressed.

transceivers have been far more complex and subtle. This has resulted in a parade of improved connection technology that has attempted to address a progression of assembly, usage, and alignment issues that have arisen over the years. Fiber connectors have tended to be redesigned to meet performance and convenience objectives, rather than the radical technology changes of copper networking.

Table 12.2 lists some of the more popular fiber optic connectors, listed in approximate order of their introduction (except for the SFF connectors, which have all been introduced at about the same time). A surprising number of the legacy connectors are still in use.

The latest issues that have demanded further modifications of connector standards are pluggability and duplexing. Prior to the -A revision of EIA/TIA-568, the SMA and ST connector types were in common use. The SMA is a threaded single-fiber connector, whereas the ST is a bayonet (push and turn to lock) single-fiber connector. Both are rather like

TABLE 12.2   Fiber Optic Connector Types

Connector	Style	Assembled cost	Attributes	Originator
Biconic	Conic ferrule, threaded	High	Good fiber alignment	AT&T
SMA	Cylindrical ferrule, threaded	Medium high	Smaller size, standard RF hardware	Amphenol
ST	Cylindrical ferrule, bayonet	Medium	Small size, quicker connection	AT&T
SC	Tapered square, push/pull	Low	Quick-connect	NEC
568SC	Dual tapered square, push/pull	Low	Duplex, quick-connect	NEC
OptiJack	Dual ferrule, modular clip-in, plastic	Low	Duplex, quick-connect, SFF	Panduit
LC	Dual ferrule, modular clip-in, plastic	Low	Duplex, quick-connect, SFF	Lucent Technologies
MT-RJ	Dual cylindrical ferrule, modular clip-in	Low	Duplex, quick-connect, SFF	Alcoa/AMP /Siecor
WF-45	Dual ferrule-less V channel, modular clip-in	Low	Duplex, quick-disconnect, SFF	3M

miniature, optical versions of the Type N and BNC connectors (respectively) that we are familiar with from the early days of Ethernet.

Unfortunately, the SMA and ST connectors had to be mated to corresponding chassis or patch connectors individually. Even if the connectors were somehow marked or color coded, it was quite possible for the connectors to be reversed, mixing up transmit and receive, and causing a link failure. As all fiber interfaces consisted of two connections, a connector type that could be paired was needed. The initial solution was the SC connector, which was designed as the preferred connector in the TIA/EIA-568-A revision to the standard. The SC connector had the advantage of a push-to-mate/pull-to-unmate connection that allowed two connectors to simultaneously plug into an interface. When two SC connectors are linked by a clip, which aligns the pair and allows them to be plugged as a unit, the assembly is designated 568SC.

The SC connector also continues the trend toward quick-termination fiber connectors. Such connectors typically use a combination of no-polish, crimp-on construction to speed termination time to as little as 60 seconds per connector. These quick-termination connectors drop the time-consuming step of epoxy-bonding the fiber to the connector by using a crimp-on system, somewhat similar to coaxial cable-connector assembly. The main difference is that the fragile glass fiber absolutely must not be deformed during the crimping process, or it may fracture or break.

In addition, some of the new connector implementations now use a preterminated connector arrangement. As you may know, one of the most time-consuming steps in conventional fiber optic connector assembly is the polishing of the end of the terminated fiber that pokes through the tip of the fiber connector. One style of preterminated connector has a built-in prepolished fiber stub at the tip of the connector. The bare fiber strand is inserted through the back of the connector and into a special "matching fluid" that closely matches the index of refraction of both fiber pieces. The result is that the usual deflection of the incident light rays that would occur at an angled fiber end is essentially absent. Instead, the beam passes into the matching fluid and then into the fiber stub with virtually no refraction, eliminating a very time-consuming step from the assembly process. An alternative process uses an index-matched optical-quality sphere to minimize the light dispersion at the end of the fiber and direct it to the prepolished fiber stub. However, there is sometimes a price to pay in using the quick-termination fiber connectors. A standard cut-and-polish fiber termination normally has a lower connector loss than the quick-termination type.

Small Form Factor (SFF) Connectors   The small form factor (SFF) connector is a very new type of fiber optic connection. Although the SC connector met the push-pull and duplex criteria, it was a large connector that required two adjacent slots in a modular faceplate. Patch panels had to be completely redesigned to fit this connector, and it prevented the high-density arrangement needed for fiber blades (option cards) in switches, hubs, and routers. The concept of the SFF connector was put forth to remedy this size problem.

The SFF connector is generally defined as a fiber optic connector that terminates two fiber connections within the form factor of a single insert in a modular jack plate. This arrangement generally complies with the requirement to align a duplex fiber connection. The new TIA/EIA-568-B.3, *Optical Fiber Cabling Components Standard,* drops the requirement for a specific connector on the basis that this will encourage innovation and that the market will determine usage. The connector can be of any configuration that meets certain requirements and does not have to be exactly like the copper modular (RJ-type) connector. Figure 12.5 shows some of the more popular SFF connector types in contrast to the conventional 568SC connector.

There are two major requirements that must be met by any SFF connector that complies with TIA/EIA-568-B.3. First, the alignment of the optical components must meet the applicable FOCIS[5] document. Second, the connector must allow the two-position A/B orientation of the stan-

---

[5]FOCIS, Fiber Optic Connector Intermateability Standard, ANSI/TIA/EIA-604-3, sets standards for optical alignment and loss for single mode plugs, multimode plugs, and adapters.

Figure 12.5
Four of the SFF fiber optic connectors are shown with two older styles. Shown (left to right) are the 568SC, ST, LC, MT-RJ, OptiJack, and WF-45.

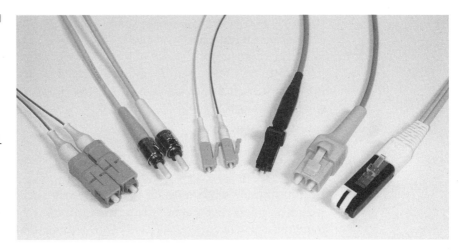

dard to be maintained. This orientation is explained in detail and illustrated in Chap. 8. The basic concept is that the A/B orientation is reversed at each adapter interface in such a way that the transmit/receive relationship is maintained absolutely throughout any sequence of cable and patch cord interconnections. The system automatically creates a "crossover" between transmit and receive at the equipment interfaces at each end of a run. The SC connectors and adapters must be oriented in a particular manner to preserve this relationship, and they are still often used to illustrate the arrangement. SFF connectors automatically orient themselves because each fiber is permanently placed in its location in the connector. However, this does make pair reversals harder to remedy, as one cannot simply unclip and reverse the SCs.

The SFF connectors may also include a latch, as with the 8-pin modular RJ connectors. The use of a latch implies that the A/B orientation will be maintained, because the plug can be inserted only one way. On the other hand, it is up to the connector/adapter manufacturer to indicate the A/B connection polarity to the installer, since the whole fiber interconnection system of the TIA is organized to create a consistent reversal at each adapter.

Connector Color Codes   TIA/EIA-568-B.3 requires a different color for connectors and adapters according to the fiber mode. Multimode components are required to be beige, while single mode components are required to be blue. This can eliminate an enormous potential problem in identifying which type of fiber is present. Fiber mode mismatch can cause a 4 to 6 dB loss (one way) and can cause a system to fail or to be unreliable.

This relationship is so important that we have repeated it in Table 12.3. Installers and network managers should ensure that they use the proper color code to indicate fiber mode on all connections and patch cords. However, there are no special keying or connector differences to prevent mating single mode to multimode if that is really what you try to do. But you should easily be able to see the color mismatch, if the color code is followed.

Optical Fiber Bandwidth

In multimode fiber, the most significant factor that limits link distance is called *optical bandwidth*. This term is a bit of a misnomer, as what is meant is the capacity for information transmission. As with many of

TABLE 12.3

Connector Color
Code for Optical
Fiber

Fiber type	Connector color code	Adapter color
Multimode	Beige	Beige
Single mode	Blue	Blue

the optical parameters, there are significant differences between the commonly used coaxial cable transmission terms and their corresponding optical-lingo meanings. The actual theoretical bandwidth of a signal that can be carried on a fiber strand is extremely high, on the order of 50 to 300 GHz for short lengths. In practice, a system's bandwidth is determined by the wavelength, the optics, the electronics, the connecting hardware, the fiber, and the circuit length. Figure 12.6 shows some typical system bandwidth versus cable distance curves for single mode and multimode fiber.

As you can clearly see, there is a vast performance difference in system bandwidth between multimode and single mode fiber types. Multimode fiber, as the name implies, allows the propagation of many light modes, which gives rise to chromatic and modal distortion. This distortion of the light signal severely limits the information capacity of multimode fiber. The minimum performance standards are shown in Table 12.4. The useful bandwidth of a particular cable length can be inferred by dividing the stated bandwidth-distance product by the actual length in kilometers. Thus, from the table, a 100-m fiber run of 50/125 μm can be seen to have a (500 MHz·km)/(0.1 km) = 5000 MHz effective bandwidth. On the other hand, a 220-m run of 62.5/125 μm will have a 727 MHz effective bandwidth, which could cause a real problem for Gigabit Ethernet. Refer to Chap. 13 for a complete discussion on the fiber cable lengths that are supported by Gigabit Ethernet.

The system designer faces something of a dilemma in specifying fiber optic cabling for use to a workstation outlet. On the one hand, multimode fiber is less expensive and supported on more current types of workstation/hub equipment. However, single mode fiber has an information bandwidth that is vastly superior, and might be usable for many more years. In addition, now that most of the equipment optics for high-speed networking are laser-diode-based, the assumed cost advantage of multimode-friendly LED sources is essentially nil.

So, the problem is whether to specify less expensive, more common multimode fiber to the desktop, or to use single mode fiber. The single mode fiber has the opposite disadvantages, but will have a significantly higher bandwidth for future network growth. However, single mode

Figure 12.6
Optical system band-
width.

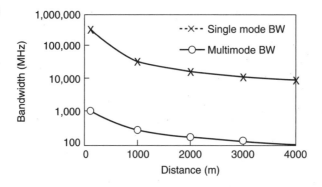

TABLE 12.4

Information Capaci-
ty "Bandwidth" of
Multimode* Opti-
cal Fiber

Modal type	Core/cladding diameter, μm	Operating wavelength, nm	Bandwidth-distance product, MHz·km
Multimode	50/125	850	500
Multimode	50/125	1300	500
Multimode	62.5/125	850	160
Multimode	62.5/125	1300	500

*Single mode optical fiber is not rated for mode-dispersion bandwidth, as it propagates essentially a sin-
gle mode.

fiber alone is incompatible with much current networking equipment.
The answer that many are realizing is that you can do both. The standard specifies a minimum of two cables to each work area outlet, one of
which may be fiber. However, it does not prohibit placing more than
two cables to a work area. Virtually all modular cover plate designs allow
at least four module inserts. So, the optimum solution for today and the
future is to place two copper cables and two fiber cables, one single
mode and one multimode, in each outlet box. If you use the new SFF
fiber connectors, and Cat 5e or Cat 6 copper modules, you should be set
for many application/years to come.

VCSEL Laser Source Multimode Problem

A popular laser source for high-speed networking is the vertical-cavity
surface-emitter laser (VCSEL). This semiconductor diode combines high

bandwidth with low cost and is an ideal choice for the gigabit networking options. In a single mode fiber environment, the laser source works flawlessly. Its highly collimated beam of light is easily aimed right down the center of the narrow single mode core and propagates in essentially a single mode of transmission, with all the attendant advantages.

However, as we just covered, much of the installed fiber in the local network world is actually multimode fiber. Back when we were using LED sources for the lower-speed technologies, such as 10BaseFL and 100BaseFX, we simply aimed the LED at the transmit end of a multimode fiber and let things happen. In effect, the excess light that did not fill the core was lost, but the light that did enter the core of the fiber propagated along the core in all modes (thus the term *multimode*). The method of applying the source light to the fiber is called the *launch,* and applying more light than is needed is an *overfilled launch.*

Unfortunately, when multimode fiber was made in the past, little attention was paid to minor defects in the glass core. If multiple modes are launched into the fiber, a tiny aberration here and there causes little harm, because all the other modes (rays) of light get through just fine. At most, these defects might cause a small increase in cable attenuation. Nothing to be worried about, unless you launch with a laser source!

Remember that a laser is an extremely narrow beam of coherent light, on the order of 1 μm from most VCSELs, that is launched precisely down the center axis of the fiber core. This means that any minor defect may deflect a substantial amount of the available light energy, causing an excessive amount of attenuation. This may, in turn, cause a link failure if the system cannot tolerate an extra 2 to 10 dB of loss. Worse yet, minor deflections of the fiber, which might be caused by mechanical movements or even temperature changes, can cause the link problem to be intermittent.

Two solutions exist to what has been dubbed the VCSEL problem. One is to get better fiber. Certainly, if you are installing new cable, this is the best option. You must be sure that the fiber cable you specify is rated for laser operation, as well as extended bandwidth (the bandwidth-distance product). Such fiber is often called *gigabit-ready* or *laser-compatible.*

If you have existing multimode fiber, you may have to use a short *conditional launch adapter* cable to spread the single mode laser source to multiple modes. The bad news is that the adapter may add from $50 to $100 to the cost of your installation. If you suspect that there may be a problem with your older multimode cable, you may still be able to use

1000BaseSX, as many of the transceivers for this technology include built-in mode conditioning. Check with your equipment manufacturer.

Standardized Fiber Optic Lan Cabling

The basic principle of the standardized cabling infrastructure is predictability. A strong measure of network integrity and reliability can be realized by utilizing a standards-based cabling system. Much has been said about the value of providing a structured cable plant that meets recognized standards for copper cable installations. A standards-based installation for fiber optic cabling is just as important. Manufacturers of networking equipment depend on the cable standards of the TIA and the operating standards of the IEEE, among others, for the compatibility and performance parameters of their products. The cabling designer and network manager depend on these standards as well.

With an inside cable plant that is designed and installed according to the standards, you can be assured that the equipment you purchase will operate properly to the limits specified. Actually, the two types of standards work in concert to bring the performance of the network up to the necessary levels. The TIA committees carefully consider the existing and planned LAN technology innovations when they create their cable standards. Likewise, the IEEE standards committees consider existing and planned cabling standards and the ability of emerging networking technologies to use these widely deployed cabling plants.

Initially, the network technologies that were to be supported required modest performance parameters to support networks with 10 to 20 MHz bandwidths. However, that speed target has moved rapidly upward until we find the standards now supporting link testing limits of 100 MHz, 250 MHz, and, in the future, even 600 MHz! As the cabling standards have moved forward, the ability to place these higher-speed networks on the cable plant has advanced as well. Fiber optic cabling is an important part of cabling technology and adds to the wealth of options available to the network technologist.

Although fiber optic cabling has an inherent bandwidth advantage over copper in many applications, it is still subject to standards of installation and operation. From the connectors to the very fiber itself, every feature of a fiber link must be appropriate for the application and for the equipment with which it will be used.

For this reason, we will spend some time describing the two types of fiber optic cabling standards that are currently in use. These standards are the structured cabling system and centralized cabling.

Structured Fiber Cabling—TIA Standards

The tried-and-true method of fiber connection is the structured cabling system conceptualized in TIA/EIA-568-B. As you recall, this standard allows for both single mode and multimode fiber, although the earlier -A version really encouraged the use of multimode in the horizontal structure and single mode in the backbone. Multimode fiber and the lower-speed LED optics have traditionally been less expensive than single mode fiber and laser optics. As a result, much of the fiber that has been installed in conformance with the structured wiring pattern is multimode fiber.

Multimode fiber in the structured environment works equally well for Ethernet, Fast Ethernet, and Gigabit Ethernet. This is quite an accomplishment, when you think about it. Part of the reason for this high-speed capability is the 90-m horizontal runs of structured cabling. This short distance limit is really dictated by the attenuation and crosstalk parameters of twisted pair cable, when tested to 100 MHz. It is no small irony that fiber advocates can thank the use of Category 5 copper cable for this relatively short run, and the consequent longevity of multimode fiber.

As you can see from Fig. 12.7, backbone cabling runs are made from a central distribution point, such as the main cross-connect (MC) shown, to the telecommunications rooms (TRs) throughout the building. At each TR, a fiber hub or switch distributes the network signal to individual horizontal fiber runs, one per workstation to be served. An optional multiuser telecommunications outlet adapter (MUTOA) may be placed in the horizontal run, between the TR and the work area outlet/connector (WOA). As explained in Chap. 11, open-office wiring practices allow the use of the MUTOA.

In the strict sense, the run from the MUTOA to the workstation is supposed to be a user cord that plugs directly into the workstation. However, the designer may wish to use a fixed WOA to protect the longer fiber run back to the MUTOA. This is not permitted for copper cables, primarily because each termination and interconnection adds significantly to the crosstalk problem. With good fiber terminations, only a few tenths of a decibel of additional link loss are added, and the additional interconnection at the WOA is generally insignificant. In addition, the potential for damage to the duplex-fiber user cord is

Figure 12.7
A typical structured
fiber layout.

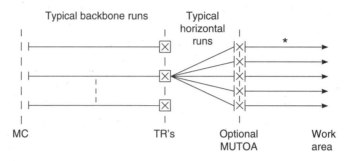

Key:		
⊢ Termination	MC = Main cross-connect	
⋉ Interconnect or cross-connect	TR =Telecommunications room	
⊠ Hub equipment	MUTOA = Multi-user telecom. outlet adapter	
➤ Work area telecom. outlet connector	Opt = Optional	

* These runs are technically user cords if MUTOA is used and may connect directly to
workstation. Otherwise, they are a continuation of the horizontal cable.

greater than with the more robust copper cord, and it would be tedious
and costly to replace a damaged user cord all the way back to the
MUTOA.

Centralized Fiber Cabling—TSB-72 and TIA/EIA-568-B

Fiber optics can support distances far in excess of those at which copper
pairs fail to work. In recognition of this fact, it is often possible to make
a fiber run all the way from a central equipment room or MC directly
to the work area. This topology is shown in Fig. 12.8. In this arrangement, the structured approach is abandoned for the fiber connections,
and is replaced by centralized fiber cabling.

Centralized fiber has two primary advantages. The cost of each run is
presumably less because the system eliminates at least two connectors for
each run, along with associated termination costs, and the cost of a
patch panel in each TR. In addition, a centralized fiber hub may be
used, presumably at lower cost (and better port usage) than would be the
case with the hub in each TR. The centralized hub also simplifies
administration and management.

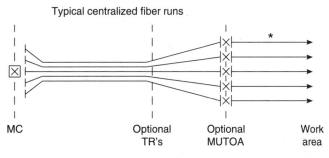

Typical centralized fiber runs

MC   Optional   Optional   Work
                       TR's   MUTOA   area

Key:

├─ Termination   MC = Main cross-connnect

|✕| Interconnect or cross-connect   TR = Telecommunications room

☒ Hub equipment   MUTOA = Multi-user telecom.
                                                    outlet adapter
► Work area telecom.
    outlet connector

* These runs are technically user cords if MUTOA is used and may connect directly to
  workstation. Otherwise, they are a continuation of the horizontal cable.

Figure 12.8
A typical centralized
fiber layout.

Centralized cabling was first described in TSB-72, *Centralized Optical Fiber Cabling Guidelines.* This telecommunications bulletin supplemented TIA/EIA-568-A to add design parameters for the use of centralized fiber runs. The -B revision of the standard incorporates this information.

In practice, the centralized fiber cables are routed through existing telecommunications rooms, even though there is no termination or equipment hub there. The reason for this practice is that the TR is frequently the location of the interfloor ports that physically connect to the MC or other point of consolidation. In addition, if a blown-in fiber system is used, the multicell tubes may be run to the TR, where they are broken out into individual tubes and routed to each work area.

Alternatively, loose-buffer multifiber cables may be run to each TR and spliced onto individual work-area fiber runs. This method avoids the additional bulk, cost, and loss of an interconnection (patch) panel. The splices are allowed in fiber runs, unlike copper runs, as they add insignificant signal degradation to the optical path. The TR splice method may be used in blown-in fiber systems, as well as in a hybrid system, which would use loose-buffer cables to the TR and blown-in fiber tubes to the work area.

As with the structured system, you may use multiuser outlet (MUTOA) consolidation for open offices. As we said before, all fiber systems are inherently interconnect systems, rather than cross-connect systems. From a practical standpoint, it matters not where the interconnection occurs, so the MUTOA takes the place of an interconnect in a TR, in the centralized fiber layout.

Centralized fiber cabling is a viable option for any fiber cable installation. Care should be taken that the total run length is within the operating parameters of the networking technology you are trying to support. For example, certain fiber runs may be limited to 220 m for Gigabit Ethernet, so you would want to stay under that length.

If centralized fiber does have one limitation, it is the ability to handle future increases in network bandwidth. Remember that the current crop of legacy fiber in the horizontal cabling is miraculously able to support gigabit because of the short horizontal limit of structured cabling. Newer laser-friendly multimode and single mode cable is still an overkill at 90 m for gigabit, but it might be just right at 10 gigabit operation. A measure of future-proofing can be obtained by at least running your centralized fiber through a TR. Then, if you absolutely must have shorter fiber runs to support a future technology, you can cut and terminate the fiber runs within the TR and place your hub equipment there.

Standard Fiber Optic Network Links

Fiber is used for a wide variety of network connections. Many of these are described elsewhere in this book, including Chaps. 4 and 13. However, some of this information is repeated in Table 12.5 to illustrate the wide variety of purposes to which fiber optic cabling can effectively be utilized.

Fiber Optic Equipment Considerations

Fiber to the Desktop

The watchword in the fiber cabling industry is "fiber to the desktop." In the early days of networking, this was seen as an ability to offer almost unlimited bandwidth to the user (or at least to the user's workstation, as

TABLE 12.5

Fiber Optic Networking Standards

Standard	Speed	Media	Distance
10BaseFL	10 Mbps	Multimode	2000 m
100BaseFX	100 Mbps	Multimode	400 m
100BaseSX	100 Mbps	Multimode	300 m
ATM/OC-3	155 Mbps	Single mode	N/A
ATM/OC-12	622 Mbps	Single mode	N/A
1000BaseSX	1 Gbps	Multimode	220–550 m*
1000BaseLX	1 Gbps	Multimode/ single mode	400–550 m* 5 km
1000BaseSLX	1 Gbps	Single mode	10 km

*Distance for multimode fiber depends on the core/cladding size, bandwidth-distance product, and operating wavelength.

few of us have much bandwidth whatsoever). The most popular option for this desktop fiber originally was OC-3 at 155 Mbps. This connection was primarily suggested because the 10BaseF options were not considered high-speed and the 10 Mbps data rate could easily be supported on Cat 3 wiring. However, a more practical problem that existed with OC-3 was that it primarily utilized ATM, which was a rare protocol in the local area and not very well supported by software or hardware manufacturers. FDDI was also available as a desktop option, but at an interface cost that frequently equaled or exceeded the cost of the equipment it was connecting.

100BaseFX and any of the 1000BaseF Gigabit Ethernet options now offer very viable desktop fiber network connections. Of course, not many users currently need the speed of gigabit, but certainly the network manager (and the boss) do. Equipment interfaces are the switch/hub fiber connections and the network adapter cards in workstations, servers, and routers. All these types of equipment now offer fiber interface options. The cost of the interfaces is a matter of supply and demand. The LED optics used with 100BaseFX are relatively cheap. Consequently, the adapter cards and hub interfaces are within reach of the high-end copper counterparts.

Gigabit Ethernet is somewhat in transition, as copper cabling options are emerging. Although the cost of fiber and copper Gigabit Ethernet interfaces will be about the same initially, you can expect the price of the copper interface to drop considerably. Even though the innovative VCSEL laser diodes that are used with Gigabit Ethernet fiber are much cheaper than the older technology, they still represent a significant cost penalty over copper transceivers.

The installation efficiency and cost of centralized fiber cabling may actually be a compensating factor in considering fiber to the desktop. In very large installations, the savings of using centralized fiber over structured copper may be sufficient to dictate its use. In addition, there are inherent advantages, such as interference immunity, security, and higher bandwidth future technologies that may give fiber the upper hand.

Copper to Fiber Conversion

It is practically impossible to entirely avoid the use of copper interfaces. Many types of equipment, such as printers and routers, may offer only copper interface options. It may be possible to use switch/hubs with both types of connections. The desktop workstations and servers are connected normally via fiber, and the incompatible equipment is connected to a switch/hub with an appropriate copper interface.

In cases where the hub-conversion method is impractical, you can simply use a fiber to copper converter. You have several options on fiber to copper conversion. At 10 Mbps Ethernet, you can convert from fiber to the AUI, coax, or 10Base-T connections. The AUI is convenient because the interface includes a power source for a transceiver, which can be used to power the converter. In all other cases, you can expect to use one of the boxy wall-plug transformers and a power cord to the converter.

Pay particular attention to the type of fiber optic interface on your converter. The 100 Mbps fiber optic interface is now available in either FX or SX versions, and the wavelengths are totally different. Likewise, 10BaseFL and -FB are available, although the FB variety is rare. Gigabit Ethernet fiber interfaces are likewise available in several varieties of wavelength and mode, so pay attention to the needs of your equipment and/or fiber cable when making a choice.

In addition to the copper to fiber converters, fiber mode converters are available (Fig. 12.9). Because of the common connectors, nothing physically prevents the interconnection of single mode and multimode fibers. However, the connection will cause an excessive transmission loss

Figure 12.9
A fiber optic mode converter. (Courtesy of Transition Networks, Inc.)

if the optical direction is from multimode to single mode. There is still an unexpected loss from single mode to multimode, but it is not as great. Consider that the transition from a large 50 μm fiber core to the tiny 4 to 8 μm single mode core obviously wastes most of the incident light, unless you are fortunate enough to be using a laser source that has inadvertently positioned the beam in the center of the multimode core.

Mode converters typically put a pair of wide-bandwidth transceivers back to back in a box. One set of optics is single mode and the other is multimode. In between the interfaces is the proper electronics to repeat the signal to the opposite interface. For this reason, you can expect fiber mode converters to cost several times more than a simple fiber to copper converter.

Fiber Optic Installation Practices

This section will briefly describe some of the installation requirements for fiber optic cable. Optical fibers vary widely in construction, unlike their copper counterparts. This is due in part to the fact that a complete fiber circuit requires only two strands—one pair (although not twisted, we hope). Thus the simplest fiber link could be two fiber strands with as little as the primary coating protecting the fiber. Conversely, a fiber link could be the same two fiber strands, but with tight buffer, strength members, thick jacketing, and even armor or imbedded steel cable. The first cable type would need several types of protective routing devices, while the second type would need virtually none.

Fiber Cable Protection

The degree of physical protection a fiber link requires depends on two factors: construction of the fiber cable and location of the cable run. In a fiber distribution panel, very little is needed in the way of physical protection for the fibers. What is needed is some type of fiber management and fiber storage. On the other hand, fibers that are used for horizontal or backbone runs do need more protection. These fiber runs will require the use of moderate jacketing, trays, and raceways to provide the appropriate amount of additional physical protection.

Blown-in fiber places relatively unprotected strands inside plastic tubes that have been preinstalled between the telecommunications room and the work area. The tubes are technically *raceways*. They must meet all the flammability and smoke requirements of any plenum-rated device, if necessary. Likewise, the enclosed fibers must be appropriately rated for the installation location.

Another type of protective fiber raceway is called *innerduct*. Innerduct is used in applications where several fiber cables must be protected (Fig. 12.10). The raceway is generally a ribbed construction to allow it to easily bend at a fairly large radius, without collapsing the interior, as a smooth tube would. The ribbed construction also provides mechanical strength against crushing. Innerduct may be placed inside larger structures called *ducts* that are basically round or square pipes that run between buildings. The innerducts are used to subdivide each section of the larger duct, and also to provide additional protection of fiber cable that is run through the duct.

Innerduct is an obvious choice for riser cables between building floors or between telecommunications rooms on the same floor. Innerduct is also very useful for the horizontal run between a TR and a consolidation point. An innerduct may also be used in a cable tray to isolate fibers from copper cables. A major advantage of innerduct is that it can be installed before any fiber is run. So, you can simply route the innerduct to wherever the fiber needs to go and later pull the fiber in and terminate it. This prevents possible damage to pulled fiber during new construction build-out. As with all cabling components, innerduct must meet all the proper ratings to be placed in plenum space.

Cable trays are used to help organize and protect cabling infrastructure in horizontal runs. They may be used in any size installation, but are often found in large buildings. They serve to protect all types of cable, but are very useful as an alternative to cable "hangers" that place hundreds of bends in cable runs. The cable lies flat in the tray and is

Figure 12.10
Innerduct is used to
protect fiber runs.

easily pulled straight across long distances without the need for special
devices such as pulleys. Additional cable is easily placed in the trays, as
needed. Cable trays are virtually mandatory in medium to large telecommunications rooms. Trays are available in plastic or metal construction,
and must meet the same plenum requirements as cables. Corners and
bends will require the use of a cable pulley when cables are initially
installed.

Fiber Terminations

The physical termination procedure for optical fibers varies enormously
with the particular type and brand of fiber connector you are using.
Accordingly, we will not cover all of the options here. However, there are
some common concepts that we can cover briefly, so that you will
understand what is required for this complex operation.

All fiber termination requires the preparation of a bare fiber for
assembly within an appropriate connector. The outer jacket and fibrous
plastic strength members must be removed if you are using a jacketed
cable. In this type of cable, the outer jacket is slit around its circumfer-
ence and slipped off the end of the cable. Next the Kevlar fibers are
combed to one side and cut off at the edge of the remaining jacket. The

fibers are extremely fine, and a high-quality cross-cutter must be used to trim them back.

The buffer-coated fiber (or fibers) must now be prepared. The tight buffer is removed with a tool that is essentially a wire stripper. The buffer-stripping tool is often a fixed-opening tool that is specifically designed for the diameter of the bare fiber (core and cladding), so that the fiber is not nicked when the buffer is removed. Any nick will severely compromise the flexural strength of the silica fiber and may result in spontaneous breakage. The rule of thumb is that such a break will always occur at the last possible moment, so that the installer will have to start totally over. So the lesson is that no nick is a good nick, or nothing can truly be done in the nick of time.

The fiber is now trimmed to the proper length, usually with a tool that purposely scores (yes, nicks) the fiber, and then breaks the unwanted portion away from the remaining fiber. The breaking process is called *cleaving* the fiber, and it should leave an orthogonal break at the fiber end. This brings some basic rules of optics into play. First, if this break is not precisely 90°, the light will exit the fiber at a slight angle. Second, the break is never exactly 90°. The cure in a quick-termination connector is to place the fiber end into a matching fluid, so that the off-axis break has virtually no effect. In a so-called field termination (slow-termination) connector, the fiber end is inserted through the connector ferrule so that it protrudes slightly from the end of the ferrule. That fiber end is then polished down until it is exactly flush with the end of the ferrule, and consequently precisely square with the fiber.

Final assembly is done by inserting the fiber into the connector and securing the fiber mechanically (Fig. 12.11). Older connector types, such as the SMA and the ST, generally require the application and curing of epoxy to secure the fiber to the connector. Newer quick-termination connectors of this generation use mechanical crimping in place of the epoxy. Both quick-termination and field-termination methods are available for most of the new-style connectors, including the SC and SFF connectors. An individual connector is designed either for quick or field termination, so you must obtain the appropriate connector for your proposed application.

You should be aware that there might be a significant difference in the expected connector losses between the two termination methods. For example, Lucent Technologies, one of the SFF manufacturers, offers a 0.2-dB loss expectation in their quick-termination LC connector, and achieves half that (0.1 dB) with the field termination. So, field termination offers better potential results, but may require more training,

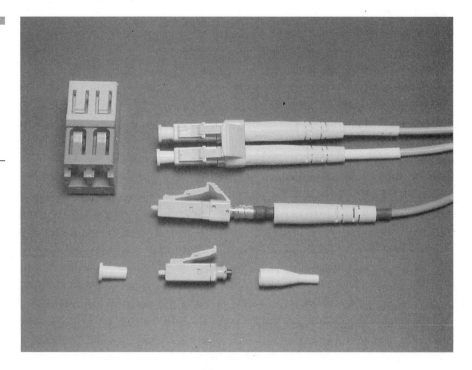

Figure 12.11
Fiber optic cable terminations, showing unterminated LC connector components, assembled simplex and duplex connectors, and an LC adaptor.

special equipment, and take longer to install. This is not to say that an inexperienced installer cannot achieve higher losses than the norms, so specific manufacturer training is quite appropriate.

A horizontal or backbone fiber run should be terminated in the same style of connector at either end, unless there are compelling reasons to do otherwise. In most cases, fiber cabling runs should terminate in a patch field in the telecommunications room. The patch panel to equipment cord is the appropriate place for a cable to adapt between different connector types. For example, an SC equipment connector could easily be interconnected to an SFF-connector patch panel with an appropriate adapter cord. Likewise, the same type of transition could be made at the user cord.

The SFF connector types offer so many varieties that an abundance of adapter cables will be made available. In addition to adaptation to the legacy connectors (which would include the 568SC), cords must be available to transition between each of the popular SFF types. It would be wise to try to minimize the number of combinations of connectors by evaluating the types of connectors that are generally available on the networking equipment you will be using. The higher-density devices will gravitate toward one of the SFF connectors. Thus, hubs and switches in the TR will tend to use an SFF connector to achieve maximum connection density. On the workstation NIC, the connector density is not an issue, so the decision to use an SFF connector will be at the manufacturer's discretion. Although there is a battle among manufacturers to control the fiber connector type in the horizontal cable plant, it is not at all clear that network managers will make equipment decisions solely on the basis of connector type. Since horizontal runs rarely terminate directly in the network equipment, an adapter cord can easily be used, no matter what connector is needed.

As the number and types of fiber optic connectors diverge, proper installer training is very important. Many of the connector designs are distinctly nonintuitive with respect to assembly. In order for the connector to operate according to its specifications, all of the installation practices must be controlled right down to the silly millimeter. Connector designs are very different from one manufacturer to another. Installers should be trained on the connector brand they are installing, and network managers should require training and certification of installers and contractors. While this is not rocket science, it is certainly not simple, and the quality and longevity of your fiber cable plant may well depend on the quality and consistency of the installation.

Fiber optic cable installation requirements call for the storage of extra fiber at the points of termination. The reason for this is simply to allow for future connector changes and for the re-termination of the fiber due to potential breaks after installation. The extra fiber must be stored in a manner that protects it from damage. At the work area outlet, the cable could theoretically be coiled up in the reassessed outlet box, if one is used. However, LAN wiring is often done with outlet rings that are totally open to the wall space, as illustrated in Chap. 7. A safe alternative is to use one of the special fiber outlet boxes that mount to the surface of the wall and contain a means for storing and protecting the fiber strands, as shown in Fig. 12.12, as well as in Chap. 7.

Fiber Optic Testing

Once installed, fiber must be tested to determine proper operation. Simple connectivity may be determined by a simple source and detector arrangement, much like the basic continuity test of copper cable. The simplest source is a flashlight (possibly with a fiber cable adapter), and the simplest detector is the human eye (Fig. 12.13). Following the safety practices in the next section, be certain you know that a low-power

Figure 12.12
Fiber outlet with
storage for extra
fiber.

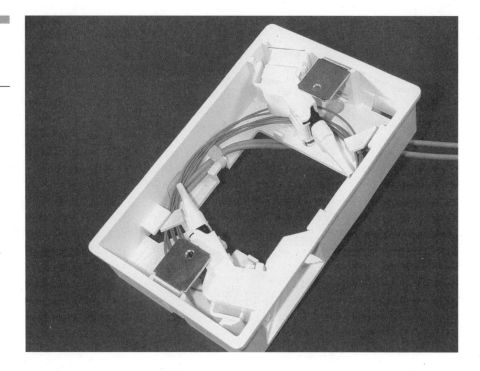

visible light source is being used for this test, and even then don't look directly into the end of the fiber.

However, as with copper cable, a much more complex test is needed to determine proper operation of the fiber link. This is accomplished with a fiber optic cable scanner, as shown in Fig. 12.14. In some cases, the scanner may be an adapter module that is used with a copper cable scanner. However, since copper scanning is inherently more complex and expensive, and since different cable crews may be involved in installing fiber optic cables, you may want to purchase a separate fiber optic tester.

All fiber optic cable testers for field use have a local and remote unit. The local unit sends light of a known intensity down the cable under test. The remote unit senses the absolute magnitude of the received light and calculates the loss. A pass/fail indication is made, and may be a function of the length of the cable. The remote may be totally passive (to the installer) and simply report the results to the local tester. The two units may be interchangeable, or at least have dual displays, so the installer at the remote end knows what is going on. Many of the tester pairs have mike/headphones and an optical "order wire" so the two installers can talk when connected to the same cable.

Figure 12.13
Testing a fiber optic
cable with visible
light.

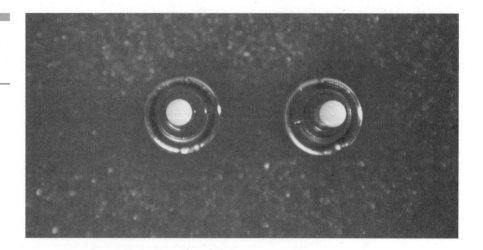

Figure 12.14
Fiber cable tester.
(Courtesy of
Fluke Networks.)

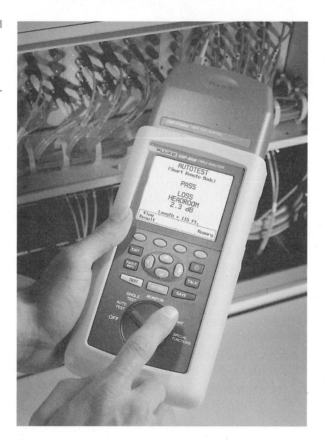

Testing installed fiber cable is a matter of measuring the link loss. In general, if the link loss is within budget for the standard (and within stricter link budgets for certain applications), your cable is fine. You should really test at the appropriate wavelength for the application, and the tester should use time-domain reflectometry (TDR) to ensure that the cable length is within prescribed lengths. Some more sophisticated testers may ensure that return losses (a measure of reflections) are within limits, but if you use appropriate cable and connectors this should not be a problem. A really sophisticated test would be to measure the link's bandwidth-length product and would catch cables that had insufficient bandwidth for gigaspeed operation at extended distances.

Most fiber optic testers use the "overfilled launch" testing method, where the entire fiber core diameter is flooded with light, usually from an LED source. This method is effective in dealing with 50 and 62.5 μm core diameters. In testers that use a VCSEL diode source, a mode dispersion technique is used to avoid the mode blockage that can occur in some multimode fibers. You should be aware that this testing technique can mask some problems that may be experienced when you actually hook up the cable run to a piece of equipment. Be aware of the specifications and limitations of your particular tester.

That takes care of the basic pass/fail testing of a fiber optic link. Troubleshooting a marginal or failed link is a different matter. One of the problems you encounter with a failed link is determining where the failure is located, with respect to the cable ends. This issue is a problem whether you are troubleshooting newly installed cable or cable that has been installed and operating for some time. Your tester may be able to function as a field TDR, in addition to the basic loss measurements. If so, it is simple to determine the location of a fiber break or a botched connector installation. Knowing which connnector is bad, or where the fiber is damaged saves a lot of repair time. Fiber breaks are permitted to be spliced because splices have such inherently low loss. Bad connectors should simply be replaced, so be thankful that extra fiber was left at the point of termination.

Safety Considerations

Fiber optic technology has safety aspects that are quite different from those of copper cabling. As long as the proper procedures are followed, fiber optic installations should be as safe as any type of cabling. Eye protection in the form of safety glasses should be used for all fiber optic assembly work.

Chemicals

Virtually all types of fiber cable termination use some type of chemical. The older epoxy-style connectors use a harsh, two-part resin and hardener combination that can harm skin, clothes, and eyes. Some of the newer field-termination connector designs use a cyano-acrylic composition that is much like a thickened version of "superglue." It is harmful to all of the above, and in addition can "weld" skin together when in its preactivated state. The cyano-acrylics are activated by water vapor, and are soluble in acetone (which many fingernail polish removers contain). You would be wise to keep a supply of chemical wipes (a sort of heavy woven tissue) and acetone solvent handy when you are using these products. If any of this superglue contacts the eye or the lid, seek medical attention immediately. You should not attempt to remove the material yourself.

In addition, the normal procedure is to use an alcohol wipe to clean the bare fiber after removal of the tight buffer. Alcohol, of course, is flammable. In addition, the type of alcohol most commonly used for a cleaning solvent is poisonous when ingested, and is harmful to the eyes.

Glass Fiber Safety

Another aspect of fiber optic safety is care with the fiber strands that are the object of our interest. Although they are very flexible, optical fibers are still made of glass (silica) and can fragment, fracture, chip, and pulverize. Tiny fragments of the glass can fly into the air (and the eye), splinter in the skin, and even be inhaled. Consequently, you should "handle like glass," as the saying goes. For the most part, fiber connector assembly is uneventful, but you should always wear your safety glasses, use the proper tools, and assemble the components on an appropriate surface. In the case of glass fiber, the most appropriate surface is a no-slip plastic or rubberized mat of a dark color. It is much easier to see the fiber (and any shards) on a black mat than on a normal light-colored table.

Use proper lighting and position the work surface at a convenient height. Place any chemicals safely off to the side and out of the way when they are not being used. Keep all liquids and glues capped when not in immediate use. Stay away from all sources of flame or sparks.

In the unlikely event that you get a glass splinter in the skin, it is best to use plastic-coated tweezers to remove the splinter. Using regular metal tweezers can cause the splinter to break at the skin surface, which results in a more difficult (and painful) problem. Always refer the removal of any glass particles in the eye to a qualified medical practitioner.

Laser Safety and Classes of Operation

Many optical transceivers that are used in optical networking use laser diodes for their light sources. The wavelengths of light that are used are in the infrared range, which is beyond the range of human sight. However, light in these wavelengths could still be harmful if sufficiently intense. All of the optical sources that are recommended for fiber LAN operation are operated at Class 1, at power levels below that considered hazardous. However, other classes of laser operation exist and you should be aware of their relative dangers.

In the United States, ANSI Z136.2 defines four classes of operation for lasers (Table 12.6). Classes 1, 2, and 3 may be used in optical communications systems. Class 4 lasers have power levels that are very high and are primarily used in medical and industrial applications. They also include some types of military applications. The optical power increases with the class.

Class 1 laser sources are generally considered inherently safe, as they have power levels that are very low. Class 2 low-power sources use visible light radiation, and should never be viewed directly. Class 3 are medium-power sources such as optical amplifiers and pump lasers. They may be visible wavelengths or not, but they should never be viewed directly and they require eye protection. Laser diodes are quite capable of producing Class 3 intensity.

Laser safety goggles are available that tune out specific wavelength ranges, while allowing enough other visible light through that you can see to work. You must choose protective goggles or glasses with a sufficient amount of attenuation at the laser's wavelength. These glasses are rated in decibels of attenuation at a particular wavelength. Be careful, as the common optical density ratings, such as OD-1 or OD-2 indicate

TABLE 12.6

Laser Safety Classes
of Operation*

Class	Relative power	Typical source	Safety characteristics
1	Low	LED and VCSEL	Safe—when operated normally
2	Low—visible	Laser diodes and ultrabright LEDs	Caution—do not view directly without eye protection
3	Medium	High-power laser diodes, optical amplifiers, and pump lasers	Danger—eye protection required
4	High	Medical and industrial lasers	Extreme danger —eye protection and physical protection required

*Classes of operation defined in ANSI Z136.2.

orders of magnitude, referenced to 10 dB, so an OD of 1 would be 10 dB and an OD of 2 would be 100 dB—quite a difference.

For European standards, you should consult IEC 825-1 and IEC 825-2. These standards cite levels of operation that differ from the U.S. standards.

In addition to the safe classes of laser operation mentioned above, you should be aware that some Class 1 laser sources are theoretically capable of operating at levels that might be harmful to the eye. In addition, some people are more sensitive to very bright point-source light, and can suffer brief effects that interfere with normal vision. If you have ever had a laser pointer flash in your eyes, you may have experienced some of this problem.

Fiber Optic Future

The future of optical fiber is secure. Fiber cable clearly provides an extreme amount of potential bandwidth, particularly in the case of single mode fiber. At present, the utilization of this incredible bandwidth is limited primarily by the speed of available optics.

Currently, 10-Gigabit Ethernet is on the horizon, and transmission beyond that is theoretically possible. One factor that will influence the introduction of these bandwidths is the need for speed. Another factor is certainly the ability of the existing fiber plant (if such exists) to support these advanced speeds. Future equipment interface implementations may actually borrow the multiple pair technique from copper cabling. Another viable (if expensive) implementation could use multiple wavelengths to multiply the effective bandwidth of the fiber.

In the end, it is all a matter of cost. If the demand is there, the volume will increase, and the price will fall. The 100 Mbps links were initially very expensive, but their price has rapidly fallen to that level once occupied by 10 Mbps. The 1000 Mbps links will definitely do the same, as in fact they already are. Fiber is definitely in our future.

CHAPTER **13**

Gigabit Cabling Technology

Chapter 13 Highlights

- Fiber Gigabit Ethernet
- Copper Gigabit Ethernet
- Wiring standards
- Planning gigabit installations
- Testing gigabit cabling
- Future of gigabit cabling

This chapter will detail the special considerations that you will encounter when providing cabling to support gigabit speeds. As you may know, the most popular gigabit-class local area networking topology is the gigabit variant of classic Ethernet. ATM at OC-12c is another topology that is now available, although it is not exactly gigabit speed. Nevertheless, it requires much the same cabling infrastructure.

Gigabit refers to a data transfer speed of 1,000,000,000 bits per second, or 1 Gbps. The corresponding Ethernet technology meets this definition exactly; it operates at 1 Gbps. Technically, ATM at OC-12c is at 622 Mbps data rate, so it is about two-thirds gigabit speed. In fact, the standards revision TIA/EIA-568-A (and now TIA/EIA-568-B as well) incorporates wiring technologies that accommodate both of these gigabit-class speeds. This standard specifies the provision of two cables to each workstation outlet, one of which may be fiber. Although it is up to the cabling system designer to include the fiber, the provision of multimode fiber makes it easy to implement either of the common gigabit-class technologies, Gigabit Ethernet or ATM/OC-12c. But, what about copper? Well, thanks to the tight specifications of this standard, we are now using Cat 5/5e and higher cabling for gigabit copper networks for both Ethernet and ATM-622. Network speeds go up by a factor of ten about every five years (Fig 13.1)!

Gigabit Ethernet

Gigabit Ethernet is formally specified in the supplement to the base 802.3 standard known as IEEE 802.3z, which includes the fiber implementations, and in IEEE 802.3ab for twisted-pair copper. It is sometimes referred to as 1000Base—X. The "X" signifies a physical layer based on the ANSI X3.230 Fibre Channel[1] standard, in the same manner that 100Base—X fiber modes were based on the FDDI physical layer.

The choice of Fibre Channel signaling was made simply to expedite the introduction of the Gigabit Ethernet technology. At the time of first consideration of expanding Ethernet operation to the gigabit range, Fibre Channel existed as an accepted standard. More importantly, integrated circuits existed to implement the signaling, so it was very time- and cost-effective to utilize the same basic circuitry. The signaling rate was modified to allow for an 8-line by 125 Mbps interface to achieve a data rate of

---

[1]Fibre Channel is spelled according to the European convention, as it is expected to become an international standard.

Figure 13.1
Time-speed curve.

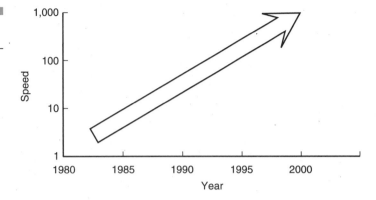

exactly 1000 Mbps. The actual medium-independent layer of Gigabit Ethernet is the 8-bit-wide Gigabit Media Independent Interface (GMII), corresponding to the MII in Fast Ethernet and the AUI in standard Ethernet.

Gigabit Ethernet is topologically similar to Fast Ethernet (100—Base—x). All workstations connect directly to a switch or hub, and collision domains are tightly controlled to allow the collision-detection mechanism to operate properly. As a matter of fact, shared-medium operation of Gigabit Ethernet is virtually absent, as Gigabit Ethernet switches are used in most hub-type applications. In essence, the switch is an OSI Layer 2 bridge that operates much like its lower speed, multiport counterparts.

Gigabit Ethernet was originally specified as a fiber optic topology, because the bulk of installed cabling at the time was only Category 3. Cat 3 simply was not capable of carrying the higher symbol rates needed to support 1 Gbps data. Today, all of the newly installed cabling is Cat 5 or above, which makes a copper implementation possible. However, most of the gigabit switches and routers on the market still use the traditional fiber connections. Fiber is also much more flexible in terms of the distances covered and its ease of use. Most any multimode fiber link will support Gigabit Ethernet at more than double the copper distance. Furthermore, for proper operation at gigabit, enhanced Cat 5 or better cable and connectors must be installed to the very highest standards to be usable, and even then the distance is limited to only 100 m.

We should mention that a lesser-known copper standard, 1000BaseCX, exists that uses twinax over a distance of 25 m. This standard is not often seen, but exists to allow the short-range interconnection of two Gigabit Ethernet devices without the need for expensive laser optics.

Fiber Optic Standards for Gigabit Ethernet

Three fiber-optic standards are available for Gigabit Ethernet. The two primary standards use a range of fiber modes and transceiver wavelengths and are intended to support link distances from 220 m to 5000 m. A third standard is in development, and equipment manufacturers are already offering this option. All require laser optics, because of the higher bandwidth requirements for gigabit data rates. Each fiber standard is intended to support an increasing link distance.

1000BaseSX, LX, SLX, ELX, ZX   Table 13.1 shows the three fiber optic standards, along with the types of optical fiber and the specified operating distance. The three standards are 1000BaseSX, 1000BaseLX, and 1000BaseSLX (an emerging standard). Within the gigabit speed range, they are often referred to simply by the suffix letters SX, LX, and SLX. The three suffixes actually refer to the wavelength of light that is transmitted. As we saw in Chap. 3, the wavelengths commonly used for fiber optic transmission are 850, 1350, and 1550 nm, with the last two being used primarily on single-mode fiber. So, you can say that the SX on 850 is the "short" wavelength, the LX the "long" wavelength, and the SLX the "superlong" wave mode. Conveniently, the letters also correspond to the relative transmission distances of each operating mode. Thus, SX is short distance, LX is longer distance, and SLX is superlong distance. Convenient, isn't it? The new ELX and ZX standards are currently proprietary, but offer extreme distances.

TABLE 13.1

Gigabit Ethernet Fiber Optic Link Parameters

IEEE 802.3z fiber mode	Wavelength,* nm	Core/cladding diameter, μm	Mode†	Bandwidth, MHz/km	Range, m
1000BaseSX	850	50/100	MMF	400–500	500–550
1000BaseSX	850	62.5/125	MMF	160–200	220–275
1000BaseLX	1300	50/100	MMF	400–500	550
1000BaseLX	1300	62.5/125	MMF	500	550
1000BaseLX	1300	10	SMF	N/A	5000
1000BaseSLX‡	1550	10	SMF	N/A	10,000
1000Base ELX/ZX	1550	10	SMF	N/A	70,000

*IEEE 802.3z specifies a range of values, such as 770–860 nm for 1000BaseSX; however, the accepted nominal wavelength is shown. It should be noted that the shorter wavelength actually represents a higher light frequency and is associated with a greater attenuation.
†MMF is multimode fiber; SMF is single mode fiber. See Chap. 3 for details.
‡Standards for 1000BaseSLX are under review and may change in the future.

Understanding Fiber Performance at Gigabit Ethernet Speeds
Several factors affect the bandwidth and range of Gigabit Ethernet fiber
links. Most of these were discussed earlier in Chap. 3; however, they gain
special importance at these much higher data rates. Chief among these
factors is the bandwidth of the transmitting optics that "launches" the
beam of light along the fiber. In lower-speed transmissions, inexpensive
light-emitting diodes (LEDs) are used. However, LEDs do not have the
speed to support gigabit transmission bandwidths, so more expensive
laser diodes must be used.

Another problem that affects link performance is the spectral width
of the transmit optics. An ideal light source for fiber optic transmission
would have zero spectral width; it would produce a pure light of just
one wavelength. In practice, these solid-state light sources produce light
that is distributed across a band of wavelengths, approximated as a bell-
shaped curve, with the peak at the desired nominal wavelength. An LED
has a relatively broad spectral width of about 150 nm. That is, a device
that operates at 850 nm actually produces light *near* 850 nm that may
range from 775 to about 900 nm. A typical laser diode source has a spectral width of about 5 nm. See Fig. 13.2.

The special width is important because of a phenomenon called *chromatic dispersion.* As you know from looking at a prism, light of different
colors is refracted differently when passing through glass. The fiber core
of a multimode graded-index fiber will refract the different wavelengths
of light accordingly, and the components of a transmitted wave front
will arrive at the destination receiver at slightly differing times. At the
high data rate of Gigabit Ethernet, this signal distortion can make the
signal unusable as the distance increases. For standard LEDs, with a very
wide spectrum to begin with, this usable distance is far too short to be
used for a gigabit connection.

Figure 13.2
Spectral width of
laser diode and light-
emitting diode
sources.

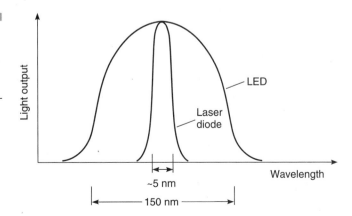

Laser diodes have a much narrower bandwidth, on the order of 5 to 10 nm, and therefore experience much less of the chromatic dispersion phenomenon. Consequently, laser optics have a much greater range when used at gigabit speeds. This range is generally true for both multimode and single mode fiber. However, multimode fiber may present another challenge for gigabit users.

A great deal of multimode fiber has been installed for future high-speed network use. The expectation was that fiber, with its inherently high bandwidth, would be quite suitable for gigabit networks when the networking hardware became available. Unfortunately, this is not necessarily the case. The problem occurs precisely because we are required to use laser sources to achieve any effective distance at gigabit speeds, because of the chromatic dispersion problem.

A laser source launches a coherent beam of light that propagates down the fiber core in a single transmission mode, regardless of whether the fiber is single or multimode. The operation is rather complex, but you can think of it as a single ray of light (or wave of energy) that bounces back and forth on the fiber core walls. A step-index fiber, such as single-mode fiber, operates a lot like this. However, in a graded-index fiber imperfections in the glass core may cause fatal variations in the refraction of a single coherent light mode. If the single mode of light encounters one of these imperfections, the attenuation for that mode will be greatly increased and the transmission may be totally lost. An LED, on the other hand, has incoherent (multiple mode) light output, so the loss of a single mode has very little effect on the amount of light at the receiver. However, the unpredictable nature of the problem with many multimode fiber cables makes this potentially a severe problem for Gigabit Ethernet and other gigabit speed services.

This phenomenon is known as *differential mode delay* (DMD), and it varies greatly from fiber to fiber among different manufacturers. Unfortunately, the problem had little effect on the LED transmission optics that were traditionally used with multimode fiber, so millions of meters of fiber were installed with no specification for DMD. The problem does not occur in single-mode fibers, because they propagate only one mode. However, a cure exists for legacy multimode fiber to enable the gigabit operation. The solution is a short length of specially constructed fiber cable that has been designed to "spread" the laser optic's single mode into multiple modes before launch down the fiber.

This length of cable is called a *conditioned launch adapter* (Fig. 13.3), and it may be used for LX transceivers to assist in operation over multimode

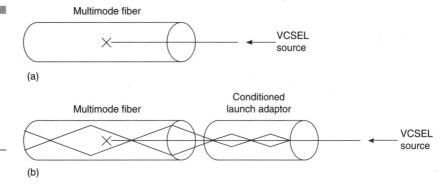

Figure 13.3
Conditioned launch
adapter. (a) Without
adapter, glass flaw
blocks narrow
modes; (b) adapter
(mode spreader)
allows most modes
to bypass glass flaw.

fiber. LX operation can be over either single or multimode fiber, so the external adapter must be used. The conditioned launch mode-spreading function can be contained within the short wavelength transceiver for SX operation, because SX only uses multimode fiber. Many manufacturers recognize this problem and now provide a type of higher-quality multimode fiber that minimizes the glass imperfections that cause DMD. You can generally recognize these advanced fiber cables by their claims to operate well at gigabit speeds with the "vertical-cavity surface-emitter launch" VCSEL laser diodes. These VCSEL optics are relatively inexpensive and have been a key component in providing cost-effective gigabit optics.

The final factor that affects gigabit performance is fiber attenuation. In Chap. 3, we saw that the attenuation of fiber ranges between 1.0 and 3.5 dB for multimode fiber at 1300 and 850 nm and between 0.4 and 0.5 dB for single mode fiber at 1550 and 1310 nm. This attenuation affects the distance one can successfully transmit on a fiber, because the light output power of the laser transmitting optics is limited for safety reasons, and because the receive power must be sufficient for signal recovery by the receive optics.

Copper Standards for Gigabit Ethernet

In the "old" days of Category 3 cabling, the concept of Gigabit Ethernet operation over twisted pair or any other type of copper cabling was widely ridiculed. It was well known that the operation of higher-speed networks required higher frequencies to be launched down the copper pairs, and the combination of crosstalk, attenuation, and return loss clearly made such operation impractical if not impossible. At least this was true for the copper cable technology of the day.

However, great advances have been made in identifying the phenomena that reduce cable operating distance and frequency range. As soon as these phenomena were discovered, the corresponding channel parameters that produced them became obvious, and the race was on to produce even better cable and components. Initially, these advances led to the wide use of Cat 5 cable, with component and link testing to 100 MHz. Topologies such as Fast Ethernet actually only require fundamental frequency bandwidths to about 33 MHz (with spectral distribution to three times the fundamental), because of the complex encoding of data signals used. In addition, 4 pair transmission methods, such as 100BaseT4 and 100BaseVG-AnyLan pioneered the use of the two unused pairs to increase the data rate far above the 2 pair range.

Advanced cable constructions, such as Category 5e, Category 6, and even Category 7 (ISO Classes D, E, and F) now allow technologists to push the limit far beyond the 100 and 155 Mbps limits that were generally accepted for ordinary Cat 5 only a few years back. These and other factors have made Gigabit Ethernet signaling over copper cables practical and desirable.

1000BaseT   After much debate and testing, the IEEE 802.3ab standard for Gigabit Ethernet over twisted pair copper cabling has been released. This technology is just emerging, but we will see an increasing availability of gigabit over copper in the future. The standard, called 1000BaseT, implements a complete gigabit data interface using Category 5e-grade 8-pin modular (RJ-45 style) connectors and cables. The distance supported is the standard 100 m link, and conforms to most modern structured cabling designs. All four pairs are used, in order to keep the frequency requirements very low. The operating specifications are shown in Table 13.2.

This ability to use standard hardware is extremely important in the ultimate acceptance of this technology. In addition, the topology and connecting hardware requirements of Gigabit Ethernet are essentially the same as those of 100BaseTX, although all four pairs are used, as with 100BaseT4. That means that any link that can support Fast Ethernet can probably also support Gigabit Ethernet. Common sense, however, says that these links are a little nearer the marginal limits when operated at this higher data rate, and corresponding higher operating frequencies. We will cover this in more detail in the section "Wiring Standards to Support Gigabit Speeds," which follows.

TABLE 13.2

Operating Specs
for 1000BaseT

1000BaseT standard	IEEE 802.3ab
Medium	Copper UTP or ScTP
Number of pairs required	4 pairs
Pair usage	Bidirectional
Operating speed	1000 Mbps
Distance limit	100 m
Propagation delay	548 ns
Delay skew	50 ns

The important thing to realize is that network speeds are moving at a rate that defies Moore's law.[2] The jump to gigabit occurred much faster than many had expected, and the jump to 10 Gbps will occur very quickly as well. In fact, 10 Gbps operation to over 300 m has already been demonstrated. However, in all fairness, advances in networking technology have not really occurred linearly, as they are totally dependent on standards implementations, and as such are not incremental at all. This means that a 9 or 10% increase in a possible network operating speed will never be implemented, because enterprise users deploy new networks in tandem, unlike workstation and server technology. It is quite possible to have one workstation that operates at a processor speed of X and another that operates at a much improved processor speed of $X + 30\%$. Yet a network that envisioned one connection at 100 Mbps and the next one at 130 Mbps would be deemed patently ridiculous. Or would it?

As it turns out, we describe a network more or less like this in Chap. 14. It uses wireless transmission as the medium, which is much more tolerant of mixed speeds than a wired network would be. The technique that is used for wireless is to always send the header at the lowest (common-denominator) data rate, and then allow what follows to be speeded up, super-encoded, or encrypted. But such a technique could not possibly be used with a wired network...or could it?

---

[2]Moore's law states that computing speeds will double approximately every 18 months. It is generally recognized that the jump from 100 Mbps to 1000 Mbps has occurred at approximately double this rate (double would have been to 200 Mbps), in concert with the increase of bandwidth requirements on the Internet. Moore's law is named for Gordon Moore, a founder of Intel Corporation, and who first proposed it.

1000BaseCX   Another copper standard has existed since the early inception of Gigabit Ethernet, 1000BaseCX. As the "X" implies, this is simply a non-fiber-based implementation of the Fibre Channel technology. This standard is intended to make use of the less expensive direct-copper connection methods, rather than the pricier fiber optic methods. The CX mode uses a type of cable known as twinax (see more detail on coaxial and twin-axial shielded cables in Chap. 3). As you can see from Table 13.3, the distance limitations are extreme.

However, this is not seen as a problem, because the CX connection was envisioned to be used only in the telecommunications closet, and 25 m would be a relatively long distance to connect in that environment. In reality, CX implementations are rare, as development of less expensive gigabit optics has greatly increased the cost associated with a fiber link.

ATM Gigabit Standards

Several methods are available for operating ATM in the local area environment, whether direct to workstations and servers, or as a backbone technology that might run between floors or buildings on a campus. As you know, ATM is a signaling protocol that is independent of data rate and media. As a result, much of the information that is available for ATM covers wide area implementations, such as those for ATM over T1 or DS3. The ATM-155 UNI and ATM-622 UNI interfaces are capable of operating both in the local and wide area, and a LAN can certainly be implemented in native ATM networking technology.

Unfortunately, in the LAN, most of the equipment connections are Ethernet-oriented. Operating a virtual LAN connection, such as Ethernet or Token Ring, over an ATM network requires a number of additional protocols that adds a tremendous complexity to a local network. This technology is called LAN emulation (LANE). In addition, 155 Mbps and 622 Mbps optical interfaces are fairly expensive, both for the hub and the workstation. For these reasons, the market has been much more open to the Gigabit Ethernet implementations than to the comparable ATM speeds.

TABLE 13.3

Gigabit Ethernet
Copper Link
Parameters

IEEE 802.3z fiber mode	Cable type	Number of pairs	Range, m
1000BaseCX	Twinax or quad	N/A	25
1000BaseT	Category 5, 5e, 6, or above	4	100

A viable alternative to LANE is to simply operate the workstation/server connections in a native ATM mode. As the ATM technology becomes more widely available, ATM network adapter cards with software drivers for popular operating systems are becoming available. In the future, it may be practical to implement gigabit-speed networks over our structured cabling networks.

ATM-155 over fiber uses the OC-3c physical layer. This connection has been available for some time, and can be supported by standards-based fiber cabling. With the proper workstation network interface cards, OC-3 to the desktop is practical. In this application, the ATM network runs in native mode, alleviating the requirement for LAN emulation.

At this point, the most copper development has been done for 155 Mbps ATM. In fact, a method now exists to support the non-return-to-zero (NRZ) signaling of ATM-155 over Cat 5 cabling. The ATM Forum provides a standard for a copper physical layer that allows for the same link parameters as the Channel defined by TSB-67 and other supplements. Unlike Ethernet, the copper ATM connection uses the 1-2 pair and the 7-8 pair. However, standard structured cabling connects all four pairs, so there is no problem in utilizing standards-based cable of the proper category. This means that a properly installed Cat 5 cable of 90 m, 10 m or less of patch and user cords, and a maximum of four Cat 5 connectors comply with the requirements for ATM-155 over copper.

ATM-622 uses the OC-12c physical layer at 622 Mbps. At this time, implementations for this technology are fiber optic–based. However, techniques for operating over copper are in the testing phase. It is reasonable to assume that the 622 Mbps signaling can use the same techniques that Gigabit Ethernet uses, as the transmission frequencies should be no greater. Keep in mind that the effective data rate and the maximum signal frequency may not be in step. In all of these very high data rate technologies, symbol encoding and signal splitting are both used to reduce the maximum signaling frequency that must be transmitted.

ATM generally uses standard data rates that are multiples of a lower-order data rate, Optical Carrier 1 (OC-1). As each standard rate is set, a judgment must be made to determine which multiple is optimum. Often, the data rate chosen is a result of a careful analysis of operating requirements and media performance. Thus, a rather unusual multiple of 3 was chosen for the 155 Mbps specification.[3] However, this rate was well within

---

[3]The OC-3 rate, at 155.52 Mbps, is three times the base rate of 51.84 Mbps, OC-1. The OC data rates are frequently referred to by the number before the decimal in the exact bit rate, e.g., 155 or 622.

the frequency and bandwidth capabilities of the fiber optics and the fiber media to allow operation at the desired distances.

The OC-3 rate is in many ways considered the base rate of the synchronous optical network (SONET), STM-1, and subsequent higher rates are at multiples of four from that base. So, the next speed chosen was 4 times 155 ≅ 622 Mbps, or a multiple of 12 from OC-1, thus OC-12. A multiple of 24 would technically yield a rate of 1244 Mbps, or a little above Gigabit Ethernet, but it is rarely employed. Instead, the next higher rate that is commonly used is again 4 times greater, 2.488 Gbps for OC-48. OC-192 then is about 10 Gbps, and OC-768 (the highest rate currently in vogue) is about 40 Gbps. See Table 13.4.

Wiring Standards to Support Gigabit Speeds

It is fortunate that so much modern cabling infrastructure has been installed in the past few years. Had we not been aggressively installing Category 5 copper links and multimode cable to the work area, it is likely that technologies to support gigabit speeds would have diverged. As it turned out, the vast high-speed cable plants have acted as both enabling and guiding forces to focus the introduction of Gigabit Ethernet and other gigabit speed networking technologies.

The most widely deployed cabling standard is TIA/EIA-568-B. This standard has provided a structured approach to cabling the workplace,

TABLE 13.4

Sonet Optical Carrier Rates

Synchronous transport level (electrical)	Optical carrier level (optical)	Carrier rate (actual, in Mbps)	Data rate classification
STS-1	OC-1	51.84	52 Mbps
STS-3	OC-3	155.52	155 Mbps
	OC-12	622.08	622 Mbps
	OC-24	1,244.16	1244 Mbps
	OC-48	2,488.32	2.4 Gbps
	OC-192	9,953.28	10 Gbps
	OC-768	39,813.12	40 Gbps

and has been a very significant factor in "normalizing" the wiring infrastructure in commercial buildings. The standard continues to evolve to incorporate new developments. Over the past few years, many enhancements have been made to the science and art of high-speed data wiring. The TIA's TR-42 User Premises Telecommunications Infrastructure standards group has been quick to recognize advances in the field and has implemented study groups to define standards and practices for new technological developments, including gigabit networking.

Gigabit Standards for Fiber

The fiber specification in TIA-568-B allows the use of either 62.5/125 μm or 50/125 μm fiber and single-mode fiber in horizontal and backbone cabling. However, the standard contains minimal specifications for bandwidth, requiring 160 MHz · km at 850 nm and 500 MHz · km at 1300 nm, and does not get into the complexities of laser-launched multimode optics. Fiber does not have the luxury of splitting the gigabit signal among multiple fibers, as with copper, so the bandwidth component is quite critical. Some manufacturers are offering higher-bandwidth fiber to support gigabit applications. You will typically find this fiber promoted as gigaspeed, to emphasize its capabilities.

The link budget for a 100-m horizontal fiber link is very modest, considering the length. For example, at 850 nm, the standard requires a maximum cable attenuation of 3.75 dB/km, which translates to about 0.34 dB loss in the 90-m horizontal run. It turns out that the losses of the connectors and adapters are much more significant in the overall fiber link performance. In recognition of this, the cabling standard permits a 0.75-dB coupling loss for the SC connector termination, and a total 1.5-dB loss for a cross-connect to another fiber. Again, this might result in a link loss of as much as 7.87 dB for the fiber equivalent of the copper cable Channel, with a 90-m horizontal drop and user and equipment patch cords totaling 10 m length. The fiber would contribute 0.37 dB, the six SC connectors 4.5 dB, and the two coupling adapters a total of 3.0 dB. See Fig. 13.4.

In practice, the link loss is far less. A typical connector termination loss is 0.15 to 0.20 dB, and coupling losses are about 0.5 dB. Redoing the calculations using the minimum values, you could expect the link loss to be only 2.27 dB. As a matter of fact, Gigabit Ethernet over fiber functions very well with a 3.5-dB loss budget.

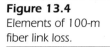

Figure 13.4
Elements of 100-m
fiber link loss.

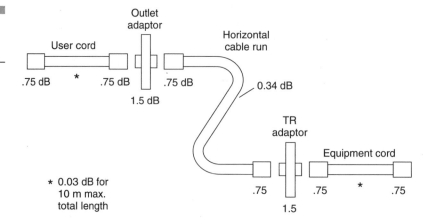

In most applications, Gigabit Ethernet is used as a backbone technology. TIA/EIA-568-A[4] allowed slightly different requirements for the backbone links, permitting both multimode and single mode fiber. In this application, multimode fiber is used much the same, although the bandwidth, mode dispersion, and transmission attenuation are much more significant issues. These factors are the primary parameters limiting the range of multimode fiber at gigabit data rates.

Here the differences of using the recommended TIA/EIA-568-A 62.5/125 μm fiber and the more expensive 50/125 μm fiber are clear. The 62.5 fiber is generally limited to a range of about 250 m, whereas the 50 μm fiber can extend over twice the distance using 850 nm optics (see Table 13.1). One must use more expensive 1300 nm optics to use the 62.5 μm fiber at 500+ m. For this and other reasons, the TIA/EIA-568-B standard allows the use of both 50/125 μm multimode and single mode fiber in horizontal and other applications.

Gigabit Ethernet Fiber Optic Connectors Since the TIA/EIA-568-A standard was finalized, the preferred connector for horizontal and backbone fiber runs has been the 568 SC connector, an assembly of two SC fiber terminations linked together into a pluggable module. However, that arrangement has an outlet footprint about 1½ times as wide as the modular RJ inserts used for copper connections. The only way to use the 568 SC in a wall plate is to decouple the modules so they plug into a nonadjacent plate opening, or to redesign the plate and perhaps the snap-in modules so that a two-wide insert can be accommodated.

---

[4]TIA/EIA-568-B allows single mode fiber to be used in horizontal, backbone, centralized, and open-office applications.

Naturally, fiber connector manufacturers responded by beginning to develop connectors that allowed duplex fiber mating within the standard snap-in module space that was designed for the RJ connectors. In a departure from past standardization actions, the TIA in TIA/EIA-568-B took the unusual step of declining a specific recommendation for the small form factor connector, as it is called. Rather, they specify only the connector and mating performance characteristics and maintain the transmit-receive A/B orientation of the connection. The result is that several different duplex fiber connectors are in use, each supported by up to 10 manufacturers. You can find more information on this issue in Chap. 12.

Gigabit Standards for Copper

TIA-568-A defined cabling standards for both twisted pair copper and fiber optic cabling. The minimum standard that is considered usable for gigabit networking is Category 5. However, if you are planning on using Cat 5 installed cabling that was tested before Addendum 5 to TIA-568-A, you should use the additional testing requirements of TSB-95 to verify the cable for gigabit performance levels (as described later in this section).

The original specifications for Cat 5 components (connecting hardware and cable) were developed with the view that 155 Mbps was about the limit for the technology. Also, it was assumed (rashly, in hindsight) that most implementations would use only two pairs of the 4 pair cable, as was the case with 100BaseTX. So the test parameters based on a 100 MHz limit tested two pairs at a time, to allow for the expected signaling structures. This allows for a third harmonic component of a 33.3 MHz fundamental signal, essentially within the requirements for the 100 and 155 Mbps technologies.

In order to place gigabit rates on the same type of cable, all four pairs are required. The net effect is that the rate is split by four and through further encoding reduced to allow operation within the 100 MHz–tested Channel. But just barely! It is a little risky to place the reliable operation of your network on an infrastructure that just meets the minimum criteria for operation on day one. This means that there can be absolutely no room for any aging of any of the components, and no deterioration of the connections for the life of the installation. That borders on the absurd.

The solution has been to provide a series of enhanced-performance categories or levels to make certain that installed wiring could meet the needs of Gigabit Ethernet, and other technologies. So-called enhanced Cat 5 cable became available almost immediately. To sort out the various claims, the TIA began studies to supplement and revise the TIA/EIA-568-A standard. At the same time, Anixter, the wiring component distributor that began the system that resulted in the TIA's component category definition, set up a new levels system for Cat 5 cable. This system recognizes that "better" can be quantified (and, no doubt, priced as well). See Table 13.5.

The most obvious difference in Cat 5 cable performance is seen through crosstalk. A cable with a lower NEXT (and FEXT, the far-end measurement) will naturally perform better at all the frequencies. As a matter of fact, the point at which a link ceases to function for a particular frequency range is where the attenuation and the NEXT are equal. For early constructions of Cat 5 cable, this point was engineered to just over 100 MHz.

So, let's look first at several of the critical component parameters at gigabit speeds. Then we will discuss the advances beyond Cat 5 that can get you reasonable performance margins at gigabit and beyond.

Attenuation   As a signal travels along a cable, the resistance, capacitance, and inductance of the copper wires act to reduce the amplitude of the signal. Some of the energy may be lost through pure resistance,

TABLE 13.5

Category 5 Extended Performance Levels for Cable*

Parameter	Units	Level 5	Level 6	Level 7
Test range	MHz	1–200	1–350	1–400
PS-ACR ≥ 0dB†	MHz	140	165	280
PS-ACR ≥ 10dB	MHz	80	100	180
Attenuation ≤ 33 dB	MHz	200	200	240
Connecting hardware				
Test range	MHz	100	100	100
Attenuation	dB	0.4	0.4	.02
PS-NEXT	dB	N/A	40	50
NEXT	dB	40	42	54
Return loss	dB	N/A	18	20

*These cable and connecting hardware performance levels were promulgated by Anixter Inc. to define advanced quality levels of Cat 5 components in structured cabling.
†The 0 dB level of ACR is generally considered the bandwidth of the cable.

and some may be coupled out of the cable pair to adjacent pairs or to other conductive objects. In addition, any variation in the magnitude of resistance, capacitance, or inductance along the pair or at a termination point can cause a mismatch that serves to further attenuate the signal (among other things).

Attenuation in a cable is frequency dependent. The higher the frequency, the more the attenuation. This effect is particularly noticeable at the extremely high frequencies that make up the gigabit signal. Attenuation is measured in dB in 1 MHz increments up to the maximum range of the cable.

Advanced cables maintain very tight control on the geometry of the cable to minimize attenuation up to 100 MHz, 250 MHz, and beyond. Accordingly, the attenuation at 100 MHz progressively improves for categories above Cat 5, as it must to achieve reasonable levels at the higher frequencies.

Near- and Far-End Crosstalk   In reality, it is the ratio of signal to interference (presumably the most significant component of noise) at the far end of a link that is critical. At the far end, however, signal components exist of both the local near-end crosstalk and the distributed far-end crosstalk (including the transmitting connector's NEXT and the pair-to-pair coupling along the cable).

In gigabit signaling, all the pairs are in use, so the situation is really complex. It turns out that it is particularly complex to test the parameters that cause performance to degrade at gigabit rates. This testing is described in detail in Chap. 16, but it seems a little like a day at the races. In the lead is attenuation, followed by NEXT, FEXT, ELFEXT,..., then comes return loss, impedance, and, close behind, ACR. Well, the resulting scramble for the top technology spot keeps us all betting on next year's network foals.

ACR—Attenuation to Crosstalk Ratio   The difference in dB between the measured attenuation and NEXT at a particular frequency is commonly referred to as the attenuation to crosstalk ratio (ACR). The ACR is given in dB, as are the attenuation and NEXT of a cable. Since both those values are converted from actual power levels, the ratio between those two power levels can be expressed in dB. Thus the ratio is developed by simply subtracting the two measurements. For example, an attenuation of 37 dB and a NEXT of 40 dB yields an ACR of 3 dB. One could say that, in essence, the remaining transmitted signal would be about 3 dB more than the potential interference from near-end crosstalk.

Beyond Category 5: Cat 5e, 6, and 7 Enhanced Cat 5, or Category 5e, has now been canonized by the TIA. This is somewhat in recognition of the Cat 5 levels program, which has been used to spec cable and connecting hardware that exceed the Cat 5 standard. For that matter, you can certainly now specify components that even exceed Cat 5e. Categories 6 and 7 are well along the standards road, and will soon be a requirement for advanced network technologies. All of these have analogs in the ISO standards as Class D, E, and F cable definitions. Taking a look at Table 13.6, you can see that there are substantial performance advantages to the higher cable categories. For more information on these new performance categories, refer to Chaps. 5 and 12 and the individual component sections in Part 2.

Using the advanced cabling categories will clearly enhance performance. However, the standard states that you should be able to operate Gigabit Ethernet over simple Category 5 cabling systems. The issue for most wiring designers is how much of a performance margin should be

TABLE 13.6

Comparison of Category 5, Category 5e, Category 6, and Category 7 Channel Performance Limits

Parameter	Category 5 Class D	Category 5e Class D	Category 6 Class E	Category 7 Class F*
Frequency range, MHz	1–100	1–100	1–250	1–750
Positive ACR frequency, min., MHz	100	100	200	600
Attenuation, † dB	24.0	24.0	21.7	20.8*
Return loss, † dB	8.0	10.0	12.0	14.1*
NEXT,† dB	27.1	30.1	39.9	62.1*
PSNEXT, dB	‡	27.1	37.1	59.1*
ACR,† dB	3.1	6.1	18.2	41.3*
ELFEXT,† dB	17.0‡	17.4	23.2	f/s*
Propagation delay, ns	548‡	548	548	504*
Delay skew, ns	50‡	50	50	20*

*Cat 7 parameters are approximations from preliminary information.
†Specified at 100 MHz for installed link.
‡Requirement for Cat 5 added by TSB-95 to certify installed cable for gigabit operation.

specified. At this point, Category 6 specs are preliminary, and to an extent you use them at your own risk. The situation was similar for the original Category 5 specification, where the parameter of delay skew was found to be critical for 4 pair network link implementations. Those who specified Cat 5 cable when only Cat 3 was really required for their current application have been rewarded, now that most network hardware needs Cat 5. Fortunately, the hardware for Cat 3 and Cat 5 is completely interchangeable, although a mixed use downgrades the link performance expectations to the lower category.

Backward Compatibility of Categories 5e, 6, and 7   The new performance categories present a more serious problem. You will recall that the major performance hit in a Category 5 link turns out to be the crosstalk due to the connectors. The 8P8C, RJ-45 style, terminations have a relatively poor score for NEXT and FEXT, which always show up as the major contributors to return loss and impedance mismatches. The Category 6 designs adjust the conductor geometry within the RJ form factor to bring these degrading performance parameters an absolute minimum. This allows a positive ACR out to 200 MHz, rather than the 100 MHz range of Cat 5.

This performance holds true as long as you are using all Cat 6 hardware, which is designed to compensate for reactance differences between the plug and jack. Unfortunately, preliminary testing revealed that Cat 6 hardware does not always give sterling performance when mated with legacy Cat 5 components. In some cases, the link performance may be degraded *below* even Category 5. Needless to say, this is a shocking result. We have depended on being able to use a compatible jack, plug, or patch cord of a higher category with no penalty. However, Cat 5 and Cat 6 may not mix, although the connections may give mechanical compatibility and have DC electrical continuity.

This problem appears to lie only with the *backward compatibility* of Category 6 *connecting hardware*. So any cable that meets Cat 6 standards should work fine in a Cat 5 application. There is total compatibility between Cat 5e and Cat 5 components, although performance may not stretch to the enhanced level.

Category 7 presents an entirely different problem, as both cable and connecting hardware are expected to be quite different from the lower categories. The Category 7 cable type and connectors have been generally defined, although this is still early in the standardization process. The connector is different from the RJ-style, and the cable is no longer UTP (unshielded, twisted-pair). Note that modified RJ-style connector designs are under study, but may present another backward-compatibility issue, as with Cat 6. In order to provide an operating range well above the lower categories, Cat 7 uses four individually shielded pairs with an overall screen and the non-RJ hardware uses a rectangular shielded connector. Any reverse compatibility here will be through the use of adapter cables that can translate the connector types, rather like those the fiber systems have been using to adapt the SC to older ST and SMA connectors.

For more information on advanced performance categories, refer to Chaps. 3, 5, and 16.

Testing Legacy Category 5 Installations   The specifications for Category 5 were developed before all the engineering for gigabit speeds was done. At that time, transmission was primarily unidirectional on only two pairs of the cable, and life was good. The main parameters that caused link failure were attenuation and crosstalk (NEXT), and the other cable parameters were usually all right , if those two requirements were met.

The unexpected use of all four pairs for gigabit, in a split-signal, bidirectional manner, brought out the problem of the differing velocities of propagation between the pairs. A common cable manufacturing practice is to vary the pitch of the twisting, in order to minimize crosstalk between pairs. This results in each pair having a slightly different twists-per-inch, and therefore the unwanted coupling between pairs (the source of crosstalk) decreases. However, this practice also causes the pairs to vary in their respective velocities of propagation—how long it takes for a signal to traverse a length of cable.

If we split the gigabit signal among four pairs, as Gigabit Ethernet does, the four signals will arrive at the remote end at slightly different times. This phenomenon is called *delay skew.* There may be no way for the remote receiver to reassemble the signal, as the bit times may virtually overlap. The only way that this parameter can be controlled is to specify it. This initially occurred in Addendum 1 to TIA/EIA-568-A, and has become an additional specification recognized in Addendum 5 for Cat 5e and in later standard decisions for Cat 6 and Cat 7. See Table 13.7.

What do you do if you have legacy Cat 5 cabling that was never tested for gigabit operation? The answer is provided by the TIA's bulletin, *TSB-95 Additional Transmission Performance Guidelines for 100 ohm 4-pair Category 5 Cabling*, which covers additional specifications for testing installed Cat 5 to gigabit performance standards. TSB-95 adds requirements for equal level far-end crosstalk (ELFEXT) and return loss (RL), and includes the propagation delay and delay skew requirements that were added to TIA/EIA-568-A. More information is given in Chap. 16 about testing new and existing cabling, as well as what to do if it fails a test parameter.

TABLE 13.7

Cable Propagation
Delay and Delay
Skew Limits

Frequency, MHz	Delay (max.), ns/m	Velocity of propagation (min.), %	Delay skew (max.), ns
1	5.70	58.5	45
10	5.45	61.1	45
100	5.38	62.0	45
1000	5.04	66.1	20*

*Inferred from Cat 7 proposed specifications, which actually specify to 600 MHz.

Open offices may present special considerations at gigabit speeds. The addition of a multiuser outlet or a consolidation point in the middle of the horizontal run makes the absolute performance characteristics of the wiring components even more critical. The TIA's recommendations are in TSB-75, and are more fully described in Chap. 11, "Open-Office Wiring."

Planning a Gigabit Installation

Before you can move forward with a Gigabit Ethernet installation, you should plan your network topology. This process will allow you to determine what network infrastructure standards you need to provide, including the network cabling and connecting hardware. The first step is to determine a layout of the network. From this step you will be able to determine the distances, speeds, and types of media you will use.

Next, you must determine any special parameters that should be verified, such as a fiber optic loss budget for runs longer than 100 m. Finally, you must specify the components and the installation criteria so that you can properly estimate costs and bid out the project. Figure 13.5 can be used as a guide to understand the cabling distance limits for different types of cabling, under IEEE 802.3z and 802.3ab.

Layout Topology

The first step is to determine a topological layout for the gigabit network. Whether you are using Gigabit Ethernet or ATM technology, this will be an important guide to the facilities you must provide. We will cover the process for Gigabit Ethernet, because it is in much wider use and the process for ATM is essentially the same.

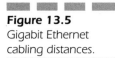

Figure 13.5
Gigabit Ethernet
cabling distances.

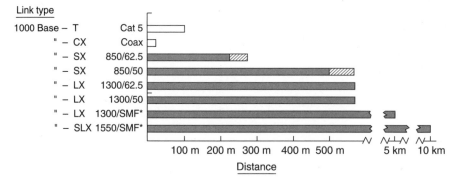

★ SMF = Single mode fiber. All multimode fiber is shown with wavelength/core diam.

As you know, Gigabit Ethernet can be used for either work area or backbone connections. Chances are that you currently have an installation that uses Fast Ethernet (100 Mbps) switches with either 10 or 100 Mbps network interface cards (NICs) in workstations and servers. Typically, these devices are cabled according to TIA/EIA-568-B cabling standards, and operate at either 10BaseT or 100BaseTX. Let's say you want to migrate your network to gigabit, and intend to continue using the Fast Ethernet NICs to the workstations, but you want to add gigabit NICs to the servers.

This is very viable migration strategy. Using this method, you can use gigabit switches in the backbone of the network (Fig. 13.6), and Fast Ethernet to the desktop. This will eliminate most of the bottlenecks from your network topology, while allowing you to preserve your existing infrastructure. Keep in mind that this essentially allows a gigabit-connected server to service as many as 10 of your 100 Mbps workstations at the same time. If you are still using some 10 Mbps workstations, a gigabit backbone would allow a throughput gain of almost a factor of 100 into the server.

An additional bonus from using Gigabit Ethernet technology is that it is engineered to provide throughput of around 90%. This is quite a bit above the normal utilization limit of 30 to 40% on 10 Mbps Ethernet (although it is possible to adjust some of the network parameters to achieve more than double this for discrete links).

You may also want to consider gigabit to the desktop. The need for speed is acute in certain enterprise environments. Surprisingly, many applications have very critical data-throughput needs. For example, it is now possible to do both voice and video communications from a desktop workstation. Both of these applications require a large amount of

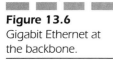

Figure 13.6
Gigabit Ethernet at
the backbone.

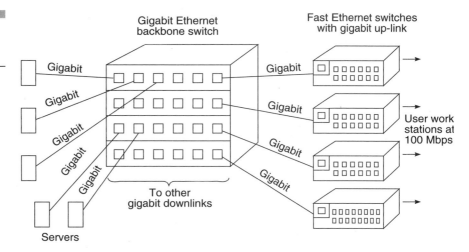

throughput, and both are very time-critical. Adding gigabit at all levels
of your network can help minimize the quality-of-service issues that
degrade these applications. In addition, anyone working with video or
graphics editing or retrieval will need to consider a gigabit network
solution.

Specifying Gigabit Cabling

A cabling specification should address a number of issues. A typical
description of the cable and installation should include the specific category of cable and connecting components, as well as a narrative description of how the installation is to proceed. In addition, the method of
testing the installed links and the express warranty of the installation
should be stated.

For Gigabit Ethernet, in a new installation, you should spec in at least
Category 5e, as it includes additional parameters that are important for 4
pair full-duplex operation at 1000 Mbps. Require that all components
meet the appropriate category's performance criteria.

Remember that if you specify Category 6 components (for both cable
and connecting hardware) you need to consider the backward-compatibility issue. The easiest way to handle this is to simply use only Cat 6 components from a single manufacturer. This is simple to do if you are installing
a new cable plant. If you are the contractor, insist that the contract recognize that proper operation is guaranteed only with user patch cords that
match the Category 6 specifications of the outlet/patch manufacturer.

Manufacturer's Warranty Programs   You may want to consider one of the manufacturer's installation warranty programs. These programs warrant performance of the components and installation for a period of 15 or 20 years, as long as an installer who is manufacturer-certified is used. As we covered in earlier chapters, these warranty programs require you to use specific brands (and material types) of connectors and cable to qualify. Generally, they guarantee that performance of the Basic Link (Permanent Link) will continue to meet minimum requirements for that category for the period of the warranty. The use of a certified installer will presumably result in the use of proper installation practices. If a link is later shown to fail those same performance measurements, the manufacturer will pay to have the failed component replaced at its expense.

In a few cases, these programs are offered by manufacturers' partnership, for example when neither makes both cable and connectors. Be aware that this may produce interesting (and unanticipated) results if a competing component vendor acquires one of the partners.

Although including a program warranty clause in your contract may produce a warm fuzzy feeling, you know quite well that network requirements change much faster than once in 15 years. By that time, you will very likely need totally new cabling and hardware. On the other hand, the warranty essentially extends the standard manufacturer's defect period from a norm of 1 year to 15 or 20 years. The cost to the manufacturer is negligible, as it is highly unlikely that a component that can last a year will spontaneously fail in 5 or 10 years, and that is virtually the only way performance can change. The benefits to the manufacturer, however, are enormous, as it "guarantees" you will buy their components to the exclusion of all others.

Testing Gigabit Cabling

Most of the information you will need to know on testing installed cable is contained in Chap. 16. However, it might be of benefit to concentrate here on a few issues that are particularly applicable to cable installations used for gigabit networks. It is also useful to know some of the gigabit networking issues that may cause a link failure that is not the fault of the cabling.

Gigabit Limitations

The first item in the "not-the-cable" category is Gigabit Ethernet's repeater limitation. Unlike older technologies, particularly 10 Mbps Ethernet, you cannot stack repeater hubs back to back. You will recall that you were allowed to place up to four repeater hops on an Ethernet segment. Fast Ethernet limited that to two, and Gigabit Ethernet limits that to one.

Gigabit Ethernet (as well as other gigabit-class technologies) has a very finite distance limit, depending on the medium. Part of this limit is due to the actual losses in transmission relative to noise (copper, particularly) and part is due to the physics of the transmission (fiber dispersion, for example).

Loss Measurements

It seems self-evident, but you must use test and measurement equipment that is appropriate for the gigabit range. Sure, you can use simple wire map and optical continuity testers during initial installation. But to ensure operation at gigabit speeds, you must use cable scanners that are capable of measuring all the parameters for the category of links you are installing. In this case, you should be testing to Cat 5e or higher, commensurate with the category of components you have installed.

For twisted pair, the measurements fall into four areas: attenuation, crosstalk, impedance, and propagation delay. Attenuation and impedance problems cause direct signal losses. Crosstalk, such as NEXT and ELFEXT, causes a lower signal-to-noise level, and indirectly affects signal range. Finally, differences in propagation delay degrade 4 pair performance. All these areas are critical in gigabit applications.

In fiber links, the essential parameter areas are attenuation and bandwidth, as they are the primary factors that limit range. Other "soft" parameters may radically affect fiber link performance. For example, multimode fiber may have to be mode-conditioned to minimize the effect of normal fiber manufacturing characteristics when used with the laser diode sources of gigabit technologies.

If you have a multimode fiber link that appears to pass testing, but will not work in the application, the first place to look is at differential mode delay problems. Fiber testing is normally done with an OFL (over-filled launch) source, which fills the fiber core with multiple modes.

DMD causes a problem when the core is illuminated with a laser source, which uses only a few modes. This additional parameter is difficult to test for, so a simple troubleshooting technique may be to keep a mode-conditioning patch cable handy. If the link works with this special patch cable, you will need to provide one on a permanent basis.

Another place to look on a fiber with excessive losses is a possible mode mismatch. The connectors on single and multimode fiber are the same, yet there is a substantial loss coupling from one to the other. A mismatch that is less disastrous is a core-diameter mismatch. As we covered in the chapter on fiber optic cabling, 50 and 62.5 µm core fibers may be coupled, but with a dB loss that may be unacceptable in a particular link.

Cable discontinuities can be found by time-domain reflectometry techniques. This applies to either copper or fiber. A substantial mismatch, in impedance or in refraction, will clearly show up on most cable scanners with a graphic display. You will be able to determine the distance from the scanner to the cable defect. Be sure to subtract out the length of the test cord when you hunt for the problem. Ordinarily, this test will direct your attention to a termination point, such as cross-connect, patch, or outlet. Infrequently, you will find a problem within a cable run, often where the cable has been severely bent and straightened out improperly.

Finally, lost performance can come from older fiber that just does not have sufficient bandwidth to operate with gigabit at a given distance. For example, Gigabit Ethernet can operate to 275 m at 850 nm on 62.5/125 µm multimode fiber that has a bandwidth product of 200 MHz · km. However, the link is limited to an expected range of only 220 m with 160 MHz · km fiber with the same core/cladding specs.

Workmanship

Frankly, at lower speeds, you can be fairly certain that any link that could pass the simple tests would also pass the full cable scan, as long as workmanship practices were reasonable.

However, at gigabit, you must be perfect. A single pair (our favorite is the brown/white) that is not twisted up to the point of termination can cause the link to fail. Fortunately, the workmanship practices are not particularly different from what they were for Fast Ethernet. You just must actually follow them with gigabit. You could get away with a few cable kinks, some sharp bends, and careless placement of cable, but not with gigabit. It uses absolutely every ounce (or gram) of margin in a Category 5 cable, and it cuts you no slack.

Be kind to your installers, though. Four or five years ago, cable installation techniques were simpler. Many installations were accomplished with Cat 5 components before all the testing criteria were determined. As a result, many installations were tested only to Category 3, although higher-performance components were used. In most of these installations, operation to Fast Ethernet speeds is fairly well assured. But too much new information and practice has appeared for anyone to have provided for gigabit performance at that time. This is the reason the TSB-95 was created: to recognize that additional testing must be done on legacy Cat 5 installations to ensure operation at gigabit.

The Future of Gigabit Cabling

We all need to be concerned with the issue of future-proofing the network infrastructure. The best assurances we had from both manufacturers and standards bodies several years ago was that Category 5 and multimode fiber would support several iterations of technology. However, many observers had concerns about the future viability of copper cabling for data rates above 155 Mbps. Most were enthusiastic about fiber's role in this upward speed spiral. In fact, it seems that the reverse may have been true.

We are now placing gigaspeed networks on solidly installed Category 5 cable, while our multimode is struggling to maintain very modest backbone distances. Of course, one could argue that the multimode fiber is performing quite well in the horizontal run, but many were encouraged to place only cheaper multimode cables in the backbone without realizing that there would be a future shock. If the backbone in a particular site includes multiple buildings, the 220-m distance can be exceeded very quickly.

Flexible Tubing

In Chap. 6, we discussed a way to allow for future changes in cable technology by running flexible tubing to each workstation outlet, rather than hard-wiring cable. Admittedly, this method is somewhat more expensive than the conventional cable-in-place method. However, it does have a distinct advantage if and when new cables must be installed to meet future needs. With the tubing run, the old cable is simply removed and new cable pulled in. None of the ceiling tiles and work areas are disturbed, it takes very little time, and it can be easily done one workstation outlet at a time. When it comes time to recable, the costs are distinctly lower.

For fiber runs, an interesting technology called *blown-in fiber* can be used. This method connects fiber distribution panels with workstation areas by small plastic tubes. The tubes in a particular work area are gathered into a large tubing to bring them back to the distribution point. For more information on this technique, consult Chap. 12, "Fiber Optic Techniques."

Quality Counts

As for copper runs, we are increasingly finding that better performance can be achieved with better quality cable. Well, that should be obvious. However, we now know more about what makes a better quality cable and connector, and can measure the relative performance of such components. This involves choosing wiring components with a higher ACR (attenuation to crosstalk ratio). Such components have lower transmission loss, lower return loss, and lower near-end and far-end crosstalk parameters. From the information in earlier chapters, you should be able to choose wiring components that meet your current and anticipated needs.

Keep the data-rate-doubling rule in mind when you pick out a cable. Remember that it is much less expensive to spend 20 or 30% more on an advanced cable construction now, than to spend 100% more to redo your cabling when the future comes slamming in the door. Many in the industry are predicting a 5-year, factor-of-10 cycle in networking speeds. That would mean that the gigabit network of today would be replaced by a 10 gigabit network in only five years. Considering that the average cable plant is expected to last 15 to 20 years (and many manufacturers offer such warranties), it is likely that you will have to upgrade your network long before the cable becomes physically unusable.

The Next Step—10 Gigabit

It is foolish to assume that we will be satisfied to halt network development at 1 Gbps. In fact, products that up the data rate to 10 Gbps are already emerging. For local networking, this portends a move to 10 Gigabit Ethernet. It has been shown to operate at over 100 m on multimode fiber, and will be a clear backbone for the gigabit networks of the near future. The 10 gigabit speed is OC-192, so the ATM and SONET technologies will logically advance to that level soon. In fact, some OC-192 systems are already emerging.

Another promising technology, particularly in the wide area, is dense wave-division multiplexing (DWDM). This technology multiplexes different wavelengths of light onto a single fiber to effectively multiply the fiber bandwidth over a hundredfold. Optical networking and switching are becoming increasingly more practical, and more affordable.

The message to take away is that technology is on an incredible spiral to faster speeds and more capability. Gigabit is not the beginning of this spiral, and neither is it the end.

CHAPTER **14**

Wireless LANs

Chapter 14 Highlights

- Wireless basics
- Spread spectrum technology
- Wireless LAN standards—802.11
- Planning a wireless system
- Troubleshooting wireless LANs
- Plan for the future, 30–100 Mbps

Wireless networks are an exciting new addition to the portfolio of the LAN manager. Many times, the requirements of the physical network are best served by using a wireless connection to the workstation. The applications where a wireless bridge should be considered fall into two general categories: need and value.

The wireless applications that need portability are the most obvious. For example, consider the use of networking in a hospital environment. Portable computers and other network devices can be mounted on a portable cart that is wheeled to each patient room for monitoring and care administration tasks. Instead of connecting the cart to a workstation outlet in each room, a wireless network interface card (NIC) can be used to interconnect to a wireless bridge in the hospital. With the inclusion of portable power on the cart, the staff can utilize a completely portable wireless workstation that may be transported to any room in the hospital. Clearly, the needs of this application are for the flexibility and portability of a wireless network.

The same needs-based analysis may be applied in other circumstances. For example, portable or hand-held devices can be used in stores and warehouses to perform inventory and order fulfillment functions. It is hard to imagine the computer terminal on a forklift being tethered to a network by a cable!

Wireless networks may also be justified by their value, either through a direct cost analysis, or through their relative value to the user. As an example, consider the case of an older building that must be wired for the network. Many older buildings are totally without the drop ceilings of modern office space. In some instances, using conventional wiring can be done only through extensive (and expensive) routing of cable through older wall structures, masonry, crawl spaces, and attics. The only other wired alternative, the use of surface raceways, may detract from the appearance of the rooms and hallways. Only a short time ago, one of these wired options was all that was available.

However, one can now utilize a wireless network system to connect workstations to network resources. This can be done at a comparable cost of conventional wiring, in many cases, without conventional wiring's disadvantages in the older structure. Historic structures and architectural treasures can be preserved without forgoing the benefits of modern technology.

Another example of the value-oriented decision is convenience. You could easily connect to the corporate network at the conference table, in the cafeteria, or in the lobby with wireless networking. In an educational setting, the instructor could carry a laptop computer into the classroom and immediately connect to the network from the lectern or the desk. Registration, security, information desks and more can easily be connected in a moment to a wireless network. The flexibility and convenience is enormous, thus justifying the slight extra cost of wireless over wired connections. And who wouldn't want to sit in an easy chair while fully connected to the Internet? So the complete value equation includes more than just cost.

Structural Advantages

In some areas, many buildings were constructed or were remodeled with asbestos materials before the health hazards of the material were known. Now that the danger of asbestos contamination is fully understood, laws have been enacted to require the removal or encapsulation of the asbestos in existing structures. Many building owners have chosen to merely encapsulate or block access to existing asbestos-covered ceilings, pipes, and flooring. However, even minor remodeling of the building, such as may be necessary to place network cabling, may require abatement or removal of the asbestos. Treated as a very hazardous substance, asbestos can be extremely expensive to abate. Consequently, a more cost-effective solution for older buildings may be to use a wireless network method, rather than incur the expense of abatement.

In a typical older college dormitory, for example, a network cabling job might cost about $200,000 to provide dual-jack network outlets in each dorm room. If asbestos is present that would be disturbed by the wall penetrations and wire placement, complete asbestos removal would be required in most circumstances. A typical estimate for this work could easily come to another $200,000. However, if wireless networking were used, the cost might be merely $300,000. Then, obviously, the project cost could be reduced by 25% by using a wireless network. In addition to the immediate cost savings, the entire network infrastructure could be reused, with practically no additional cost, when the older dorm was replaced with a new structure in a few years.

Wireless network component costs are dropping dramatically. In fact, typical wireless interface cards have dropped to nearly half the cost in only a couple of years. The price of providing a wireless network may soon be almost as low as a wired network. However, all wireless networks still require conventional cabling to the wireless access points distributed throughout the workspace.

Wireless Basics

Before we get into the operating complexities of interconnecting network devices wirelessly, it is important to garner a few facts about how radio devices work. In many ways, wireless equipment is all around us. From pure audio-style radio, to broadcast television, to cellular telephones, we use wireless technology every day. But to use wireless networks, we need to understand a little more detail about how wireless transmission and reception work.

Radio-Frequency Operation

The transmission of information through the air (or through space) is actually done by the generation of electromagnetic (EM) fields, on which is placed content in the form of sound, coded pictures, or data. These fields act a little like three-dimensional waves on a lake, because they propagate in all directions, reflect off walls and other objects, and are affected by the medium they pass through. Indeed, if you can imagine a wave on a lake passing through an area of swamp grass, you can get an idea of what happens when radio waves are attenuated by passing through a plasterboard wall, or even a cement structure.

Electromagnetic transmission occurs at a particular base frequency that in the simplest case is called the *carrier frequency*, or just the *operating frequency*. Typical operating frequencies in the United States are 0.540 to 1.600 MHz for AM radio, 88.1 to 107.9 MHz for FM radio, and 850 to 950 for cellular phones, and "900 MHz" two-way radio. Most of the police and fire departments operate in three bands, 150 to 175 MHz, 450 to 475 MHz, and "900" MHz. Television operating frequencies range from 54 to 800 MHz, microwave ovens are typically around 1250 MHz, and microwave radio and satellite links operate in various bands to over 10 GHz. Many radar systems operate in a couple of bands around 1.3 GHz (C band) and around 9 GHz (X band). Similar operating frequencies are used in other countries, with worldwide coordination done through the International Telecommunications Union (ITU).

The current frequency ranges for wireless LAN (WLAN) networking devices are in the 2.4 GHz and 5.2 GHz frequency bands. These are existing bands designated for industrial, scientific, and medical applications. Most of the current WLAN equipment uses the 2.4 GHz band, while some of the newer high-speed equipment is targeted for the 5.2 GHz band. This is a low-power, unlicensed operation that minimizes potential interference by limiting the distance that the signal can travel. An older wireless networking standard exists in the cellular band, but is not widely used.

Modulation

The information we transmit is placed on the carrier wave by a process called *modulation.* A wide variety of modulation schemes exist, but they basically are variations of either amplitude or frequency modulation. The AM broadcast band is the simplest example of amplitude modulation. The carrier is simply increased or decreased in intensity by the amplitude of an audio signal. In contrast, FM broadcast transmission allows the frequency of the output to vary in proportion to the audio signal's amplitude.

Wireless networking uses a variation of these modulation methods to allow the base or carrier frequency of the transmission to hop around among a select number of discrete frequencies, while encoding the data varying both the amplitude and phase of the signal. This encoding is similar to the encoding and modulation methods we use for high-speed modems, with the addition of this frequency hopping. The formal name for this jumping-bean transmission is *spread-spectrum* operation. We will cover more on the reasons for using this technique a little later.

Line-of-Site Concepts

One of the most important rules in the operation of wireless networking is that the technology operates only in a straight line, with a relatively unencumbered path. This is called *line-of-sight* operation, because you can usually operate the connection satisfactorily between two points if you can see between them, within obvious range limits (Fig. 14.1).

In the upper frequencies of radio transmission, the electromagnetic waves tend to be incapable of penetrating dense objects, particularly those with any surface conductivity. In other words, air and glass are no problem, steel and other metals are, and everything else in between is...well, in between. Many materials will cause an attenuation of the signal. Plasterboard and plaster are mild attenuators, cement and cinderblock walls are bad, steel enclosures are worse. The use of foil-covered wallboard, metal wall studs, reinforcing bars, or mesh greatly increases the attenuation of the wireless signal.

Figure 14.1
Line-of-site operation.
Radio-frequency sig-
nals at ultrahigh fre-
quencies are blocked
by obstructions.

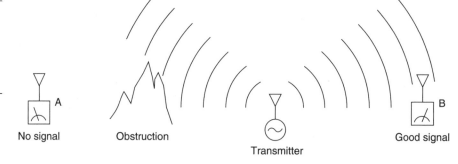

Reflections of a signal may occur because of highly conductive objects in the area of the transmitter or receiver. For example, a signal might bounce off a metal door, a steel enclosure, or even a parked truck. If the signal travels to the receiver both via a direct and a reflected path, an interference problem called *multipath interference* can occur. If the two signals are of similar strengths and arrive at the receiver out of phase, it may be impossible to recover a usable signal. We commonly see this phenomenon on a broadcast television signal as a ghost image when the antenna is misaligned or when large reflective objects are nearby.

Antenna Operation

Antennas operate by capturing the transmitted electromagnetic field and converting the energy to an electrical signal of the same frequency. All of the information that is packed on the field is preserved for recovery by the demodulation circuitry in the receiving device. As we deal with very tiny signals in wireless networking, the antennas must be as efficient as possible.

The typical wireless network interface card (W-NIC) contains a small antenna that is usable in locations with adequate signal strength. However, if the distances, surroundings, or line-of-sight attenuation dictates, you may need to add a supplementary antenna (Fig. 14.2). Such antennas are often called *range extenders*, after their function.

Antennas may be either directional or omnidirectional. An *omnidirectional* antenna is an antenna that transmits equally in all directions. The only truly three-dimensional omni antenna is the theoretical point-source, or *isotropic*, antenna. Such an antenna, if perfectly matched to its source, would convert all its applied electrical energy to electromagnetic energy, and launch the EM waves in all directions. The converse for

Figure 14.2
Gain antenna (range extender) with wireless NICs.

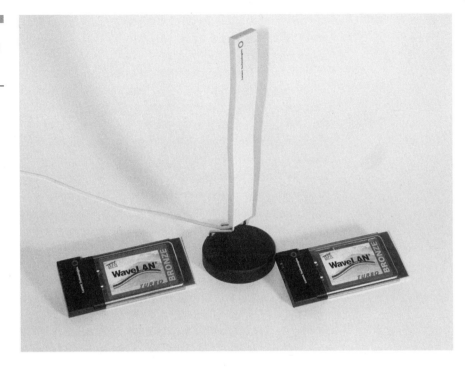

receiving EM energy would be true, as well. From any direction, in three dimensions, the transmitted/received energy would be equal. We rate real-world antennas in comparison to this theoretical model, in decibels above or below the isotropic model, or dBi. Thus, a perfect isotropic antenna would be rated as "unity gain," 0 dBi.

In order to optimize the transmission of the wireless network signal, we assume that most of the devices we want to access will be roughly along a plane parallel to the ground—the floor of the building, for example. We can make an antenna that does a better job of transmitting side to side than up and down. When we measure the received signal strength from side to side (horizontally) for such an antenna, we find that it is greater than we would have expected from an isotropic antenna. This type of antenna is commonly called a *gain antenna*. A gain antenna that has equal gain in all horizontal directions (360°) is what we usually mean when we describe an antenna as omnidirectional. Omnidirectional gain antennas typically have gains of 3 to 6 dBi.

The directional antenna, on the other hand, concentrates its gain in a single direction, both vertically and horizontally. Common types of directional antennas are the Yagi and the parabolic dish. Each type produces a much narrower gain lobe of concentrated EM signal, which can almost be called a beam of energy. Directional antennas for this band usually range from around 12 to 15 dBi, although parabolic dishes can go as much as double that.

The Yagi antenna (properly the Yagi-Uda Array, so named after its inventors) consists of array of parallel conductive elements (Fig. 14.3). The rear-most element is called the reflector, followed by the driven element (to which the signal is applied), and then a series of oddly spaced conductive elements that "point" in the direction of greatest gain. In most cases, you will see Yagis in the microwave frequency ranges (above 1 GHz) encased in a plastic cylindrical covering made of the same material as a radome, transparent to the radio waves.

It is interesting to note that all practical antennas exhibit a phenomenon called *effective area*. Effective area is sort of a "capture" area of the antenna. Thus an antenna with half the effective area would capture half the electromagnetic signal. For a dipole antenna tuned to the frequency of operation, the effective area is inversely proportional to the frequency of operation. For our wireless networking subject, this means that the antennas used for 5.2 GHz will have about half the effective area of those we use at 2.4 GHz. This accounts for a received signal that is about half that of the 2.4 GHz case, so we can expect that there will be

Figure 14.3
Yagi antenna and radome.

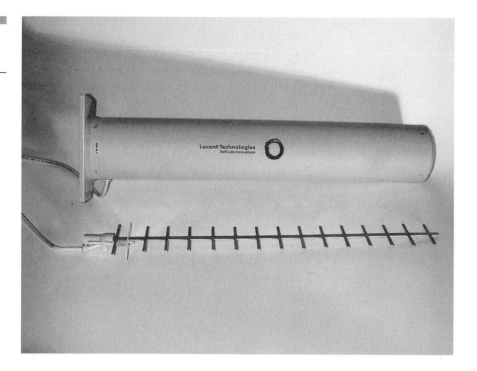

about a 3 dB lower signal.[1] The net effect of all this is that we can expect 5.2 GHz networks to be more limited in range for the same transmitted power and that we may have to use higher gain antennas to compensate.

So-called *parabolic* antennas (really paraboloid-reflector antennas) are a special case. The actual antenna is a small dipole or a feed horn at the focus of the parabolic dish. The dish essentially captures extra signal, thus artificially increasing the effective area, and then reflects/concentrates it onto the feed point. Dish antennas are capable of truly awesome gain, and the increase in effective area is proportional to their size. Practical dish antennas for 2.4 GHz are 12 to 18 in in diameter, and can produce about 15 to 18 dB of gain, which is not that much better than a multielement Yagi. Dish antennas are also more obtrusive, harder to mount, and more affected by wind. Still, if you could use a 3-m dish...wow.

Interference Sources

All radio transmission is subject to interference from a variety of causes. With the tiny signals involved in the unlicensed low-power transmission of 2.4 GHz and 5.2 GHz wireless networking, the most significant issues are the strength of the available receive signal, relative to other innocuous signals. Unfortunately, other types of RF equipment operate at nearby frequencies and all digital equipment produces unwanted or spurious signals. For example, a microwave transmitter, microwave oven, radar, or certain industrial equipment may produce interfering signals. For that matter, other wireless devices operating in the same (or mathematically related) frequency ranges may interfere with a wireless network.

The unlicensed nature of wireless networking technology also means that the user must tolerate legal interference sources. This means that one must carefully design and site the network so as to avoid or minimize interference problems.

Spread Spectrum Technology

The term *spread spectrum* comes from the transmission method, whereby the transmitted signal is spread out among many frequencies, rather than

---

[1]A decibel (dB) is simply 10 times the log of the power input/output ratio, or 10 log (P_i/P_o). Thus a power ratio of 2:1 equals approximately 3 dB. Likewise, 4:1 corresponds to 6 dB, and 10:1 corresponds to 10 dB. This transforms all the gain/loss calculations into simple addition and subtraction.

based at just one carrier frequency, as with conventional AM and FM transmission. Spread spectrum operation was originally developed for military use, where radio and radar transmissions needed to be difficult to detect and relatively immune to detection or jamming. This method can put the transmitted signal level just above the level of ambient noise and uses a mathematical algorithm to allow the receiver to know where to look for each signal increment.

Spread spectrum has many advantages in wireless networking applications. It is inherently low power, and the transmission technique rejects most random interference. Multiple devices can share the same set of frequencies, simply by hopping to them at different times. The current level of electronic technology makes the complex RF and digital circuitry small and affordable.

Frequency-Spreading Variants

Several types of spread spectrum standards are available. The two most common are direct sequence spread spectrum (DSSS) and frequency-hopping spread spectrum (FHSS). Direct sequence (DSSS) is the method used for current 2 Mbps and above wireless networking implementations at 2.4 GHz.

In order to receive the transmissions, the discrete transmit frequencies and their exact sequence must be known. Also, a method of synchronizing the beginning of the sequence must be allowed. Any system that meets the international standards will use compatible methods of generating the sequence and recovering the transmitted data.

Resistance to Interference

Spread spectrum transmission is inherently tolerant of random interference. If at any one instant an interfering signal is present on one of the hopping frequencies, the encoding of the data will allow the recovery of the information. This is very much the case with pure noise interference. At any given time, a noise pulse may be present on a particular discrete frequency, but it is unlikely that it will reoccur on subsequent hops to that frequency.

The operation of the NIC is such that if the circuitry determines that a particular frequency is unavailable, because of a constant interfering signal, the frequency can simply be avoided, thus lowering the data throughput. Standard operation allows for graceful fallback of the operating data rate to allow for these conditions.

In addition, some wireless manufacturers offer advanced signal processing techniques to reduce the degradation caused by multipath interference.

Wireless LAN Standards

IEEE 802.11 Standard Operation

The IEEE 802.11 committee was formed to implement standards for operating Ethernet networks over wireless connections. The standard, as usual, embraces several competing modes of operation, while lending interoperability and order to future developments. The Wireless LAN Interoperability Forum and the IEEE both participate in bringing forth operating and protocol standards for this service.

IEEE 802.11 networks operate at 1, 2, and 11 Mbps nominal data rates. The 11 Mbps rate, added in the IEEE 802.11b supplement, is often called a 10 Mbps link, to compare to the wired Ethernet. However it is better to use the terms 11 Mbps wireless or high-rate mode, to differentiate it from the 10BaseT Ethernet standards. In fact, there are other significant differences between wired and wireless Ethernet.

The 802.11 wireless Ethernet method uses a collision-avoidance technique, rather than the collision-detection technique of wired Ethernet. To accomplish this, the wireless stations are far more aware of each other than are wired stations, and they either cooperate to sequence their transmissions, or they cede control to a server. We will cover more on that subject in a moment.

Operation of an IEEE 802.11 network is in the 2.4 GHz Industrial, Scientific, and Medical (ISM) band, according to a fixed channel plan. Operation in the 5.2 GHz ISM band is already beginning, and the standard for that mode will soon be finalized. Modulation types used are FH (frequency-hopping), DS (direct sequence), and ODFM (for 20 Mbps and above in the 5.2 GHz band). This operation is summarized in Table 14.1.

The standard also provides for infrared Ethernet links, although they have a more restricted application and are not covered here, as they are not general-purpose wired-LAN replacements.

Another important feature of IEEE 802.11 is that it allows for two types of wireless architectures, ad hoc, and client/server (Fig. 14.4). The ad hoc operation allows any two or more stations with wireless network interface cards (W-NICs) to communicate, much in the same way as peer-to-peer operation occurs on a conventional Ethernet network. The

TABLE 14.1

IEEE 802.11
Operating Modes

IEEE 802.11 mode	Frequency band, GHz	Data rate,* Mbps	Spread spectrum mode
802.11 FH	2.4	1	Frequency hopping (FH)
802.11 DS	2.4	1, 2	Direct sequence (DS)
802.11 HS	2.4	1, 2, 11	DS and ODFM
"	5.2	1, 2, 10, 20	DS and ODFM

*Data symbol rate, theoretical 100% transfer rate.

only difference is that one of the W-NICs becomes an ad-hoc (tempo-rary) controller, to sequence the transmissions of all stations. Remember, the operation is "collision-avoidance," which implies a synchronization between stations that does not occur on regular Ethernet. The main drawback to this operation is that there is no logical point for intercon-nection between the wireless and a wired Ethernet LAN, and there is no central control point to apply a rigorous discipline to the interchange.

In a client/server network, a single device, called an *access point server,* (Fig. 14-5) controls network operation and regulates access to the wireless medium (Fig. 14.4(b)). Typically, client/server WLANs feature better and more consistent throughput. A wireless station may not "see" another peer and might cause unwanted interference in an ad-hoc network. However, the access point can detect these conditions and manage the network in a more orderly fashion. In addition, the access point can adjust the speed of operation individually for each station, and can be equipped with a gain antenna to extend the effective range of all stations on the network, with-out the necessity for all end points to have gain antennas.

Some access points feature multiple W-NICs and can bridge between indoor WLAN networks and outdoor point-to-point network connections to other buildings. For example, you might want to interconnect two buildings using IEEE 802.11 hardware, in addition to supporting wireless workstations in either building. A dual W-NIC access point could be equipped in each building, with an omnidirectional range extender anten-na for the local workstations, and a directional Yagi outdoor antenna for the remote building link. At the remote building, either wireless stations or interconnection to a remote wired Ethernet could be accommodated.

1, 2, and 10 Megabit Operation The most sterling feature of the IEEE 802.11 implementation is that it is backward-compatible. That is, the very latest W-NICs can interoperate with the earlier 1 and 2 Mbps cards.

Figure 14.4
Types of wireless
architectures. (*a*) IEEE
802.11 ad hoc
WLAN; (*b*) IEEE
802.11 client/server
WLAN.

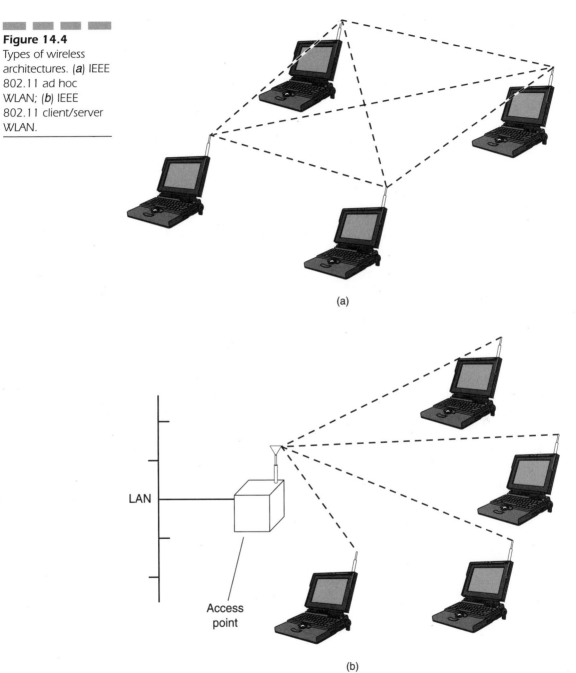

(a)

LAN

Access
point

(b)

Figure 14.5
Access point server
(shown with cover
removed and exter-
nal gain antenna).

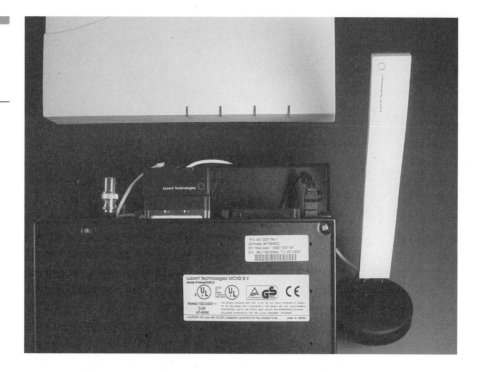

Figure 14.5
Access point server (shown with cover removed and external gain antenna).

As a matter of fact, this feature is exploited to extend the range of the WLAN at all speeds by allowing the cards to downshift as necessary to maintain transmission integrity.

Basically, 1 Mbps operation is the legacy speed. It uses frequency-hopping spread spectrum (FHSS), and provides the longest link range. The range is a function of the data rate, primarily because the higher the data rate, the less robust the signal, and the shorter the effective range. FHSS uses a two- to four-level gaussian frequency-shift keying (GFSK) modulation method. FHSS can also operate at the 2 Mbps rate.

In order to accommodate higher data rates, either differential binary phase-shift keying (DBPSK) or differential quadrature PSK is used for DSSS. This mode accommodates 1, 2, and 11 Mbps operation, as shown in Table 14.1.

2.4 GHz Channel Plan   It is important to realize that WLANs actually operate at particular "channels" within each band. For the 2.4 GHz band, these channels occupy the range from 2.400 to 2.483 GHz, a total bandwidth of 83 MHz. In the United States, the FCC plan allows 11 channels for DSSS, while the ETSI allows nine of these in Europe, and Japan allows only one, which is not common to the other two band plans. Thus, channels 1 and 2 are used only in the United States, but channels 3 to 11 are common to both the United States and Europe. Other countries use variations of these plans. The 11 megabit mode spreads across multiple channels; three sets of channels are accommodated within the 2.4 GHz band.

By contrast, FHSS uses 79 frequency channels in North America and Europe, with 1 MHz spacing, allowing 26 collocated networks. Other countries use this same standard, or additional variations that conform to local frequency coordination schemes. In Japan, for example, 23 hopping channels are provided.

Transmit power for both FHSS and DSSS can be a rather ample 1 W [or 4-W effective radiated power (EIRP)] in the United States, although a transmit power of 100 mW or less must also be supported. Europe limits this to 100-mW EIRP. As with many other types of shared-band communications, regulatory agencies require the transmitter power to be the lowest necessary to maintain communication.

HiperLAN Operation

In addition to the IEEE standard, other wireless networking standards exist. One of these is HiperLAN, promulgated by the European Telecommunications Standards Institute (ETSI). HiperLAN is intended to provide large-capacity, multimedia connections of up to 24 Mbps. It uses gaussian modulation shift keying (GMSK) to provide voice, video, and data capability. With the increasing emphasis on multimedia networking, this should be a technology to watch (pun intended).

Wireless ATM is not far behind. A new standard called HiperLAN-II intends to utilize the Asynchronous Transfer Mode (ATM) protocols over wireless networks. ATM has inherent advantages in sequencing time-critical traffic, such as voice and video, over stochastic media. Using these methods, it should be possible to guarantee the quality of service in the wireless networks that is needed to enable multimedia applications.

SWAP WLAN Operation

The Shared Wireless Access Protocol (SWAP) has recently been created by the Home RF Working Group. The increasing power to "unwire" computing and communications devices in the home and small office is producing this and other proposals. SWAP is to use FHSS at up to 2 Mbps in the 2.4 GHz ISM band to provide voice and data operation for a relatively low cost. An important feature of this and other localized wireless technologies is to prevent "roaming" between network cells. This is very understandable, as you would certainly not want your neighbor remotely typing on your keyboard, or worse yet, changing your TV channels. On the other hand, such a feature could make for great entertainment on slow nights.

Bluetooth WLAN Operation

Another wireless innovation that has some interesting potential is called *Bluetooth*. The Bluetooth technology is intended for short-range use as a cable replacement. For example, one might have a wireless keyboard, mouse, phone, remote control, or stereo speaker, all connected to their respective hosts by this low-cost technology.

Bluetooth also uses FHSS at 2.4 GHz, but at only 1 Mbps signaling rate, for the time being. Separate synchronous and asynchronous links are defined. The synchronous mode is connection-oriented with slot-reservations and data paths that are symmetrical. This makes it function essentially as a time-division multiplexer. Symmetric data rates from 108.8 kHz to 432.6 kHz are supported. The asynchronous mode, in contrast, is connectionless and packet-switched with a polling access method that allows a range of data rates from 57.6 to 721.0 kHz.

The combination of simple frequency-hopping technology, a low bandwidth 1 MHz symbol rate, common 2.4 GHz electronics, and adaptable power levels will allow this technology to become a very low-cost, low-power, single-chip solution. Bluetooth is endorsed by almost every major semiconductor, computer, and consumer products manufacturer. The long-term implication is that this will be a very common method to connect a wide range of devices within a 10-m distance. Increased power levels can jump the range up to about 100 m.

For those of you who are moderately curious, the name for this technology is taken from Harald Blaatand II, King of Denmark from 940 to 980 A.D., and otherwise known as Bluetooth. No doubt this illustrates the engineer's capacity both for intellectual endeavor and for obscure reference.

Enhanced Features

As with all network hardware implementations, it is always possible to advance the state of the art beyond the state of the standards. Of course, one would be naïve to not expect that some manufacturers would use advanced features to market advantage.

For example, some of the features that enhance operation of a wireless network are private networks and encryption. The concept of the private network comes from the fact that all IEEE 802.11 networks are designed to be inherently "open." That is, they allow a new device to join the network, merely by identifying its presence. This is totally in concert with normal Ethernet operation, where anyone can plug into an Ethernet jack and gain access to the physical layer of the network. However, higher layers are not necessarily open, either through the limited station isolation provided through an Ethernet switch, or through VLAN (virtual LAN) network privacy.

Private networks simply implement a crude method of V-LAN through a network "name" and associated password. This can be a significant advantage, as the casual user cannot simply jump onto a wireless LAN and perhaps use sniffer technology to penetrate further.

Encryption methods may also be an enhanced feature that is important to a network administrator. While conventional wired networks do indeed emit tiny signals that can be picked up by an intruder, at least they do not intentionally broadcast network traffic. But your sensitive data, from correspondence to personnel information, passes along any network, and now can be broadcast to the nearby world through the wonder of WLANs. The only way to deal with this real or perceived threat is to encrypt all data passing over the WLAN. Many encryption standards are available, and this fact can lead to incompatible "enhancements" between manufacturers.

The enhanced features may be an important consideration in certain wireless networks. Nevertheless, these features do not prevent operation in accordance with approved standards, as long as they are turned off.

Wireless Link Types

Before we get into planning the wireless network, let's look at the types of wireless links. Regardless of the type of WLAN technology, there are basically three types of links, of which two are closely related (Fig. 14.6).

Figure 14.6
Three basic types of
wireless links: (*a*)
indoor point to
multipoint.

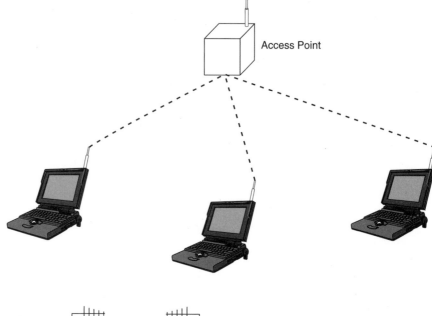

Access Point

Figure 14.6
Three basic types of
wireless links: (*b*) outdoor point to point.

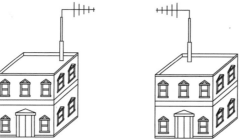

Indoor Point-to-Multipoint LANs

The majority of WLAN implementations are made within a building
(Fig. 14-6*a*). Invariably, these are multipoint connections, although rarely
a point-to-point link is made to span a large distance in an industrial
plant or an aircraft hanger. These indoor point-to-point links are sometimes best accomplished with infrared links, as none of the disadvantages normally associated with outdoor infrared links are present.

The typical indoor LAN serves to link wireless workstations to a corporate LAN backbone. Access point servers are distributed throughout
the building to provide wireless coverage at all the needed locations.
Often the access point servers are mounted above the ceiling tiles and
out of view, except for an unobtrusive antenna.

Figure 14.6
Three basic types of
wireless links: (c)
outdoor point to
multipoint.

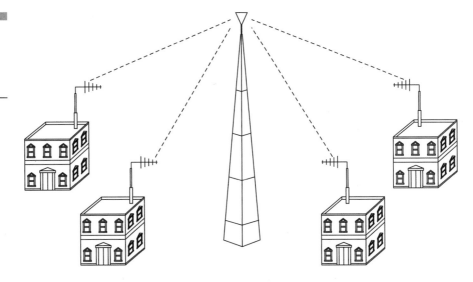

Multipoint LANs may also be implemented without access point servers, but this is normally limited to a small number of devices within a particular room, as there is no straightforward method to distribute the signal to other areas.

Outdoor Point-to-Point Links

An outdoor point-to-point connection is typically made to connect two buildings, each with an internal wired network (Fig. 14-6 b). The main differences from the indoor network are that the access point servers must use special bridging software and that weatherproof outdoor directional antennas are connected via an appropriate length of RF cable to the indoor-mounted access point server. The server is then connected to the wired LAN, and the bridge is complete.

The operation is somewhat more complex than the classic wireless LAN, because each building may have an extensive wired LAN, and the appropriate bridge tables must be constructed by either wireless server/bridge. A router function could also be employed, which would allow the buildings to have separate network addressed (IP subnets, or IPX network numbers). Bridges intended for outdoor point-to-point operation require special software on the access point server.

Outdoor point-to-point bridging can span distances of up to 20 km, if there is clear line of sight and high-gain directional antennas are used. However, it is sometimes possible to bounce signals off buildings and other structures. Other considerations are the provision of very low loss RF cable to the antennas and lightning protection for the outdoor structures.

Outdoor Point-to-Multipoint Systems

The relatively high power output in the United States lends itself to the use of these networks in outdoor point-to-multipoint networks (Fig. 14-6c). A classic multipoint transmission is a broadcast TV station. An omni-directional antenna at the transmitting tower links to directional antennas at the home. Likewise, a central access point server for a WLAN would connect to an omnidirectional antenna and each remote workstation would use an outdoor directional antenna to access the server.

With this method, a central station could typically link to user stations within 8 to 10 km, provided it had clear line of sight. As you might expect, this is an ideal method to provide high-speed Internet access to users that are within line of sight. It is also a good method to link several buildings on a campus.

Planning a Wireless System

A wireless networking system consists of one or more wireless bridges, which function as a point of access between the wired LAN network and the wireless workstations. A desktop personal computer, laptop computer, hand-held terminal, or other device can function as a wireless workstation. All that is needed is an appropriate wireless NIC, software drivers, and the wireless bridge.

Most of the planning information here will address the general-purpose wireless LAN applications. As we mentioned, another use of the 2.4/5.2 GHz technology is to operate a relatively wide area wireless link. We will describe this application in "Outdoor Operation," below.

Determining Coverage Area

It is relatively easy to determine the outdoor, line-of-sight range of a wireless LAN networking link. With standard, built-in antennas, most manufacturers claim performance at up to 500 m. However, in practice, this range is very dependent on the characteristics of the building in which you are operating, the minimum data throughput you will tolerate, and other environmental noise and interference factors.

The IEEE standards for wireless LANs provide for a graceful fallback in data rate as it gets more difficult to push the signal through. What this means in practical terms is that the data rate will fall as the distance from the bridge increases, or as any degrading interference increases. Top performing data rates are 2 Mbps for the standard IEEE 802.11 wireless systems, and 10/11 Mbps for the high-rate or "turbo" systems. The standard systems can fall back to a 1 Mbps rate, while the high-speed systems have a medium-speed fallback rate in addition to the 2 and 1 Mbps rates.

In addition, it is important to recognize that the supposed "data rates" are a representation of the symbol rate, and are proportional, but not equivalent, to the "data throughput rate." This is certainly true with any data network topology. For example, a typical maximum transfer rate for a high-performance 10BaseT NIC card is actually about 5.5 Mbps, half-duplex operation, with collision detection enabled. So when you say that an 11 Mbps wireless network achieves 4.5 Mbps throughput, that is actually pretty good.

Table 14.2 shows a coverage comparison for the various modes of operation. As you can see, distance increases with lower data throughputs. The gentle degradation in data rate allows the wireless network to continue operating, even in difficult conditions. This greatly adds to the robust nature of the technology. You can much more easily tolerate a slow network connection than a dropped one.

Situating Wireless Network Bridges

Using the numbers in Table 14.2, you can begin to plan logically where to place the wireless bridges. Figure 14.7 shows a typical floor plan and a possible placement of wireless bridge antennas. Note that it is the placement of the antenna that is critical, not the placement of the bridge!

Here we have placed two access point bridges in the two front hallways, to provide coverage of the lobby and nearby offices and portions of the modular office space. Office walls will often attenuate the WLAN signal, so this assures good coverage in these areas. A third bridge is placed at the back of the modular offices to provide coverage there, as well as to nearby offices and the storeroom. A large room (which could be an auditorium, a cafeteria, or a meeting room) has walls that obstruct the wireless signal, so bridge 4 is placed there.

TABLE 14.2

Typical Wireless
Link Coverage
Distances

IEEE speed	Throughput,* Mbps	Range,† m
High	4.3	40–125
Medium	2.6	55–200
Standard	1.4	90–400
Standard low	0.8	115–550

*Typical measured data transfer rate.
†Maximum distances shown are for open areas, minimums for indoor/obstructed areas.

Figure 14.7
Typical access point
server placement
within a building.

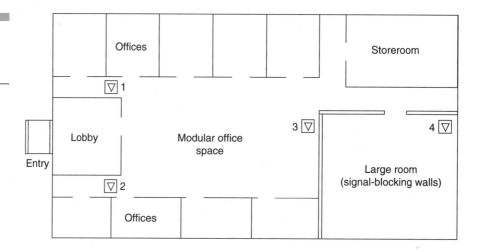

In some cases, you can use bridges with multiple transceivers and antennas to cover additional areas. For example, some of the manufacturers offer dual-transceiver bridges, so one antenna can be placed in an interior location and the other placed outside the building. In the example floor plan, it might be possible to eliminate bridge 4 by using a dual NIC bridge in location 3, with one antenna inside the large room. In addition, an external antenna could be used either for outside wireless coverage or to bridge to another building's LAN.

The exact number of wireless devices that can be covered from a single bridge/antenna depends totally on the conditions and the desired coverage area. An upper limit is about 200 devices, but you will want to limit the number of devices in the same way you would with a normal Ethernet segment. Most network managers like to keep this number well under 100 stations for a 10 Mbps Ethernet segment, depending on usage and applications.

In a crowded environment, with limited coverage because of building walls or other obstructions, you will probably want to limit the number of wireless workstations to no more than 50. Actually, this limit is not that low, as it may be difficult to get 50 wireless workstations within the 200- to 300-ft indoor range of the wireless bridge.

Using Range Extender Antennas

Gain antennas may effectively extend the range of a wireless system (See Fig. 14-2). Antennas are rated in dB over isotropic, and the higher the number, the greater the amplitude of the transmitted or received signal. Decibels are a logarithmic power ratio. A good rule of thumb is that a 3-dB increase represents a doubling of the power. Unfortunately, the available captured power decreases as the square of the distance from the transmitter, so it takes about a 6-dB increase to effectively double the range.

Nevertheless, gain antennas are a very effective means of increasing the usable coverage area of a wireless bridge. A cautious approach is to use gain antennas for every bridge, and add a gain antenna for specific workstations that suffer the slow data rates that go along with a marginal signal.

It is important to recognize that the effect of the extender antenna is bidirectional. That is, you increase both the received signal and the effective transmitted signal. With a gain antenna, you really cannot lose.

You place the antennas where you want the signal. This means that you could locate a bridge some distance from the antenna, and connect the two with coaxial cable. Keep in mind that the dB losses increase rather quickly in proportion to the length of the cable. So, you must use high-quality cable designed for operation at the 2.4 GHz or higher frequencies to keep losses to a minimum. Most WLAN gain antennas come with a 3- to 10-m segment of coax. Feel free to have a custom length made, with appropriate test results to guarantee performance. Extremely low-loss cable is available for a premium.

For that matter, you could make up your own cables, if you have the proper tools and knowledge, After all, you are capable of doing your own twisted-pair cabling. Just keep in mind that you are dealing with frequencies up to 20 times higher than Ethernet networks, so you really have to construct these cables with care. If you have access to a loss-measuring device for 2.4 GHz, you can test the resulting cable to be sure you get what you are paying yourself for.

A word about the care and feeding of coaxial cable is appropriate here. Just as with our relatively fragile Category 5 and better cable, you must be careful with the handling and routing of coaxial cable. As we covered in Chap. 3, coax consists of an outer conductive shield and an inner center conductor, separated by an insulating "dielectric" layer. Any distortion of the dielectric material, through bending or compression, will cause an attenuation of the signal. Attenuation is sort of like a negative gain. So don't crimp, bend, break, pull, or tie-wrap coaxial cables excessively. Take care of your cable and it will take care of you.

Site Survey for Wireless LANs

In our simple building example, placing the wireless bridges was fairly easy. We had a small number of offices to cover in a simple building structure. However, this is rarely the case in practice. To create a much better estimate of the number of bridges and their placement, you should conduct a site survey.

There are two ways to do a site survey, and as you might expect, the survey's cost is the most apparent difference. However, the quality of the results is the real defining factor. Be careful which you use, unless you have no problem adding additional access points and resituating existing ones to optimize coverage. If you have bid an installation, with no caveats to the buyer, adding the additional bridge or two can be very costly. On the other hand, if you are the user, and if you have a relatively small location to cover, it may be much cheaper to use the "casual" site survey and the advice of a knowledgeable installer to make a best-effort estimate of your requirements.

The two site survey methods are what we will call "casual" and "full." In both situations, the building layout is considered, probable bridge locations plotted, and the extent of wireless coverage is detailed.

The *casual site survey* is done with actual WLAN links. A wireless bridge is temporarily set up in each likely bridge location and the antenna that will be used is temporarily mounted at or near the expected permanent location. For example, if you plan to mount the bridges above the ceiling tile in a hallway, place the antenna in the approximate planned location below the ceiling grid and drape the coax down to the bridge. Remember that it is the antenna that transmits the signal into space, and the bridge location does not matter. If you plan to use the internal antennas on the bridge's transceiver card, then you really should try to get the bridge up near the ceiling. Try mounting the bridge on an 8-ft board (such as a 1 × 6 × 8 ft) and propping the board up against the wall to simulate mounting the bridge on the wall.

Now, use a laptop that is equipped with an appropriate W-NIC to test the signal strength in the desired locations with the expected range of the bridge. The client software drivers for the W-NIC generally have a test mode that displays the signal strength at each end of the link. You need to know how well the bridge is receiving the remote signal, as well as how well you are receiving the bridge's signal. Remember that walls, ceilings, and even floors are important in relation to the wireless devices. So, place the laptop on a desk where it will be used to test the link. If you doubt this phenomenon, simply go to a distance from the bridge where the signal becomes weak, and place your hand over the W-NIC's antenna. The signal will drop to unusable in an instant.

The *full site survey* is performed with sophisticated field-strength measuring equipment. A test signal is sent to a test antenna in the planned wireless bridge location and the receive signal strength is measured at various locations through the desired coverage area. The test antenna is moved as necessary to achieve adequate signal strength at every desired point. The *path loss* can be determined from the measurements, and appropriate corrective actions taken. Typical actions that increase signal strength are to use gain antennas, use secondary antennas, adjust antenna (and possibly bridge) locations, and add additional bridge locations.

It is amazing what a tiny adjustment to a bridge antenna placement can do for the received signal strength at these frequencies. For example, in a location where metallic shielding of sheet metal or foil-covered duct work is a factor, a mere 6- to 12-in change in antenna location can cause a 10- to 12-dB increase in signal strength, measured 50 ft away!

It sometimes helps to pretend that you can sense the LAN signals. Try to imagine that the radio waves are light from a bulb. The farther away you go, the dimmer the light, and the less the signal. If you move the light behind an opaque object, the light decreases almost to darkness. If the light passes through a window screen or curtains, some of the light gets through, but it is much dimmer. It is just as if you had increased the distance. That simulates the attenuation of ultrahigh-frequency signals, except that real radio waves can go right through objects, such as plasterboard, that are quite opaque to light. While this method is very useful in WLAN planning, please don't tell anybody else you are doing this, and do try to keep it to a minimum. We wouldn't want you running around acting as if radio waves are real.

Outdoor Operation

Outdoor operation of a wireless LAN has two requirements that differ from the indoor counterparts. Those are the use of weatherproof wireless antennas (see Fig. 14-3), and the concept of LAN-to-LAN bridges. The first requirement is really rather trivial, as it exactly mimics the indoor operation, except that a waterproof antenna and cable are used, lightning protection must be provided, and special antenna mounting hardware and structures are required.

As with any outdoor mountings, you must make a stable mechanical mounting, and allow for the effects of wind, rain, sun, snow, and ice. The antennas for 2.5 GHz and above are often mounted with plastic radome coverings that protect the antenna from many of the effects of weather and corrosion. You should use highest-quality connectors on the connecting coaxial cable, and protect them from the elements with a clear silicone rubber material that is designed for this purpose. The sealant is available in most stores that sell TV antennas.

While national and local codes provide the exact specifications for lightning protection of outdoor antenna systems, common sense should insist that you provide the very best protection for the sensitive networking equipment that you can. Run the drain wires to appropriate ground rods or approved structures. Lightning happens . . . don't let it be your problem.

Outdoor operation can be also multipoint, with an omnidirectional antenna, and you could certainly connect to laptop computers outside. This would technically be an outdoor application.

The LAN-to-LAN bridge, however, is generally what most vendors mean when they refer to "outdoor" operation. Point-to-point or point-to-multipoint operation often requires special operating software for the bridges, in addition to the same weatherproof antennas and mounts.

If you are connecting from a multiple-station LAN on one side of the link to a multiple-station LAN on the other side, you will need two bridges with the special inter-LAN connection software. Some bridges actually have two wireless transceiver cards so they can connect between two wired LANs on either side and, in addition, to two wireless networks on either side of the link.

The name of the game for outdoor LAN-to-LAN operation is to use directional gain antennas. If you are strictly point-to-point, which means only two end points, you can use directional antennas at both bridges. Highly directional antennas are available to give you a range of 10 to 20 km. You will need Yagi or even parabolic antennas to achieve this range, and you must still have optical line of sight.

If you want to go point to multipoint, you will use an omnidirectional antenna at the main site, and directional antennas at the remote sites. These directional antennas will be pointed at the main site, as before. The range will be reduced, because the central omniantenna will not have as high a gain as the directional antenna used in the point-to-point case. However, coverage will still be excellent. Line of sight is required, as before. This mode of operation is suitable for a small campus connection to several buildings or even a connection between users and an Internet service provider.

New options are available at data rates beyond 10 Mbps, to 30 MHz and above. Indeed, even 100 Mbps base-rate operation is available if you are willing to pay the price. We have covered wireless LAN operation here. However, operation of wireless connections for non-LAN traffic is also possible with special hardware. For example, wireless devices are available to interconnect buildings at 56 kbps, T1, DS3, and OC-3 rates. In fact, all of the older microwave links used these rates, but they required special licensing and siting, and were intended for long intercity links. Our WLAN operation, on the other hand, uses unlicensed frequency bands, lower power, and shorter distances.

Troubleshooting Wireless LANs

All computer systems periodically require maintenance and repair. If you think of the wireless network as just another computer system, you can use the same techniques to troubleshoot and repair these networks. We prefer a step-by-step approach that separates the overall system into each component part. We then determine which subset of component parts could be causing the symptoms we see. Then through an orderly process of elimination and substitution, we isolate the component that has failed and make a repair.

Determine the Scope

First in this process is to determine specifically how much of the network is affected by the problem. In the case of the wireless network, we are frequently interconnected to a wired network, which forms the backbone for our wireless system.

Ask logical questions about the network outage, even if parenthetically. Actually, if you ask the questions to yourself and then simply state the conclusion aloud, you may be seen as a wizard, rather than as a fool. So you might think, "Gee, are all the wireless devices out of service, or only those in one area?" But you would say, "Only the devices served by bridge 3 are out of service, so let's check it out first." Then you can begin to check the LAN connection to the bridge, the W-NIC cards, the antenna, the coax, and even the power to bridge 3.

Line-of-Site Problems

Many of the problems with wireless LANs are caused by simple obstructions in the path from bridge to workstation. You need to know if something has changed in that path. For example, remodeling could add walls or shielding metal to the path. Or, something as simple as a metal filing cabinet could have been moved in the way. Look for simple things first, and determine if something has changed to make the signal strength marginal.

The easiest way to determine line-of-sight problems is to imagine the radio waves' path from workstation to access point. Remember that the signals can penetrate certain types of structures, although there is always a signal decrease when this happens. Use any built-in signal level indicator in the wireless client software to help diagnose this issue.

You may add or move a range extender antenna at the workstation. However, the antennas at the wireless access point should be relocated with caution, as their position affects the reception at all workstations. In a few cases, a time-of-day problem may actually be a line-of-sight problem. For example, a delivery truck might block signal from a nearby server.

In a few cases, you may actually be able to compensate for a line-of-sight problem with a reflector made from sheet metal or foil. If all else fails, try to plot the signal path and experiment with reflectors from 2 to 4 ft square. Radio waves bounce off a reflector much like light off a mirror, so place the reflector accordingly. While almost any metal object will reflect the wireless LAN signal (or attenuate it from passing through), a highly conductive surface, such as a copper sheet, will work best. If you are skeptical, simply take a close look at many of the microwave communication towers littering the landscape. You may note that many of them actually have the transmitting dish mounted at ground level, pointed straight up to a reflector high up on the tower.

Interference

Many types of equipment can create interfering signals in the WLAN operating bands. As we mentioned before, these bands are shared with other services, and we must tolerate any interference that is deemed to be a legal use of the band.

In some cases, the interference may be significant because of the relative weakness of our desired signal (for example, from the bridge). The cure may be to add an extender antenna to the workstation, or to the bridge. In severe cases, a directional antenna with a higher gain may be necessary. Also, the application may simply require an addition of another access point bridge.

In rare cases, an improper level of interference may even be illegal. You should be able to locate the source and measure the level of the interference with appropriate signal-strength equipment. The same type of equipment that is used in a full site survey can also help locate these interference sources if you use a directional antenna for the tracking.

Typical sources of interference might be microwave ovens, microwave or radar transmissions, industrial heating and welding equipment, certain medical equipment, any digital equipment (such as a computer or a monitor), radio transmitters, and finally other WLANs. Any device that emits RF energy in the frequency bands used by your WLAN is a potential culprit, whether the emission is intentional or not. If you determine that it is improper, you can ask the owner of the interference source to correct its operation, or in extreme circumstances, you can take legal action to shut down the source.

If the interference is from another WLAN, try changing the channel you are operating on. The U.S. band plan, for example, comprises 11 channels in the 2.4 Gbps band. Most equipment defaults to Channel 1 or 3 (the lowest channel in common with Europe). Simply change all the default channel settings to another channel. Note that this must be done in all the workstation W-NICs as well as those of the access points.

This brings up an interesting point. IEEE 802.11 allows a workstation to "roam," that is, to scan for the best signal among the complement of channels. Many suppliers lock the W-NIC into a default channel, so it will not inadvertently "catch" on another network with a good signal (what we might call a *foreign network*). Although the workstation would generally not be authorized on the foreign network, the W-NIC might stay on that channel, and not be able to connect to its home network. The safest practice is to determine exactly which channel you will operate on and set the bridge and workstations to use this particular channel. Remember to check for this fault if a workstation inexplicably will not connect.

Software and Compatibility Issues

A wireless LAN is inherently more complex than a wired network. In the latter case, all you have to worry about is the proper connections at each termination point, the proper components, and the proper installation techniques. However, a wireless LAN actually takes you beyond the physical connection into the network hardware and driver software.

For example, to install a wireless LAN, you must install the W-NIC into the workstation and install the driver software, either from disk or from CD. To install the access point server, you must install conventional LAN wiring to a workstation outlet near the server, securely mount the server and antenna, and connect the network and power connections. Then you must configure the server, usually through an RS-232 or Ethernet connection to your laptop computer.

Not only is this process somewhat complicated, requiring special training for the installer, but it is subject to all the problems with software revisions and firmware compatibility that you have with any actual computer network. In order to function properly, the workstation's driver software, the W-NIC firmware, and the server's firmware must all be compatible and up to date. Moreover, you must properly configure IP addresses, RF channels, and, in some cases, network names on both the server and the W-NICs to get everything to work right (or at all).

When an addition or repair to a wireless network component is made, you must verify that the software and firmware revisions are correct in all the devices in the network, and that the new hardware has been properly configured to work with the existing hardware.

Plan for the Future

Stay Compatible

The major consideration in making your network available for future innovations is to make sure that your wireless component vendors adhere strictly to the standards. We cannot predict what turns technology will take over many years, but fortunately all the IEEE 802.11 innovations have remained backward-compatible. This means that if you have bought 1 or 2 Mbps hardware, it will still work within a faster 11 Mbps system. Consequently, the same should be true of faster networks to come.

Standards compatibility is the key here. It is unfortunately true that final standards are often not able to support early implementations. This was certainly true with early twisted pair Ethernet technology, and the same is true with wireless. As a matter of fact, early non-standards-based hardware is often discontinued overnight when a noncompatible standard is released.

30 and 100 Mbps Systems

Clearly, the need for speed of the wired network is placing pressure on wireless networking. Already, special-purpose wireless bridges that support 30 to 100 Mbps links are on the market. The next step is 30 to 100 Mbps wireless links. Manufacturers are implementing these higher speeds very quickly. As prices fall on 11 Mbps W-NICs, the need for faster links, albeit with higher prices, will appeal to both user and equipment manufacturers.

It is hoped that these new faster wireless link speeds will be available with the same accommodation for older and slower wireless cards.

LAN Wiring Management

15

Telecommunications Rooms and Wire Management

Chapter 15 Highlights

- Telecommunications rooms
- Horizontal and backbone wiring
- Accessibility and appearance
- Size requirements
- Power requirements
- Backbones and concentration

The concept of a series of interconnected telecommunications rooms is central to the structured cabling system design. The planning and placement of these telecommunications rooms is a subject sufficiently important for the EIA/TIA to devote an entire standard to it, EIA/TIA 569-A, *Commercial Building Standard for Telecommunications Pathways and Spaces.*

In this chapter, we will discuss the placement of telecommunications rooms in modern telecommunications installations. Naturally, this discussion will center around designing a wiring system for LAN use. This will impose some additional considerations on the location of our telecommunications rooms, primarily because of the distance restrictions of LAN wiring.

The telecommunications room is more broadly defined in EIA/TIA 569-A as a hierarchical system of telecommunication spaces that are used for the concentration and connection of a building's telecommunications wiring. This standard covers wiring spaces for all telecommunications activities and actually includes telephone as well as computer wiring. Because of the much lower operating frequency, telephone wiring is not subject to the severe distance limitations of LAN wiring. For example, the standard allows a total distance of 800 m (2624 ft) for an unshielded twisted pair circuit run through a combination of telecommunications rooms—far beyond that allowed for a single LAN leg, which is limited to 100 m (328 ft) without passing through a repeater or bridge. So, we will concentrate on criteria appropriate for LANs, and assume that the telephone wiring will either easily meet its requirements or be separately handled.

In this hierarchical telecommunications room system (so-called structured wiring), there are three types of wiring closets: the Telecommunications room (TR), the Intermediate Cross-connect (IC), and the Main Cross-connect (MC). Figure 15.1 shows this hierarchy. The structure does not assume that any of these telecommunications rooms necessarily contain equipment (such as hubs or telephone gear) but allows for any telecommunications room to be collocated with an Equipment Room (ER). In the traditional structured wiring installation for telephones, a cable from the local exchange carrier (LEC) would enter the building and run to the MC, which would generally also house the user's telephone switching equipment (a PABX or EKTS). This would be a combination MC/ER.

Cross-connections would be made between the LEC lines and the telephone switching equipment and additionally between the telephone switch and large multipair cables running to the ICs or TRs on each

Figure 15.1
The structured wiring hierarchy includes the Telecommunications Room (TR), the Intermediate Cross-connect (IC), and the Main Cross-connect (MC).

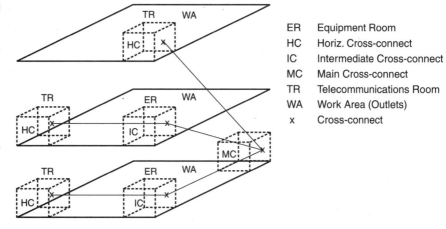

ER	Equipment Room
HC	Horiz. Cross-connect
IC	Intermediate Cross-connect
MC	Main Cross-connect
TR	Telecommunications Room
WA	Work Area (Outlets)
x	Cross-connect

floor. A large building would have intermediate cross-connect (IC) spaces where the multipair cables would be terminated and split out to other feeder cables to TRs near various office locations. At the TR, the station cables from each workplace location would be cross-connected to the feeder cables. The line for each phone would actually be connected through the TR, to the IC, to the MC, and finally to the telephone switch. In a smaller building, the functions of the various closets may be combined. In a very small building, all connections might be made in one space, a combination TR/MC/ER.

LAN wiring is implemented on basically the same scheme but with several important differences. First of all, LAN systems are distributed systems. There is no large central device similar to the PABX from which all the connections emanate; instead, the connections are made to hubs located in the TR (thus a TR/ER). Then, as we will see, the hubs are interconnected to each other and to common resources such as servers. With regard to LANs, TRs are usually interconnected with no regard to hierarchical level, since the network components are distributed. Still, the telecommunications room hierarchy is a useful model.

Horizontal and Backbone Wiring

All of the wiring from workstations to telecommunications rooms and between telecommunications rooms is referred to as "horizontal" wiring and "backbone" wiring, respectively, in the terminology of structured wiring. Backbone wiring is sometimes called *vertical wiring*. If we envision workstations on each floor of a building, and wiring closets vertically stacked with one per floor, this terminology makes sense. However, multiple telecommunications rooms are often located on the same floor, so the term "backbone" is clearer than the term "vertical"—the wiring from workstation to telecommunications room is the horizontal wiring and the wiring between rooms is the backbone wiring. If we also have additional telecommunications rooms, such as the intermediate and main cross-connect rooms, the wiring between those rooms is also called backbone wiring. This terminology holds whether the rooms are on multiple floors or are on the same floor.

The EIA/TIA standards always use the term *backbone* to refer to wiring between telecommunications rooms. Unfortunately, backbone has a very different meaning in LAN systems as compared to telephone systems. A LAN backbone is a logical subdivision of a network that has nothing to do with where the wiring goes. In fact, wiring on the same LAN backbone segment can be in two or more telecommunications rooms, or two LAN backbones can be in the same telecommunications room. However, we will stick to the wording in the standards and refer to all wiring between telecommunications rooms as backbone wire (or cable), unless we are specifically referring to that portion of a LAN system that interconnects multiple LAN hubs or bridges.

EIA/TIA 569-A recommends that the cross-connects in a telecommunications room be grouped together and prescribes a color code to further show this grouping. Table 15.1 shows the recommended colors for various terminations and cross-connect locations. For the cross-connect locations, the standard generally uses traditional colors, such as orange for the Telco demarcation punch blocks, and blue for the station cable punchdown locations. Actually, the punchblock itself is usually white, but its cover or mounting board may have the code color.

The standard also recommends that different jacket colors be used on the cables to distinguish between the usage types of the cables. This presents an interesting dilemma, as the colors in the standard bear little resemblance to the traditional colors used in commercial cabling. For example, station wire (to the workstation) has traditionally been run in beige or gray (the plenum-rated default), but the standard shows blue for this use. The color gray is used in the standard but only for second-level backbones, from the IC to the TR. Many LAN installers would agree that there is an advantage to having the LAN cables a distinctly different color from the telephone cables, but the standard assigns the same color to both types of station cables. (Advantage lost!) Of course, if the telephone cabling is simply run in a traditional color such as gray,

TABLE 15.1

Color Coding Areas in the Structured Wiring System

Orange	Telephone company demarcation	Blue	Horizontal
Green	Customer network connections	Purple	PBX, host, LAN, MUX
White	Backbone intrabuilding, 1st level	Red	KTS, EKTS
Brown	Backbone interbuilding	Yellow	Misc.
Gray	Backbone intrabuilding, 2nd level		

NOTE: It is recommended by the standard that cross-connects be grouped by color for ease of identification. It is also recommended that different sheath colors (cable jackets) be used to aid in distinction between types of circuits.
(SOURCE: TIA/EIA-569)

despite the standard, then the LAN cables may be run in the recommended blue color to distinguish them from the telephone cables. However, in a design that includes both telephone and LAN cabling, you could not simply specify that the EIA/TIA standards be used in your statement of work, because the telephone cable color would not be in compliance.

Frankly, it would be easier to just use traditional colors for all the telephone wiring. Intercloset cables are easy to recognize because they have distinctly larger pair counts and are consequently physically much bigger than the station cables. You could adhere to the color code at all points of termination or interconnection for the telephone system. Since many cabling veterans strongly recommend keeping LAN wiring completely separate from telephone wiring, the use of contrasting blue LAN station wire is appealing. You could allow a deviation from the cable color recommendation for the telephone wiring but would maintain color coding for the LAN cable.

How many horizontal cables should you run between telecommunications rooms and to the work areas? The standard states that you should run two cables to each individual work area, but is this enough? The answer to this question depends upon many factors, but the two most important considerations are growth and utility. You must determine what amount of growth planning is reasonable, and what it will cost to implement. For example, if you have large individual offices that may eventually include two workers, overwiring by a factor of two would be reasonable. The type of work that is to be done is also important. For example, some workers may eventually need two computers or two phone lines. In such cases, you should provide at least one additional station cable outlet that can be used for expansion. Generally, it is far less expensive to do a certain amount of overwiring during initial construction than it is to add outlets later.

What methods should you use to terminate the cables in the telecommunications room? The answer to this question depends upon whether you intend to install a universal wiring system, where any cable can serve your telephone or LAN systems, or install separate telephone and LAN wiring systems. Telephone wiring is quite forgiving in that you may run longer distances and make a larger number of cross-connects between the telephone instrument and the switching equipment. Modern LANs are much more restrictive in this regard. Early cabling standards allowed two cross-connects for a single LAN connection, but the latest Category 5e/6 wiring standards (as shown in Fig. 15.2) allow only one cross-connect and also severely restrict cable lengths, including that for the cross-connect wire. This is to allow for the strict attenuation, delay, and crosstalk limits of 100 MHz networks. Many experts even recommend that the workstation cables be terminated directly into a patch panel, with no intermediate cross-connects at all.

Figure 15.2
The methods of interconnection and cross-connection are shown.

Interconnection

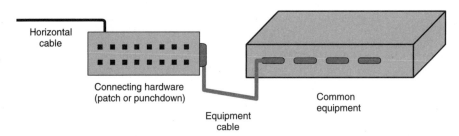

Cross-connection

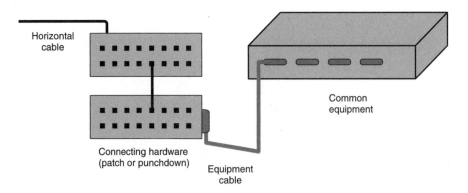

One of the reasons for this cross-connect limit is that all wiring terminations require that the pairs be slightly untwisted to be punched down at the point of termination. When the cable being used is very tightly twisted, as with Category 5e/6 cable, this untwisted portion increases crosstalk, introduces noise, disturbs the line impedance, and increases RF emissions. Too much untwist can cause a link to fail at the higher LAN data rates, especially if the cable run is near the allowable length limit. All punchdown blocks introduce untwist into a link. The extent of the problem is less with the 110-style blocks than with the older 66-style blocks, but it exists nevertheless. The effect of untwist is more disruptive as the network speed increases. Terminating the station cables directly onto the patch panel minimizes the amount of untwisted wire and increases the performance of the link.

Distance Considerations

The primary guideline in locating telecom rooms is to place them near the work areas they will serve. Advisory language in the standard says to place them in the center of the area to be served, but this central location is simply impossible in many existing buildings. The layout of the building may cause you to have to run your cables from a TR location on one side of the floor to each of the workstations. In such cases, it is critical that you predetermine the lengths of the wiring runs; if those runs will exceed the appropriate limits, additional TRs will be needed.

In new construction, the architects should be given firm instructions as to the placement of infrastructure facilities such as telecommunications rooms and equipment rooms. If you make these needs known at the inception of building design, you will be more likely to have the telecommunication spaces that meet standards.

You should give consideration to placing the telecommunications rooms vertically, one over the other, so that wiring distances between rooms will be minimal. In large buildings, multiple "utilities shafts" not only simplify telecommunications wiring but also eliminate the single point of failure in single-shaft designs. The benefits of this approach go beyond the telecommunications issues. It is also a good idea to provide an orderly system of cable ducts or sleeves, both for interfloor and fire-wall use. Proper ducting will allow workers to place UL-rated firestopping material in the ducts to enhance fire protection. Also, because many of the utility shafts are closed and reinforced concrete structures, this avoids the problem of random wall penetrations and "sneaker" cables that are forced alongside existing pipe and HVAC penetrations.

Maximum Drop Length

LAN wiring standards dictate strict distance limits for horizontal cable runs. These cable drop lengths include the entire cable from the point of termination in the telecommunications room to the workstation outlet. Standards also cover the maximum distances between telecommunications rooms. In this way, the requirements of TIA/EIA-568-B and 569-A work together, in balance with the building's architecture, to achieve a compliant wiring system for your LAN that will meet your needed performance.

The maximum length for a horizontal cable is 90 m (295 ft). This 90-m length excludes the user cord that connects from the outlet to the user workstation (or telephone instrument) and the patch cord (or jumper) that connects from the patch panel (or cross-connect field) to a hub or other device. This is a maximum acceptable length and not simply an average length or a goal. You should try to keep your wiring well below this length in most cases. At least two telecommunications outlets must be provided for each individual work area.

Although some of the lower-speed LANs, such as 10BaseT and even Token-Ring, may run to longer distances of as much as 150 m, exceeding the standard internationally is a bad idea. You cannot expect such run lengths to support 100/1000 Mbps data rates at all, and any equipment problems will be blamed on the excessive drop lengths. As a matter of fact, cable testers are required to fail a Basic Link longer than 90 m, allowing for test cables and measurement accuracy. For a complete discussion of measuring cable lengths using cable testers, see Chap. 16.

The allowable backbone cable distances between telecommunications rooms depend upon the type of cable you use as well as practical considerations. For example, the standard allows 2000 m (6560 ft) from TR to MC for fiber optic cable, but only 800 m (2624 ft) for unshielded twisted pair cable. The distances are even less between ICs and the other two types of telecommunications rooms. However, practice may dictate much smaller distances. If you wish to connect between two telecommunications rooms on a 100BaseT or a gigabit link, for example, you would have to limit the backbone run to the same 90 m you are limited to for any hub-to-workstation link. As mentioned previously, although some situations may operate with slightly longer cable lengths, they do not meet the requirements for inevitable higher speeds of 100 Mbps or more.

Estimating Drop Length

Cable drops from the proposed telecommunications room to the workstations to be served must be within the maximum allowable length of 90 m (295 ft). You should estimate these distances before finalizing your cable system design. If the estimates exceed the maximum length, you may need to add one or more wiring closets. Although there are minimum size specifications and features for telecommunications rooms (covered in a later section), you may be justified in putting in a substandard telecommunications space to get the drop lengths down.

In order to estimate the total length of a cable run, you must first know how the cable will be routed and hung. Conventional wiring is often routed through metal floor or roof supports to keep it well above the ceiling tile. Alternatively, cable may be hung with a variety of cable hangers, clips, and tie wraps from supporting structures. In some installations, particularly old work, the cable is laid on top of the ceiling tile supporting grid. Large buildings frequently have the cable placed in cable trays suspended between the ceiling grid and the floor above. This method is becoming more popular for Category 5e/6 installations where the slightest kink or sharp bend can impair the cable performance. Any combination of the methods may be used as well.

The routing methods are important because the more orderly routings inevitably require more wire. For example, routing through cable trays involves lots of right angles (orthogonal routing) instead of diagonal runs, demanding up to 40% more cable. On the other hand, laying the cable directly on top of the ceiling grid with direct routing uses the least cable (although it is the hardest to maintain and may suffer EMI problems more often). Cable trays are easiest to use during construction build-out, when the ceiling grid is not in place. They are more difficult to add once the offices are constructed or, worse, occupied. Trays carry significant weight from the many cables they contain and must be well secured in place. For that reason, cable hangers and clips are often used to minimize disruption to an existing workplace.

To estimate an orthogonal wire placement, where cables are to be bundled or laid in cable trays, you must calculate the length of run in the tray. This may involve a straight run or one with several corners. Add each leg of the run in the tray. Next, add the drop from the tray to the wall header above the workstation outlet, and the lengths of the vertical drops in the workstation wall and in the telecommunications room, and add a factor for obstacles (2 to 5%, depending upon the building). Obstacles that might be encountered are HVAC units, duct work, firewalls, columns, and pipes. The vertical drops should allow for 18 in (0.5 m) of excess wire at the wall outlet plus the distance through the wall to the height of the cable tray (not just the ceiling height). The vertical drop length in the telecommunications room should include the drop from the tray to the final termination point of the cable, including all ducts, ports, channels, and trays through which the cable must be routed. Be sure to accurately estimate the twists and turns the cable may make to "dress" the wires on the mounting board or in the equipment racks. Allow another 6 to 12 in here for termination. If your measurements are close to the limit, remember that cables are typically routed along the length of a patch panel to their termination points, which might add up to 19 in of additional length.

Diagonal wire placement is really direct routing. This means that the cable is simply run above the ceiling grid and routed directly to the workstation drop location. Routing is done either directly over the grid or through supporting structures. Sometimes wire hangers or clips are used, as are tie wraps.

The use of cable hangers and clips may allow a combination between orthogonal wire placement and diagonal placement. Bundles of cable are typically routed through cable hangers orthogonally from the telecommunications room to groups of work area locations, but individual work area drops cables may be diagonally routed from the nearest hanger.

Older buildings, houses, or buildings with fixed ceilings present special problems to cable routing. Often the cables may be run in attic or crawl spaces. Sometimes cable must be run around baseboards or even over door frames to reach the workstation location. The use of staples—a common mounting technique for telephone wires—is definitely out, even for Category 3 cable. Be careful when estimating cable lengths for these buildings. Simple detours, such as doorways, can eat up lots of cable. For example, a 10- by 20-ft room would require 10 + 20 + 17 = 47 ft of cable simply to run to an opposite corner of the room, if one 7 by 3 ft (7 + 7 + 3) doorway has to be traversed. If a drop must be run through several adjacent rooms to reach from the wiring closet to the work area, an enormous amount of cable may be required.

Accessibility and Appearance

The wiring centers of your network are key components of your LAN system, containing important networking hardware, the cable distribution points for your users, and important diagnostic and backup equipment. In addition, many important installation, configuration, and troubleshooting tasks take place there. For those reasons, you should devote the time and money necessary to maintain good telecommunications rooms and equipment rooms.

The first key to a good telecommunications room is accessibility. The wiring connections and network hardware should have their own room, out of the way and yet easy to get to. The area should be as close to the work areas as possible, since LAN workers will often have to travel between workstation locations and the telecommunications room. That means that a location down the hallway near the elevators is not as useful as a small room or office near the work areas. The room should be used exclusively for LAN wiring terminations, hubs, routers, servers, and related equipment, but never for storage.

The second key is appearance. The telecommunications room should be orderly, well laid out, and well lighted. Adequate power and ventilation should be available and designed for the specified purpose. Mounting boards on the walls should be full sheets of $^3/_4$-in plywood and well secured to the wall's interior structure. At least two walls should be plywood-sheathed, and the boards should be painted a neutral color, such as gray. Areas for specific equipment should be well marked.

If you use equipment rails or racks to mount your hubs and patch panels, they should be placed a sufficient distance from the walls to work behind them. It is not at all a bad idea to place the hubs so they can be viewed from the doorway of the telecommunications room. If you terminate the station cables directly onto a patch panel, you can use cable trays across the top of the mounting rails to route your cables. Wire management trays, mounting brackets, preassembled backboards, and distribution rings are all available to help make a neat and serviceable cable arrangement. Proper routing (or "dressing") of cables will make a neat, orderly appearance that will save maintenance time and trouble in the long run.

Category 5e/6 wiring requires special precautions in installation and routing. The cable must be treated with great care during installation to preserve its performance characteristics. This special care includes running the wire with a minimum of sharp turns, no stretching, no staples, and no tight tie wraps. Many patch panels have additional strain relief points at each cable termination to eliminate sharp bends and to help secure the cables with a minimum of stress.

The third key to a successful telecommunications room installation is security. Equipment located in the wiring closet needs to be secured. That is one reason why a separate room with a locked door is the best location for your telecommunications room. If your telecommunications room is located in a common area outside your offices, controlling access is essential. Telecommunications rooms that are within your offices still may need to be secured, particularly in a high-traffic location. Your company's mission may also dictate a need to control access to the room, because important servers are often located there in addition to the networking hubs.

Size Requirements

Telecommunications Rooms need to be large enough to provide a proper area for wiring termination, patch panels, and hubs. In some cases, other networking equipment will be located in the same area. Rooms containing network hardware such as servers, bridges, and routers are properly called equipment rooms (ERs). ERs may be combined with telecommunications rooms at any level, including main cross-connects (MCs), intermediate cross-connects (ICs), and telecommunications rooms (TRs).

The space requirements for wiring closets are clearly defined in EIA/TIA 569. The minimum size for a TR or IC is 7 by 10 ft (2.2 by 3 m). This space is designed to serve a 5000-square-foot area (500 square meters). A minimum size for an equipment room is 150 square feet (14 square meters). Table 15.2 shows the sizes for other served areas. Note that the metric conversions are greatly rounded in this standard. This is referred to as a "soft" conversion and you should use the dimension in the appropriate unit of measure for your normal building plan dimensions.

These sizes are really minimums, but should be adequate for most instances. In the case of equipment collocation with wiring closets, you will need to adjust the size of the room accordingly. Remember that you need to have space in front and back of racks, rails, and servers for routine maintenance in the wiring closet. While you may be tempted to put the recommended area into a narrow space, remember that your LAN workers, installers, and vendors will have to work in that area, and will be hampered by too small a space.

Power Requirements

Many telecommunications rooms have networking equipment that will require electrical power. Power outlets for telecommunications rooms should be on dedicated circuits. A dedicated circuit is one that serves only one outlet, or a series of outlets in one room. Standard practice in the United States is to equip each dedicated outlet location with a bright orange outlet (either a duplex or simplex receptacle) to distinguish it. The electrical circuit should be rated appropriately for the equipment that is to be connected. Some larger equipment may require its own separate dedicated outlet.

You should be sure that an adequate number of electrical outlets are provided, based on the equipment that needs to be installed. Racks or banks of equipment may need an outlet strip or a power bar with multiple outlets. Allow lots of room for growth. A good rule of thumb is to provide twice the number of outlets you think you will need, and double that number! Remember to include open outlets that can be used to provide power to test equipment.

The standard recommends two dedicated duplex outlets AC outlets for equipment power, rated at 15 A, 110 V (or appropriate ratings for your locale) on separate circuits. It also recommends convenience outlets, wall-mounted every 1.8 m (6 ft) around the telecommunications room, in addition to any power requirements for collocation of an equipment room. Also specified is lots of ceiling mounted lighting, 540 lux (50 ft-candles). Equipment rooms should have enough HVAC capacity to maintain the room at 18 to 24°C (64 to 75°F).

Another item to consider is power conditioning and backup power. Your power connections should, at the minimum, contain surge protection. In areas that are subject to power line fluctuations, you might want to consider ferroresonant transformers that automatically compensate for sags and minor surges.

True power spikes must be dealt with by the use of line surge protectors. Several types of these exist, including gas-discharge protectors, semiconductor protectors, and metal-oxide varistors (MOVs). The MOVs are the most common and the cheapest, and provide the least protection—usually found in low-cost multioutlet strips. Their primary drawbacks in such an application are their slow response time and the user's inability to determine if they have lost protection because of prior spikes.

A form of spike also occurs as a result of telecommunications wires that run outside your building. The most common occurrence results from a

TABLE 15.2

Recommended
Telecommunica-
tions Room and
Equipment Room
Sizes
(SOURCE: TIA/EIA-569)

Served zone		Telecommunications room	
Square meters	**Square feet**	**Meters**	**Feet**
500	5,000	2.2 × 3	7 × 10
800	8,000	2.8 × 3	9 × 10.
1000	10,000	3 × 3.4	10 × 11

Equipment rooms		
Workstations	**Square meters**	**Square feet**
Up to 100	14	150
101 to 400	37	400
401 to 800	74	800
801 to 1200	111	1200

nearby lightning strike. The electrical discharge may induce destructive voltages in any outdoor cabling, including telephone and LAN cabling. It is very important to provide approved protectors on all telecommunications cables at the point of entry to your building. Fiber optic cabling with no metallic members can be an important tool in eliminating lightning risk.

Another potential power problem is a power outage. Equipment that might be disrupted by a power outage should have emergency backup power supplies provided. These power supplies usually connect between the equipment to be served and the AC power outlet. One type of uninterruptable power supply (UPS) is the backup UPS. When power is cut off, an inverter generates AC voltage from power stored in batteries. The generated power is switched to the connected equipment in a relatively short time period so that the equipment sees only a small break in the voltage. Another type of supply, the true uninterruptable power supply (true UPS), constantly generates power to the connected equipment. When there is normal line power available, the true UPS uses the AC line as its source of power. If there is a break in AC power, the UPS switches almost instantaneously to the battery source, with no appreciable interruption in voltage to the connected equipment. As you might expect, the true UPS is considerably more expensive than the backup UPS.

What type of equipment should be provided with emergency power? You should provide for equipment needed during an emergency, or equipment that can be harmed or disrupted by a power outage. Equipment needed in an emergency might include the telephone system, the alarm system, the video security system, or any network servers and hubs needed in an emergency. Some servers and other computer hardware are disrupted in a power outage and often must be manually restarted or "cleaned up" before normal use can resume. Emergency power, while temporary, may eliminate the restart process completely in a brief power outage. Some UPS systems alert servers to begin the shutdown process automatically, if it appears that the outage may be sustained. It also may be necessary to provide emergency power to network hubs to assist in the shutdown process. Unless user workstations are somehow provided with emergency power, it may be of little use to provide such power to the hubs and other networking equipment.

Backbones and Concentration

LAN backbones and cabling backbones unfortunately share a common name. During most of this book, we have been referring to backbone cabling rather than LAN backbones. However, in this section, we will actually be talking about the type of logical backbones that are used in LAN systems.

LAN backbones are often used to tie sections of the networking system together, so that workstations anywhere on the network can use resources, such as servers, that are not exactly on the same portion of the network. The workstations in a work area are connected to a hub-type device in the wiring closet. Hubs may be interconnected in a variety of ways, but often they are connected to a higher level of the network, called the backbone.

In a simple repeater backbone system, as shown in Fig. 15.3*a*, several standalone 10/100BaseT hubs are connected together via a 10Base2 coax segment. This setup has the advantages of 100BaseT star wiring to the workstations, but the disadvantages of coaxial bus cabling between the hubs. If any hub or coax connection fails, the entire backbone segment is likely to fail.

Figure 15.3*b* shows another alternative. Here, the hubs are connected to a higher-level backbone hub, via 100BaseT. Because all hubs repeat the LAN signals to all ports, the servers can be located anywhere. Some variations of this scheme put each hub onto a circuit card that mounts in a hub chassis. The hub cards are interconnected at the backplane on a high-speed bus that is independent of the topology of the network. This bus serves as a type of backbone to all of the hub cards.

Figure 15.3
Different types of
backbones used in
Ethernet networks.

A. A simple repeater backbone

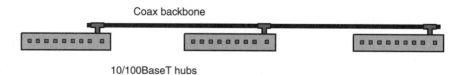

B. Multilevel hub backbone

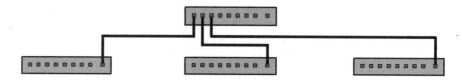

C. A bridged backbone

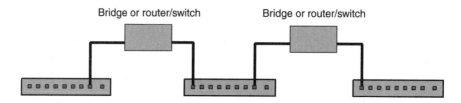

D. A collapsed-backbone router/switch

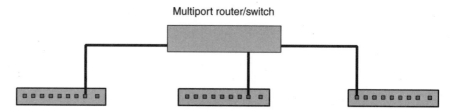

However, neither of these really tends to break up the network into smaller, more efficient chunks. Any network traffic is repeated throughout all of the connected hubs, effectively limiting the bandwidth of the entire network to a single 100 Mbps path.

Figure 15.3c shows several hubs connected via switching bridges[1] (or routers) to a backbone LAN. Notice that servers on the local hubs are

---

[1]Switching hubs function at Layer 2 of the OSI model, as do bridges. Routers and routing switches are Layer 3 devices.

available to local workstations without transiting a bridge. This effectively makes the 100 Mbps bandwidth of hub A add to the 100 Mbps bandwidth of hub B, as long as the user on one hub has no need to use the server on the other hub. If a user on either hub needs to have access to a server on the backbone, the user connects across the intervening switching bridge. Company-wide applications are often placed on servers on the backbone to give access to any user without putting traffic on other parts of the network. Thus, a user on hub A can use a backbone server at the same time that a user on hub B uses a local server. Figure 15.3*d* shows a variation using a multiport router switch.

This type of backbone hierarchy must be considered in planning the location of telecommunications rooms, equipment rooms, and the cabling between them. Repeater backbones and bridge backbones can both be supported by using the wiring standards we have covered in this book. However, the specific number of interconnecting cables between telecommunications rooms may be dependent on the way you lay out your backbone system.

Hubs (or stacked hubs) are often placed in separate telecommunications rooms. This means that the LAN backbone cabling must be run between telecommunications rooms to interconnect the hubs. These connections between telecommunications rooms are traditional backbone cabling. The actual backbone switches may be placed together in the same equipment room, or located individually at each hub that is to be connected to the backbone. In some LAN backbone topologies, a backbone ring must be formed linking each telecommunications room. Token-Ring and FDDI are examples of this backbone ring method. Unlike the telephone scenario, the LAN backbone connections do not simply combine individual station cables into larger multiplier cables. LAN backbones usually require one backbone connection for each group of hubbed workstations that is to be connected. Additional backbone cables may be run to provide backup for the primary cables or to provide closure of a ring. Backup cables or cables to complete a ring are an integral part of the backbone hierarchy.

By the way, nothing in this implies that a backbone is either fiber optic or copper, as either one may serve your needs well. Keep in mind that the allowable distances between closets in TIA/EIA 569-A exceed the distances for copper wire 10/100/1000BaseT or Token-Ring, set without reference to a specific LAN topology. However, the use of fiber optic cable can extend your cables to the limit of the TIA/EIA standard.

16

Testing and Certifying

Chapter 16 Highlights

- Certifying cable and connecting hardware
- Scanner performance
- Cable plant certification
- Testing methods
- Test equipment
- Analog versus digital testers

The major reason to specify a standards-based LAN wiring scheme is to ensure that your installed cable will meet the needed level of performance for your intended use. To specify such demands three things. First, you must make a cable plant design that will meet the desired performance criteria when the other two requirements are met. Second, you must use cable and connecting components that are certified to the proper level of performance. Third, you must employ proper installation methods so that the performance potentials of the cabling and components are achieved.

This is a long way of saying that a well-designed LAN wiring system, using the proper cable and connectors, and installed with good workmanship, will operate at the expected level of performance. Previous sections of this book have described all of these key concepts. This chapter will explore the ways in which you can verify the proper operation of your installed cable system.

We will discuss some of the ways in which cable and components are certified to a particular level of performance by their manufacturers. In addition, we will see how you can perform inspection and testing to ensure that your installed cable plant meets the required performance levels. Next, we will talk about the test equipment you will need to verify your installation. Finally, we will show how to use that test equipment to do cable certification.

Certifying Cable and Connecting Hardware

The primary standard for LAN performance certification in the United States and many other countries is TIA/EIA-568-B, *Commercial Building Telecommunications Wiring Standard.* An earlier supplement, TSB-67 1995, *Transmission Performance Specifications for Field Testing of Unshielded Twisted-Pair Cabling Systems,* provided the basic standards for the testing of installed links. The supplement TSB-95, *Additional Transmission Performance Guidelines for 100-ohm 4-Pair Category 5 Cabling,* gives added standards for gigabit operation. Several other standards bodies, including the Canadian Standards Association (CSA) and the International Organization for Standardization (ISO), have similar or coordinated standards, such as CSA-T529 and ISO/IEC IS-11801.

The TIA/EIA-568-B standard contains information regarding the performance levels of connecting hardware and cable used for telecommunications wiring. This standard contains explicit testing methods for some items and refers to documents from other organizations, such as ASTM D-4566, for other items. Cable and connecting hardware manufacturers offer components that are said to provide a certain level of performance, as delineated by the standard. The general measure of component performance is the "category" of operation. Categories 3, 5, and 5e are defined in the standard, along with their corresponding recommended uses. Please refer to Chap. 5 for an explanation of the categories. The parameters that are specified in TIA/EIA-568-B pertain to the performance of those components prior to their installation, and the performance testing of installed link components. TSB-95 allows the retesting of installed cabling to verify gigabit performance.

The actual testing of cabling components is the responsibility of the manufacturers of these items. Manufacturers may do their own testing to "certify" conformance with the standard, or they may use an outside testing facility. In some cases, manufacturers also have a recognized independent testing laboratory—such as UL—verify that the testing was performed properly. Some independent labs also offer a continuing verification or certification program so that manufacturers may further assure their prospective customers that their products meet the necessary performance criteria.

You are responsible for using only components of a proper category and for field testing the installed cable links to verify proper operation for the rated category. This process may be done by you, your installation contractor, or an independent third party to "certify" compliance at the rated category.

This means that if you expect to provide a LAN wiring system with a particular level of performance, you must use only components and cable that have been certified for that category of operation or higher. For example, for a Category 3 installation you may use either Category 3, 4, or 5/5e components, but you must use only Category 5e cable and parts for a Category 5e system.

A common misconception is that using Category 5e cable is always better than using Category 3 cable for an application that requires only Category 3 operation, such as 10BaseT. However, if you use components and installation techniques that are only appropriate for Category 3, you essentially get only that level of performance. The performance of a cabling system is only as good as its weakest link, and a category of performance is equal to the lowest certification and installation of any component. You could well argue that a higher category of cable will ensure meeting a lower category of performance. The higher grade cable may improve performance margins a bit. It will, but the only gain to you is in those instances when a lower category of cable might reveal other marginal conditions, such as interfering electrical fields or longer-than-permitted cable runs. A much better reason for using a higher category component is that you could at a later time upgrade to Category 5e/6 without repulling the wire. Be cautious, though, as this potential for later upgrade might be wasted if other Category 5e/6 installation practices are not followed.

Performance Levels

What does component certification really mean? Because recognized performance categories exist, a cable or connector that is marketed as meeting the requirements of a certain category should be tested to verify that claim. As the manufacturers are responsible for testing these components themselves, they often provide typical test results in their literature. Some manufacturers will provide detailed test reports upon request and offer product guarantees to back their claims.

In addition, independent laboratories offer a manufacturer verification testing programs that range from a one-time test report to a continuing test program with random testing. Obviously, you are more ensured of quality components with a continuing verification program than with a single test sample.

TIA/EIA-568-B outlines the performance specifications and very detailed test procedures that must be followed to certify components to a particular category. TIA/EIA-568-B.1 (Part 1) describes the essential wiring structures and link testing methods, while Parts 2 and 3 cover cable and connecting hardware standards for 100-ohm unshielded twisted pair (UTP) and fiber optic cabling components, respectively.

Pluggable connector components present some unusual challenges in reliability testing, as they must be checked for connectivity degradation. Random samples of each component are tested for degradation of contact resistance and insulation resistance in tests for durability, vibration, stress relaxation, thermal shock, and humidity/temperature cycle. This testing is to be done by the manufacturer to certify their parts to a recognized performance category. A cautious manufacturer might choose to have the certification verified by an independent testing laboratory, since a lot rides on the higher categories of operation.

Surprisingly, with regard to connector plugs and jacks, the TIA/EIA-568-B standard calls for only 200 insertion/withdrawal cycles without failure (usually defined by excessive contact resistance). Some manufacturers may specify a usable life of 500 to 2000 cycles for their connectors, although the standard only calls for 200 cycles. The point is made that these components have a finite life. Modular connectors that are used in patch panels with frequent use may be expected to exceed the standard's parameter over normal equipment life. Moreover, connectors that are used in test equipment, such as for cable certification, can exceed that number of cycles in a single afternoon! The user is only cautioned by the standards to "inspect the connectors for signs of wear." You would be much safer either to use connectors designed for high usage or to periodically replace the connectors on high-use equipment. You would also be wise to keep a usage log of your test cables and replace them regularly, as the modular plugs used on these cables have the worst degradation from multiple insertions.

Connectors must also be tested for transmission performance (attenuation and NEXT). Barring the degradation from excessive use, modular connectors have an innate contact resistance that is not significant when compared to the bulk resistance of a length of cable. The wire pins that form the connector are typically gold-plated to minimize this contact resistance. However, the modular-type connectors used in the current wiring standards do exhibit a significant amount of cross-coupling that appears as near-end crosstalk (NEXT). All of the 8-pin modular style connectors are bad in this respect, but testing is done to make sure the NEXT is not worse than allowed.

Table 16.1 shows the attenuation and NEXT performance parameter limits for Category 3, 5, and 5e components. You can see from the values that little contribution to circuit loss is expected from the circuit hardware, as most comes from contact resistance. For comparison, the DC insertion resistance is only 0.3 ohms for a connector, compared to 9.2 ohms per 100 m for the cable. However, the NEXT contribution from the connectors can be significant. For example, at 100 MHz, the NEXT loss of the connecting hardware must be better than 40.1 dB, while the cable alone must be better than 32.3 dB for a 100 m length. The entire terminated and installed Basic Link allows only an additional 3 dB (to 29.3 dB at 100 MHz) over the raw cable for all the intervening connecting hardware and installation anomalies. This is an area, by the way, where a higher "enhanced" grade cable can increase the margin of performance of a link. However, the connectors cannot be much improved and remain compatible with the standard modular design.

The low NEXT requirements of Category 5 and 5e are one of the primary reasons that the traditional screw-type wire terminals are not acceptable in those applications. You can easily observe the increased NEXT of screw-

TABLE 16.1

Connecting Hardware Attenuation and NEXT Limits of TIA/EIA-568-B

	Connecting hardware limits (loss in dB)					
	Attenuation			NEXT loss		
	Category					
Frequency, MHz	**3**	**5**	**5e**	**3**	**5**	**5e**
1.0	0.4	0.1	0.1	58.1	65.0	65.0
4.0	0.4	0.1	0.1	46.0	65.0	65.0
8.0	0.4	0.1	0.1	40.0	62.0	64.9
10.0	0.4	0.1	0.1	38.1	60.0	63.0
16.0	0.4	0.2	0.2	34.0	56.0	58.9
20.0		0.2	0.2		54.0	57.1
25.0		0.2	0.2		52.1	55.0
31.25		0.2	0.2		50.2	53.1
62.5		0.3	0.3		44.2	47.1
100.0		0.4	0.4		40.1	43.0

NOTE: The maximum attenuation values shown are specified at each frequency. Acceptable values of attenuation are a lesser (or equal) dB value than shown in the table. The NEXT loss shown is calculated from the formula in TIA/EIA-568-B. The formula is based on assumed minimum values at 16 MHz of 34 and 56 dB for Category 3 and 5, respectively. Acceptable values of NEXT loss are a greater (or equal) dB value than shown in the table, or calculated from the formula, at a given frequency.

terminal jacks with any of the more sophisticated cable scanners. (An increase in NEXT is a lower dB number, as the measure indicates how low the crosstalk signal is with respect to a transmitted signal on another pair. You might think of the NEXT loss number as the amount of isolation from an interfering signal. A higher number is better.) Connector assemblies that are designed for the higher categories normally use an integral printed circuit board with carefully routed wiring to minimize NEXT.

Screw terminals are also technically rejected on grounds of pair untwist, which must remain below 0.5 in for Categories 5 and 5e. The station cable must often be untwisted in excess of those limits to reach the screw terminals. Screw terminal connectors also have an unacceptable length of untwisted wire between the screw terminal and the actual connector body. It is these factors that lead to the unacceptable level of NEXT and impedance anomaly in screw-type jacks.

You should be aware that there are some insulation-displacement connectors (IDCs) that use similar space-wiring between the IDC block and the modular connector. Both 110- and 66-type miniblocks may be found in these types of outlet plates. These connector assemblies do not meet the needs for Category 5/5e and should be avoided. This problem comes not from the connecting blocks but from the internal outlet wiring from the jack to the block. Do not confuse these jack plates with the acceptable style that integrates the 110 block into the connector as part of a printed wiring assembly.

Cable used in horizontal and backbone runs must also meet performance standards. Table 16.2 shows the attenuation and NEXT performance parameters for Category 3, 5, and 5e cable. Note that these levels are for the raw cable, not an installed link (see "Testing Methods" later in this chapter). Horizontal cable must meet essentially the same requirements as backbone cable. The attenuation and NEXT limits are calculated from a formula in the actual standard and the values in the table merely illustrate the calculation at specific discrete frequencies. Cable attenuation is known to vary with ambient temperature. The cable must also be tested (by the manufacturer) to meet adjusted loss limits at 40 and 60°C. The maximum allowable attenuation at these elevated temperatures for Category 5 and 5e cables is adjusted by a factor of 0.4% increase per degree Celsius.

Cable must also meet other transmission-related specifications of DC resistance and unbalance, mutual capacitance, capacitance unbalance to ground, characteristic impedance, and structural return loss. The resistance and capacitance components are primarily a function of the characteristics of the copper wire and the insulation material. These values are consistent and are easily described by a single set of values for each category, as shown in Table 16.3.

Characteristic impedance is specified as 100 ohms ± 15% (or 85 to 115 ohms). However, it turns out that it is rather difficult to make a consistent measurement of characteristic impedance of twisted pair wire, because of variations in the cable's structure over its measured length. Consequently, the characteristic impedance value is derived by "smoothing" the results of measurement of a parameter called *structural return loss* (SRL). The SRL may vary rather wildly from 1 MHz to the highest referenced frequency (16, 100, or 250 MHz for Category 3, 5e, or 6, respectively). For this reason, an absolute minimum limit for SRL is imposed for each category. When the SRL value is higher than the limit, it indicates that the impedance variation is better than required for the cable pair. The limit applies to the worst pair of the cable, so all pairs must be tested.

Because the twist pitch (lay length) of different pairs is varied to decrease NEXT coupling, each pair of a cable will show a variation in values of most of the measurement parameters. Some of the standard's limits are based on the worst pair's value, while others, such as attenuation,

TABLE 16.2

Cable Attenuation
and NEXT Limits of
TIA/EIA-568-B

	Horizontal and backbone cable limits (loss in dB)					
	Attenuation			NEXT loss		
	Category					
Frequency, MHz	3	5	5e	3	5	5e
0.064	0.6	0.7	0.7			
0.150				52.7	74.7	77.7
0.256	1.2	1.1	1.1			
0.512	1.8	1.5	1.5			
0.772	2.2	1.8	1.8	43.0	64.0	67.0
1.0	2.6	2.0	2.0	41.3	62.3	65.3
4.0	5.6	4.1	4.1	32.3	53.3	56.3
8.0	8.5	5.8	5.8	27.8	48.8	51.8
10.0	9.7	6.5	6.5	26.8	47.3	50.3
16.0	13.1	8.2	8.2	23.3	44.3	47.2
20.0		9.3	9.3		42.8	45.8
25.0		10.4	10.4		41.3	44.3
31.25		11.7	11.7		39.9	42.9
62.5		17.7	17.0		35.4	37.2
100.0		22.0	22.0		32.3	35.3

NOTE: The maximum attenuation values shown are calculated from the formula in TIA/EIA-568-B. The attenuation value is per 100 m (238 ft) at 20°C. An increase of 0.4% per degree Celsius is added to the value shown for Category 5 and 5e cables. Acceptable values of attenuation are a lesser (or equal) dB value than shown in the table. Values for frequencies below 0.772 MHz are shown for reference only and are not required for testing.

The NEXT loss shown is calculated from the formula in TIA/EIA-568-B. The formula is based on assumed minimum values at 0.772 MHz of 43 and 64 dB for Categories 3 and 5, respectively. The NEXT loss of the worst pair combination, for a length greater than or equal to 100 m (328 ft), is used. Acceptable values of NEXT loss are a greater (or equal) dB value than shown in the table, or calculated from the formula, at a given frequency. The power-sum method of measurement is used for backbone cables.

require all of the pairs to be at least as good as the limit value. This intentional difference in cable geometry also contributes to some of the variations we see in the SRL.

Patch cords and cross-connect jumper wires must meet the requirements of horizontal cable with some exceptions. It is assumed that the patch cords will use stranded wire for increased flexibility. Stranded wire pairs are expected to have slightly greater attenuation than solid

TABLE 16.3

Cable Performance Requirements for Resistance, Capacitance, and Balance

Parameter	Applies to:	Conditions:	For category:	Requirement:	Mandatory?
DC resistance	Any conductor		All	≤ 9.38 ohms per 100 ohms	Yes
DC resistance unbalance	Any pair		All	≤ 5%	Yes
Mutual capacitance	Any pair	1 kHz	Cat 3	≤ 6.6 nF per 100 ohms	No
	Any pair	1 kHz	Cat 5 & 5e	≤ 5.6 nF	No
Capacitance unbalance, pair to ground	Any pair	1 kHz	All	≤ 330 pf per 100 ohms	Yes

NOTE: All measurements are to be measured at or corrected to 20°C. The table applies to both horizontal and backbone cables.
SOURCE: TIA/EIA-568-B.

copper, so an increase of 20% in attenuation is allowed over that for horizontal cable. The attenuation value is prorated for length, since patch cords are generally far shorter than horizontal cable runs. These wires need not be tested at elevated temperatures. In addition, the stranded wires must have a twist pitch (lay length) of 15 mm (0.6 in) or less.

The standard specifies a clear system of category marking for cable and connecting hardware. The category should be marked either with the words Category N (where N is the category number 3, 5, or 5e) or with a large letter "C" with the category number in the center. These markings are shown in Fig. 16.1. The category of some wiring components can be very difficult to determine once they are installed. A sticker on a plastic bag, label on the package, or note in the data sheet is not of much use. The standard requires marking, and it should be easy for most manufacturers to stamp, emboss, or label each jack, plug, or cable with the rated category number. After all, you are paying extra for the performance. The best rule is, if it doesn't say Category 5e (or whatever), don't assume it is Category 5e. Try to use marked components wherever possible.

Pitfalls and Specsmanship

The greatest danger in evaluating cable and connector hardware specifications is that the certifications may not be based upon the most current standards. The category nomenclature has evolved over a period of

Figure 16.1
Category of perfor-
mance markings for
components.

Category 3:	Cat 3	Ⓒ3
Category 5:	Cat 5	Ⓒ5
Category 5e:	Cat 5e	Ⓒ5e

years from a loose general specification to a very specific (and tough) measured-parameter specification. Originally, the progressive ratings of performance were called Levels 1 through 5. Over time, the term "Cate-gory" was introduced and eventually adopted by several standards bod-ies, including the EIA/TIA, the UL, and the ISO/IEC, which has a corresponding class A–F system for links. Now the performance parame-ters and testing methods have been further tightened, although the existing Category terminology remains. Participation in a continuous testing verification program by one of the major independent labs is the best assurance that the latest standards are used.

You may have cable or connectors on hand that do not strictly meet the current performance requirements of the categories. It is possible also that the testing was done with outdated procedures. You would be confi-dent that the latest performance standards were met only if your hard-ware and cable were certified to meet TIA/EIA-568-B. A reference to 568-A would probably suffice, since it was the revision to the original EIA/TIA 568. In some cases, previously manufactured components may reference the technical bulletins that were issued prior to being incorporated, for the most part, into TIA/EIA-568-B. Test equipment that meets the TSB-67 requirements should be fine for Category 5, but you may need to update or replace your test instrument to test for TSB-95, Cat 5e, or Cat 6.

One pitfall you should avoid is the use of components that diverge from the recognized standards. For example, other styles of plugs and jacks may have much better NEXT performance than the ubiquitous 8-pin modular hardware. Typically, these other connectors and hardware are presented as having an advantage over the standard product. This is, you see, a way to say "standard" so that it seems ordinary. However, wiring standards are not only a set of performance parameters and installation practices, but also a standard style of interface to the user equipment that allows multivendor interoperability with a high level of performance. Any item that does not comply with the standards is an exception and may be a potential problem for your network. This is, you see, also a way to hear "exceptional" that says it is noncompliant. For the time being, you may be taking a risk if you choose to use some of these better (but nonstandard) connectivity options.

What is specsmanship? *Specsmanship* is a term for the exaggeration or misleading presentation of product specifications that sometimes occurs in the marketplace. Fortunately, the existence of recognized national and international standards minimizes the occurrence of specsmanship, but it still occurs. Were there no standards, you can be sure that various manufacturers would make conflicting product claims that would be impossible for the buyer to resolve. The tough standards for telecommunications cabling make product differences clearly recognizable. All you have to do is know which standard of performance you want the product to meet, and then be sure that the products you use meet that standard.

There is a continuing controversy that involves connector performance. The controversy concerns the relatively high NEXT of all modular connectors and the measurable differences in longitudinal balance of some connector jack designs. Several types of nonmodular connectors are available that have significantly lower levels of NEXT than the popular 8-pin modular design. Some of these connectors are used for field test instruments to reduce the contribution to NEXT from the test set and are quite acceptable for testing the Basic Link. However, a compatible lower-NEXT modular connector style would be much more desirable than attempting to replace more than a decade of installed modular connectors with another connector. Several manufacturers are working on the improved products, and we're already starting to see clever plug and jack designs emerge, but the issues of longitudinal balance and resonance phenomena are more subtle. Resonance phenomena can cause short links (particularly under 15 m) to exceed the allowable NEXT performance and is related to the return loss and/or balance of the link. These are subjects for further study by a TIA task group.

Cable Plant Certification

In addition to using properly rated cable and components in your LAN wiring installation, you should plan to test your installed wiring system to the expected level of performance. Installing a cable plant or wiring system is a complex task, particularly if you intend for it to support high performance 100/1000 Mbps networks. Even 10 Mbps Ethernet and 16 Mbps Token-Ring must have station cables that meet a certain level of performance to be considered reliable.

Most LAN wiring is put to use immediately after cable plant installation. Initial operation of the workstation is a good indication that the cable is all right, but it's not positive proof that the cable is in perfect condition—particularly if your workstations are still using 10 or 16 Mbps technology, and you plan to go to higher speeds. Also, it is not unusual for some of the station drops to remain unused until some later date. How will you know if those drops will work when you need them?

Would you have a light fixture installed and not try the switch? Probably not. You need to test each drop of your cabling system as well. The best way to test cable is to use proper test equipment that looks only at the cable performance. In that way, you avoid the nagging little network problems that obscure the real issues. Cable often gets blamed for problems that are really network hardware problems and bad software configurations.

The best time to do the testing is before furnishings and people have been moved in. At this point, all outlet jacks are accessible, without the need to move desks and other furniture. Any inspections and repairs that are needed can be done quickly and without disturbing anyone. Of course, modular furniture will have to be in place because the cable must be installed in the furniture modules before it can be tested.

TIA has offered TSB-72 and changes in 568-B as a means to provide an intermediate point for interconnection to modular furniture. See all the details in Chap. 11, "Open-Office Wiring."

Testing Installed Cable

The testing of installed cable links is often referred to as "certification" of the installation. Certification is normally performed by the installer and involves testing each individual cable link to a recognized standard. The end result of the certification is a report that shows the actual measurements of each cable link and the pass or fail determination.

A dilemma initially existed in providing a method of testing and certifying installed telecommunications cable. The original EIA/TIA-568 standard specifically stated that the performance standards for cable and connecting hardware did not apply to installed systems. The reason for this was that it was known that an installed cable run had performance that was below that of cable tested at the factory. As a matter of fact, a correlation problem seemed to exist between cable tested with a factory RF network analyzer and with the field test instruments. Whether the field testers gave actual measurements of the parameters or judged the cable on pass/fail criteria, there were differences that might cause factory-certified cable to fail a field test.

Some of the differences were to be expected, as the installed cable had several wire terminations, cross-connects, and modular connections that did not exist in the factory. Also, the standard factory test was (and is) to lay the cable out on a nonconducting surface (such as a concrete floor) or loosely coiled and use special test adapters to connect to a network analyzer. Cable in the walls is bent, pulled, tie-wrapped, run along metal pipes and steel beams, placed in conduit, and installed in all manner of nonideal conditions that affect its performance. You would expect it to perform differently. Moreover, there were no standards for accuracy for field testers and their performance varied considerably. No wonder there was no correlation.

Fortunately, the situation is much improved with the newer standards, such as TIA/EIA-568-B, TSB-67, and TSB-95. The performance of installed cable is specified in detail, as is the accuracy of field testers. Even the required tests, testing methods, and method of reporting the results are specified. The correlation of field tester performance with laboratory network analyzers is also made. All of this lends a wonderful consistency and reproducibility to the testing of installed cabling that gives a real confidence factor to cable plant certification.

Certification for installed cables may be done two standard ways, Basic Link testing and Channel testing. Both of these testing methods are defined in TIA/EIA-568-B, which builds on the "worst case" link model. As we saw in Chap. 2, a Basic Link covers the horizontal cable from the workstation outlet at one end to the initial point of termination at a cross-connect block or patch panel in the telecommunications room, including the test cords for the test equipment. The test connection at the telecommunications room termination of a horizontal run may require special adapters unless the termination is directly into a modular patch panel. For example, the test cord would have to terminate in a 110 adapter plug if that type of connecting block were used.

The Basic Link and the Permanent Link, the ISO/IEC's counterpart, are very similar, except that the Permanent Link excludes the test cables. It is likely that a more sophisticated version of the Permanent Link will replace the Basic Link in the future. (See Fig. 2-8 in Chap. 2.)

A Channel includes the actual equipment cords (user cords, patch cords, and cross-connect wire) that connect to the network (or telephone) equipment at each end. Keep in mind that the word "actual" means just that. Once you certify a Channel with a particular set of equipment cables, those cables are frozen in place, as far as the standard's Channel testing method is concerned. If you change cables, to use a longer or shorter cable for example, you must recertify the Channel.

The Basic Link was designed to allow a cable installation to be certified once the horizontal cables were installed by the installation contractor. As a matter of fact, the link was originally called the Contractor's Link. Naturally, the user and patch cords were not in place at this stage and would have been impossible to test. Nevertheless, the performance of the entire link (the Channel) is ultimately what the user depends upon, so this test was also included as an important benchmark of overall link performance.

Which link test should you use? Basic Link certification is probably sufficient in most cases of new installations. It will ensure that the 90 m (maximum) of horizontal cable is performing to specifications. The Basic Link test is probably the most reasonable performance testing standard for a turnkey cable system installation, because the vendor may have little or no control over the cables used to connect to the equipment. Also, the build-out stage of construction is the best time to correct any problems that are revealed by a Basic Link test. After the network equipment is installed and connected, it will be easiest and most useful to do Channel testing. The Channel test is obviously very useful in dealing with network equipment vendors. You can rely on your cable system completely when it has been certified as a Channel.

Performance Levels

Two types of performance levels should be considered in making field tests, the first being the performance category of the cable system. You should determine long before installation what category of performance you require and then test to that level. The second type of performance is the accuracy level of the field tester you intend to use to certify the cable.

We have discussed the cable performance categories at length in previous chapters. The three categories of operation specified by TIA/EIA-568-B and related standards are Categories 3, 5, and 5e and Category 6 is close behind. Other standards, such as some of the ISO documents, characterize similar performance levels, although with some differences. For example, the ISO/IEC IS-11801 specifies identical categories for uninstalled cable and components, but uses a class system for installed links. Table 16.4 shows how the TIA categories correspond to the ISO classes. In

TABLE 16.4

Comparison of TIA/EIA Categories and ISO/IEC Classes

TIA/EIA-568-B	Category 5	Category 5e	Category 6	Category 7*
ISO/IEC IS-11801	Class D	Class D	Class E	Class F*

*Category 7 and Class F are proposed extensions to the performance models.

most instances, the TIA categories are the same or tougher in test parameters and may be substituted with no problems. If you are in a situation that needs to reference these other standards, you should adjust your testing accordingly. While most of these standards on an international level are coordinated (also called *harmonized*) to use the same procedures, nomenclature, and limits, they may also contain deviations in language and some parameters that are appropriate in that region of the world. In addition, the standards bodies that promulgate these documents revise them on differing time cycles, and they may not be exactly synchronized to implement changes at the same time that the overall international community deems necessary.

TSB-67 identified two performance levels for field testers, Accuracy Levels I and II. Accuracy Level II was the toughest, and was justified when you intended to operate your cable drops at or very near the length or performance limits, including user cords and jumpers. Table 16.5 shows the minimum requirements for the two levels. These accuracy levels were provided to ensure that the test instrument can actually measure the cable link parameters that need to be certified. Accuracy Level II was intended to closely mimic the laboratory test equipment and turned out to be more than was needed for reasonable measurements.

A third accuracy standard is now specified in TIA/EIA-568-B, called Accuracy Level IIe. It allows for measurement to Category 5e.

It is difficult to say which type of tester you need for certification. The potential problems actually occur only at or very near the acceptable performance limits. In a cable plant that uses very high quality cable and components and that has relatively short cable runs, you may never approach the limits required for your category of operation. However, if you use low-quality materials and questionable installation practices, you may see a lot of links at the performance limits. The Accuracy Level II is difficult to achieve, as its intent is to very closely emulate the performance of general-purpose laboratory analyzers costing 10 to 20 times as much. Now that TIA has endorsed Level IIe accuracy, that is really all you need to properly verify link performance. For routine use, it may be sufficient

TABLE 16.5

Field Tester Accuracy Levels

Performance parameter	Accuracy Level I	Accuracy Level II	Accuracy Level II-e
NEXT accuracy:			
Basic Link	3.8 dB	1.6 dB	2.4 dB
Channel	3.4 dB	1.5 dB	4.4 dB
ELFEXT accuracy			
Basic Link			3.0 dB
Channel			5.0 dB
Attenuation accuracy:			
Basic Link	1.3 dB	1.0 dB	1.7 dB
Channel	1.3 dB	1.0 dB	2.5 dB
Dynamic accuracy	1.0 dB	0.75 dB	
Length accuracy	4%	4%	5% (5 m)
Propagation delay			25 ns
Delay skew			10 ns
Return loss	15 dB	15 dB	25 dB
Accuracy			2.1 dB Basic Link 4.6 dB Channel
Random noise floor	$50 - k'$ dB	$65 - k'$ dB	
Residual NEXT	$40 - k'$ dB	$55 - k'$ dB	
Output signal balance	$27 - k'$ dB	$37 - k'$ dB	
Common mode rejection	$27 - k'$ dB	$37 - k'$ dB	

NOTE: The constant k' is equal to $15 \log (f/100)$ and thus varies with frequency. Performance and measurements are not required below 75 dB for residual NEXT and for random noise floor, and are not required below 60 dB for output signal balance, dynamic accuracy, and common mode rejection.

to get an Accuracy Level I tester or one that does not achieve Accuracy Level IIe on all parameters. Some testers may offer different performance levels on Basic Links and Channels, because the Channel test requires the use of the 8-pin modular connector at the test interface. As a LAN manager, you may choose to get a less expensive Accuracy Level I tester for your own casual troubleshooting use, and depend upon an installation contractor for Accuracy Level II-e tests when necessary.

Testing Costs

Proper testing costs money over and above the basic cost of installing and verifying the cable. As we will see in the next section, there are many tools that may be used to test LAN wiring. Some of these tools are fairly simple testers that check continuity and wire map (the proper pin connections). Use of such tools should be considered a normal part of cable installation and included in the price. These simple tests verify the DC electrical integrity of the link and find gross failures that would prevent the cable from functioning even in simple uses. This verification catches wiring errors and can find cases where the cable has been severely damaged after installation. Verification, by visual inspection and continuity check, once was considered sufficient testing for telecommunications cables. This is no longer the case.

The demanding cable performance that is required by modern LAN applications can no longer be met by such simple testing methods. To truly test a LAN wiring system to meet one of the more rigorous performance categories requires the use of very sophisticated test equipment. Specialized field testers have been developed to reduce the cost of testing. These testers are far less expensive to purchase than the ultrasophisticated laboratory RF network analyzers (these test all sorts of electronic networks, not local area computer networks). Yet, the field testers yield useful results that closely match the measurements to the LAN category standards, produce quick pass/fail indications, and print full reports. These field testers are also very easy to use. They can run completely automated tests, store the results, and pinpoint problem locations on the cable under test. Most of the testers will operate for several hours on battery power, so you don't even have to find a power outlet.

Although these testers provide many advantages, they are still much more expensive than the simple wiring verification tools, and they require additional skill to operate properly. Consequently, you should expect to pay extra for having your cable "scanned" with one of these field testers. The cost of the tester, the skilled operator time, the preparation of reports, and the possible retesting should be considered when budgeting for the additional cable scanning step. You should allow a budgetary cost from 5 to 15% of the installation cost per drop to certify to Basic Link standards. The cost variation is to allow for special circumstances, such as an unusually difficult and expensive per-drop cost. If the scanning is done by a separate contractor, you should allow an additional factor for problem determination or visual inspection. Scanning is best done before personnel have moved into the office space.

Although scanning involves only a brief visit to each workstation outlet, labor costs will be higher if furniture (and people) have to be moved.

If your intention is to install a full Category 5e or 6 wiring system, then you need to plan for a complete certification scan. The component standards and installation practices are simply too tough to not test completely for proper performance. If you have an existing facility that is supposed to meet Category 5 and you have not previously tested it to 100 Mbps, then a new scan would be appropriate. To recertify Category 5 links for Gigabit Ethernet, you should test to the limits in TSB-95, as the full Category 5e limits should not be required. Prior to the release of the testing procedures and limits in TSB-67, standard practice was to use Category 5 cable and components, install using the best current practices, and test using whatever cable scanners were available at that time. The older EIA/TIA 568 standard specifically excluded installed cable and connectors from its scope.

Testing Methods

The Channel and Basic Link

The installed cable link components of a horizontal link are measured either as a Basic Link or as a Channel. The difference between the two is that the Channel includes the actual equipment cords (user cords), patch cords, horizontal cross-connect, and cross-connect jumpers that will ultimately connect to the LAN equipment at each end of the horizontal link. The Basic Link includes two test equipment cords, one at each end (excluding the connectors at the tester), while the cords at the ends of the Channel must be plugged directly into the test instrument interface with no allowance for test cords. The Basic Link and Channel are shown in Fig. 16.2.

As we mentioned earlier, the reason for the two types of links is to facilitate testing at both the construction phase and the postinstallation phase. It is unusual for network hubs and workstations to be available during the construction phase of an installation. It would be time-consuming and difficult to wait to detect and repair bad cable runs at some later time. Moreover, the cable installer may have little responsibility or influence for the equipment cords. This two-step method of cable link specification allows the installed "in-the-wall" cable to be tested as soon as installation is completed. The complete link, from network hub to workstation, can be further tested during installation of that equipment.

Figure 16.2

The Basic Link and
Channel are defined
for link testing by
TSB-67. (See Fig 2-8
for Permanent Link,
which is not yet
defined for testing
by TIA.)

a. The Basic Link

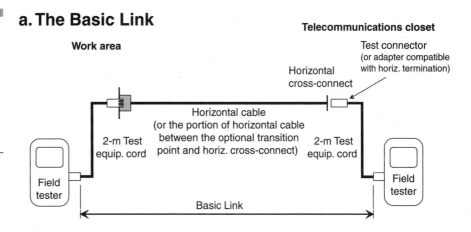

b. The Channel

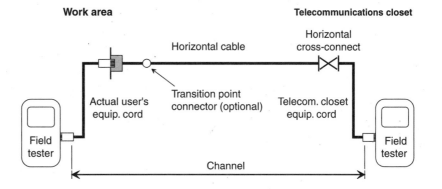

Equipment and Testing Requirements

The purpose of the standards in TSB-67 is twofold. The full title of the TSB is "Transmission Performance Specifications for Field Testing of Unshielded Twisted-Pair Cabling Systems." One explicit purpose, then, is to define cabling system performance after the cable has been installed. The other purpose, which is implied by wanting to test in the field, is to define field tester performance. Early field testers did not always produce consistent results when compared to ultraexpensive laboratory testing methods. So, the standard states which tests will be done, what measured performance levels are important, and what tester accuracy and reporting is required.

The testing that is required includes parameter measurement for the following tests:

- Wire map
- Length
- Attenuation
- Near-end crosstalk (NEXT) loss

As more is learned about the requirements of networks operating at the higher levels, additional tests may be added to the field test specifications, and others may be modified. For example, you will notice that impedance and structural return loss (SRL) that are measured on uninstalled cable (per TIA/EIA-568-B) are missing from this specification for installed links. This is not because it is considered unimportant, but because it is not yet known (or agreed) what levels are appropriate. On the other hand, the cable length, attenuation, and NEXT are currently tested. The length of the cable run is important in influencing attenuation and NEXT values, but perhaps the ratio (attenuation to crosstalk ratio, ACR) is a better indicator of overall system performance. Likewise, very high speed networks may care as much about the propagation delay as the physical length of the cable, yet transmission delay is not yet specified nor tested. These new tests could be a boon to transmission engineers but a bane to cable system designers and installers. Can you just image telling your installer to keep cable runs under 300 ns? Where do you get a tape measure that size?

Let's look individually at each of the tests that are now required.

Wire Map   The wire map tests checks for wiring errors and any significant cable faults, such as shorts or opens. It is a very good idea to check the reeled cable for shorts or opens before installation, as it will be very hard to prove that the defect was not caused during installation. You can also perform most of the other tests, as the reeled cable should easily pass them before it has been installed. Keep in mind that, if you want to check the NEXT, you will need to use a properly rated connector because its NEXT will often be significant compared to that of the cable.

The wire map test will check for:

- Continuity
- Shorts
- Crossed pairs
- Reversed pairs
- Split pairs

Figure 16.3 shows the correct pin-out/pairing and most of the common wiring errors. In the standard TIA/EIA-568-B wiring scheme, the wires are paired 1-2, 3-6, 4-5, and 7-8. The pair numbering, 1 through 4, varies slightly between the T568A and B wiring patterns. Color codes for the cable are assigned on a pair basis, so (as an example) Pair 2 will always be Orange and White but may be assigned to the 3-6 pins or to the 1-2 pins, depending upon wiring pattern. The tester has absolutely no idea what color the wires are! The technician will have to visually inspect them if the design requires that a particular pattern be used. However, if both ends of a cable or run are wired with the same pattern, the field tester will pass the cable. If not, it will show a crossed pair condition, as in Fig. 16.3*d*. The field tester also easily finds split pairs, shown in Fig. 16.3*f*, an impossible task for a DC continuity tester.

Figure 16.3
The correct pin-out/pairing and most of the common wiring.

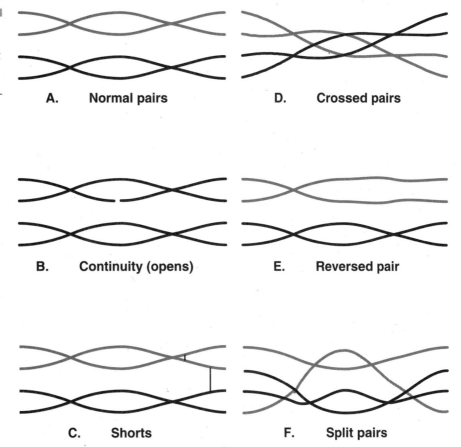

A. **Normal pairs**   D. **Crossed pairs**

B. **Continuity (opens)**   E. **Reversed pair**

C. **Shorts**   F. **Split pairs**

Length   Length measurements are made on two types of cable links, the Basic Link and the Channel. Recall that these two links differ in that the Channel includes the actual user equipment cables that will be attached to the hub, the workstation, or whatever. For the field tester, this means that the Basic Link must include the test equipment cords at each end, but the Channel must terminate directly into the tester (or somehow exclude the effect of test cables from the measurement).

The TIA/EIA-568-B standard requires a horizontal link to be 90 m (295 ft) or less, with an allowance for an additional 10 m (33 ft) for all user cords, jumpers, and patch cords. Because this horizontal link length cannot directly be measured electrically by the field tester, which must somehow connect to the link under test, an allowance is made for two 2-m test equipment cords. This means that the Basic Link is defined to have an acceptable length of 94 m. The Channel, which includes the user cords, can be as much as 100 m. From a practical standpoint, we would recommend that you use test cords that are exactly 2 m in length. If you use a shorter cord, the tester may pass a link that exceeds the limit. If you use too long a cord, you may reject links that actually are legal. Some testers require a test cord of at least 1 m. Never fear; because of individual cable variations in nominal velocity of propagation (NVP), the TSB allows an additional 10% before it declares a length failure (plus a factor allowing for the ± 1 m ± 4% tester accuracy, although pass/fail results in this zone must be marked with an asterisk).

An interesting distinction should be pointed out in regard to length. The physical length of a Channel or Basic Link is an important parameter. Physical length may be measured or determined from markings on the cable. Cable testers estimate electrical length from the propagation delay of the cable. Because propagation delay depends upon the physical characteristics, such as the twist pitch (lay length) and dielectric properties of each pair, it varies with different categories of cable and even with different lots of cable from the same manufacturer. To accurately correct for manufacturing variations in cable, it is very important to measure the propagation delay of each reel of cable installed. The cable pair with the shortest propagation delay is used.

The procedure for calibrating length on a field tester is to physically measure a length of cable from the reel to be installed and calibrate the tester through an electrical length measurement on that same sample of cable. The particular length of cable you use is not important, but it is important that its exact length be known. However, accuracy of the calibration is increased with a longer piece of cable. A length of 15 m (or about 50 ft) or longer is commonly recommended. Realistically, it would not be practical to cut and measure every reel of cable at a large job, but you should calibrate for each different cable lot, if possible. You might cut a length for one of your cable runs to use to calibrate the cable, so as to not waste the sample.

Testers are required to have a length range of at least 310 m, so that a 1000-ft reel of cable can be measured. However, until the cable on a particular reel is calibrated for NVP, this measurement should be considered only approximate. Stated lengths on reeled cable are not exact and should not be used exclusively to calibrate the NVP of a cable.

Attenuation The maximum attenuation for the purposes of testing the Basic Link or Channel has been based on the attenuation values given in TIA/EIA-568-B for horizontal cable, connecting hardware, and jumper or patch cords. It prorates the loss values, given in dB per 100 m, and applies them to the worst-case length of the link. Table 16.6 shows a summary of the allowable attenuation values for the Basic Link and the Channel at

TABLE 16.6

Attenuation Limits for Installed Cable Links, in dB

Frequency, MHz	Basic Link			Channel		
	Cat. 3	Cat. 4	Cat. 5	Cat. 3	Cat. 4	Cat. 5
1.0	3.2	2.2	2.1	4.2	2.6	2.5
4.0	6.1	4.3	4.0	7.3	4.8	4.5
8.0	8.8	6.0	5.7	10.2	6.7	6.3
10.0	10.0	6.8	6.3	11.5	7.5	7.0
16.0	13.2	8.8	8.2	14.9	9.9	9.2
20.0		9.9	9.2		11.0	10.3
25.0			10.3			11.4
31.25			11.5			12.8
62.5			16.7			18.5
100.0			21.6			24.0

NOTE: The Basic Link (sometimes referred to as the contractor's link) is comprised of the telecommunications outlet in the work area, the horizontal cable (which may include one transition, such as to under-carpet cable), the termination in the telecommunications room, and, necessarily, the test equipment cords at either end. The Channel is composed of the components of the Basic Link, excluding the test equipment cords, plus the cross-connect field or patch field to the user equipment in the telecommunications room, and the user cords at either end (which are connected to the test equipment in place of test cords). The test equipment connection is excluded in either case.

SOURCE: TSB-67.

selected frequencies. TSB-67 contains the formulas for calculating each of the components of Basic Link or Channel attenuation. The actual measurements should be made at intervals of 1 MHz or less in the appropriate frequency band, but the tester will take care of this automatically.

The attenuation test limit is for a worst case link. The actual attenuation of a measured link will be less if the cable run is shorter than allowed. However, in most testers, the measured loss is compared only to the limit for worst case. This means that the test will not directly catch cable that has greater than expected loss (based on the uninstalled cable limits of TIA/EIA-568-B), but the extra attenuation per meter should have no effect on link performance. If you wish to catch substandard cable, you will have to compare the actual loss measurement with the expected loss for that particular length of cable.

As with the specification for uninstalled horizontal cable, there is an allowance for the effect of temperature on the attenuation of the cable. The allowance is 1.5% (of the value in dB) per degree Celsius for Category 3 and 0.4% for Categories 4 and 5. To illustrate this, consider a cable installed in an attic, where the actual temperature is 30°C, rather than the nominal 20°C of the standard. The attenuation limit for a Basic Link at 10 MHz would be raised from 10.0 dB to 11.5 dB (an increase of 10.0 dB × 1.5%/°C × 10°C) for Category 3, but only to 10.4 dB (10.0 dB × 0.4%/°C × 10°C) for Category 4 or 5.

Two other environmental factors are known to cause changes in attenuation: metallic conduit and humidity. The difference expected for cable in conduit is up to a 3% increase, but the standard allows no latitude in test limits. It should be expected that long runs in close proximity to other metallic objects might also cause an increase in attenuation. You should keep this in mind in planning your design. The humidity effect is a new consideration. Some cable manufacturers have stated that the humidity effect is cumulative over time in PVC or plenum-rated PVC-compound jacketed cable and may cause cable that originally tested good to fail after a period of exposure to high humidity. The standard makes no allowance for the effects of humidity; the cable link must pass the attenuation limits regardless of the humidity. This might be an argument against using the less expensive PVC insulation or plenum-rated PVC-compound jacket and might also be a warning signal for some of the new copolymers that contain other thermoplastics mixed with the more expensive TFE compounds.

Near-End Crosstalk (NEXT) Loss   The theory behind twisted-pair wiring is that the twisting rotates the magnetic fields from the wire with the twist, and coupling to nearby pairs or other objects is minimized. However, some coupling does exist. How much coupling occurs between pairs of wires on the same cable is of particular interest. If too much signal is coupled from, say, the transmit pair to the receive pair of a connection, the receiver will not be able to distinguish the far-end signal from its own transmissions. This coupling is illustrated in Fig. 16.4.

Coupling between pairs of a cable is called *crosstalk*. Because the crosstalk between transmit and receive pairs is greatest when the transmitted signal is unaffected by attenuation, the worst case occurs at the same end of the cable where a test signal is transmitted. This is called the *near end* and yields the expression near-end crosstalk, or NEXT.

NEXT is measured in dB below the test signal. Technically, NEXT would be a negative dB number, but we turn it around by calling it NEXT loss. This means that a large NEXT number in dB is considered low NEXT, as the amount of crosstalk is much farther below the test signal. A high NEXT would be a smaller number in dB, indicating much less isolation from pair to pair. Very logical, n'est-ce pas?

NEXT must be measured in both directions on the cable link, to prevent the situation where one end of a cable passes NEXT and the other end fails. Single-ended measurements are often made in the telecommunications room, as it eliminates carrying the tester around to each workstation outlet. Excessive untwist at the connector is more often found at the workstation end, however, because the outlets naturally separate the conductors, and because the individual jack plates are more difficult to inspect. Also, the workmanship may be less because the supervisory control may not be as great as would exist in the telecommunications room. So if you only measure at one end, you are likely to miss more of

Figure 16.4
The NEXT measurement. If too much signal is coupled from the transmit pair to the receive pair of a connection, the receiver will not be able to distinguish the far-end signal from its own transmission.

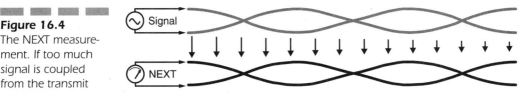

the high NEXT connections. The equipment at both ends require appropriately low NEXT to operate properly.

A related figure of merit for a cable link is attenuation to crosstalk ratio (ACR). Although not yet specified by the standards, ACR is a very good measure of how the attenuation and the NEXT can influence the performance of networking interface devices. Attenuation may be used to calculate the received signal strength of a signal transmitted from the far end. NEXT can yield the level of the interfering crosstalk signal from the near-end transmitter. The near-end receiver is thus required to cope with a signal-to-noise ratio that is basically the ratio of the two signals, the attenuated far-end transmission and the near-end crosstalk. Since both attenuation and NEXT are directly measured as part of the testing, it is very simple to ratio the two to produce the ACR figure. The higher the ACR number, the better the expected performance. Since a ratio of two logarithmic quantities is their numerical difference, the ratio is simply:

$$ACR = |\text{attenuation (in dB)} - \text{NEXT loss (in dB)}|$$

One of the things that concerns network hardware designers is that the ACR decreases steadily with frequency. This means that at 100 MHz, the ACR margin may be very small indeed. Thus, it may be of benefit to those planning to eventually go beyond 100 MHz to purchase cable with the highest ACR possible.

Reporting Pass and Fail Criteria

An important part of TSB-67 is the carefully detailed requirements for the reporting of test results. In general, the standard requires that each measured result be reported in units appropriate for the test and that a Pass or Fail determination be reported. Measurements, such as for attenuation or NEXT, that might be very close to the limits are reported with an asterisk in addition to the Pass or Fail.

A data report is simply the visual indication of the field test set. In all of the modern cable scanners, the visual display is a liquid crystal display (LCD) that also allows the display of some graphics. Of course, it is important for a permanent record of the report to be made, so the testers also allow the test data for each cable to be stored and later printed out. Figure 16.5 shows a typical cable scanning report.

Figure 16.5

A typical cable scanning report. (*Courtesy of Datacom Technologies, Inc.*)

```
                    LANcat V  CABLE CERTIFICATION REPORT #3
                            Company Name Here

Circuit ID: H7/B6-61                    Date Tested: 06/24/1996
Cable Test Standard: CAT5 Basic Link    Module Type: Modular Plug
Location: _____       Remote Module Type: TwoWay Remote
                                        Serial Number: 9510704  V2.41D1
                                        Cable NVP:  72.0%
TEST SUMMARY: PASS

Wire Map: PASS  Near End   1    2    3    6    4    5    7    8     Shield
                           |    |    |    |    |    |    |    |     open
                Remote End 1    2    3    6    4    5    7 .  8
```

Length: PASS	Limit: 308ft		
	Pair	Length	Comment
	1,2	3ft	PASS
	3,6	3ft	PASS
	4,5	3ft	PASS
	7,8	3ft	PASS

Attenuation:	PASS	1,2	3,6	4,5	7,8
		PASS	PASS	PASS	PASS
Attenuation	dB	1.9	1.8	2.0	2.3
Limit	dB	19.8	20.0	21.6	20.1
Margin	dB	+17.9	+18.2	+19.6	+17.8
Frequency	MHz	86.4	87.3	100.8	88.2

Local NEXT:	PASS	12/36	12/45	12/78	36/45	36/78	45/78
		PASS	PASS	PASS	PASS	PASS	PASS
NEXT	dB	43.7	44.5	76.5	36.4	45.5	43.2
Limit	dB	31.0	29.3	60.0	29.4	32.2	29.3
Margin	dB	+12.7	+15.2	+16.5	+7.0	+13.3	+13.9
ACR	dB	41.9	42.5	76.5	34.4	43.8	41.9
Frequency	MHz	79.2	100.1	1.1	99.0	66.8	100.1

Remote NEXT:	PASS	12/36	12/45	12/78	36/45	36/78	45/78
		PASS	PASS	PASS	PASS	PASS	PASS
NEXT	dB	42.6	41.5	83.5	34.9	45.4	43.8
Limit	dB	29.5	30.9	60.0	29.3	32.9	29.3
Margin	dB	+13.1	+10.6	+23.5	+5.6	+12.5	+14.5
ACR	dB	41.5	39.9	83.5	32.9	43.9	42.5
Frequency	MHz	97.6	80.3	1.1	100.1	60.5	100.1

```
Operator: _____       Date: _____

Comments:
```

The wire map tests are reported simply as Pass or Fail. The length measurement is performed on all pairs of the cable under test, but the pair with the shortest electrical delay is used for the Pass/Fail determination and its electrical length (in feet or meters) is reported. An additional 10% in length is permitted to allow for uncertainty in the NVP of the cable.

The tester must make a determination of attenuation Pass/Fail based on a calculation of the allowable attenuation at each frequency of measurement. If the cable fails, the attenuation at the frequency (or highest frequency) of failure is reported. Measurement steps are to be no greater than 1 MHz intervals up to 100 MHz for Category 5, or the maximum frequency for lower categories of cable. Attenuation values of less than 3 dB are reported but are not used in a Pass/Fail determination. An optional determination of the attenuation per unit of length is calculated for cables longer than 15 m and identified if they are greater than that expected. However, higher than expected attenuation per unit length values do not cause a failure to be reported, but might be a cause for concern.

The NEXT loss measurement is made in steps of 150 kHz from 1 to 31.25 MHz, and 250 kHz from 31.25 to 100 MHz. For the NEXT test, the measurement of the worst pair combination is used. The tester may report either the actual NEXT or the NEXT margin, although NEXT margin is required for failure reports or when the measurement is within the tester's accuracy from the limits. NEXT margin is a value of much significance to cable installers because it indicates when marginal installation practices may be corrected by retermination or rerouting. Often NEXT can be reduced by maintaining the twist of the pair as close to the point of termination as possible.

Test reports must indicate whether the cable under test has passed or failed each requirement. Because of the difficulty in establishing an absolute measurement near the performance limits for the cable link, a cable may be falsely accused of failure or improperly passed. The standard takes a clever way of dealing with this problem—essentially including the tester's accuracy as if it were a part of the cable under test and allowing an additional 10% of the horizontal length limit to allow for NVP uncertainty.

A Fail indication, thus, is a sure indication of a parameter that is outside the limit. A Fail* indicates that the tester measured a value that is outside the test limit but by an amount within the tester's accuracy. A Pass indication is assumed to be adequately below the limit, with the accuracy of the instrument taken into consideration.

However, a Pass indication that results from a test measurement that is within the accuracy level of the actual limit is marked with an asterisk (Pass*). The link is still considered to pass the test but is a handy visual clue to which links may be at or near the absolute limits. Prudent installers might want to take a second look at these links to see if shortening the length slightly or more carefully controlling untwist might produce a full Pass. Because the test-reporting procedures allow an additional length variation of 10% in measurement of the electrical length, to account for NVP variations, you should really respect the length limit of the tester.

Occasionally the source of a Fail indication will be the "short link" problem described earlier. Certain links under about 15 m may exhibit NEXT failure due to the resonance phenomenon. These links may actually be "cured" by replacing the run with a longer length of cable. The evaluation of such links is under study and will, no doubt, be the subject of a revision to TSB-67 or TIA/EIA-568-B at some future time.

Test Equipment

The proper use of test equipment is one of the most important aspects of a successful LAN wiring installation. Test equipment may be subdivided into two types: basic troubleshooting equipment and diagnostic/measurement equipment. Basic troubleshooting equipment consists of devices that check continuity, DC resistance, and simple wire maps. Diagnostic/measurement equipment is used to measure cable parameters, determine compliance with standards, and diagnose specific cable problems.

Which type of test equipment should you have? The answer to that question depends on the type of work you need to do. If you do any amount of cable installation or troubleshooting, you need most of the basic test equipment items described below. If you need to test or certify cable to cabling standards, you will need a field tester capable of meeting the accuracy requirements of the category you are testing. Fiber optic cable requires special test equipment, even for routine tests. For a fiber optic installation, you should have fiber attenuation measurement equipment, at the least, and you may need more sophisticated scanning equipment to locate fiber faults.

How much money should you spend for test equipment if you are the network manager? A good rule of thumb is to expect to spend 2 to 5% of the value of your networking equipment (including workstations and file servers) for test, analysis, and monitoring equipment. As an alternative, you could allow 10% of the value of just the cable, hubs, and bridges for test equipment. The point is that you need to have a reasonable capability to test and troubleshoot your network. Without the proper equipment, you are flying blind. You will suffer from periodic cases of "equipmentus substitutis" (as evidenced by lots of running around while carrying computers, hubs, and cables). To avoid this syndrome, simply obtain the necessary test equipment to simplify and speed your troubleshooting. You also need a certain amount of test and monitoring equipment to verify the health of your network and to spot trends that may have a negative effect on performance in the future.

The rest of this chapter is devoted to the essential test equipment you will need and the methods to use in employing the more sophisticated testers. In addition to the physical layer equipment we show here, you may also have need for LAN protocol analyzers that deal with the higher layers of networking. These protocol tools will help you answer the age-old question, "OK, if it's not the cable, then what is it?"

Continuity Testers (Voltmeter, Continuity Tester)

The most basic item of test equipment that should be in your toolbox is a good electronic voltmeter. This voltmeter should be able to read DC volts, AC volts, and ohms. It should be durable and have an easy-to-read display. A "hold" function is also nice, as it allows you to make a measurement and later read the display. This is great for those times when you are deep inside a punchdown block or in an area with low light. Many voltmeters also have a continuity function with an audible beep (some of us consider this a mandatory requirement).

Voltmeters can make quite a few basic tests. For example, you can check continuity and DC resistance on a new reel of cable. If you know the resistance of a full reel, you can easily estimate the remaining length from the DC resistance value. With test adapters, you can check continuity of a jack, path panel, or patch cord. You can make a few simple adapters, wrap-around plugs, and break-out cables to help you use the meter for continuity and voltage measurements. You can also check for potentially dangerous voltages that may exist on cables run between telecommunications rooms or between buildings.

A type of voltmeter that is particularly handy in the field is the probe-style digital voltmeter. This type of meter, shown in Fig. 16.6, has the entire meter circuitry and display in one oversized probe. The other probe is connected via a short wire to the probe/meter. With this type

Figure 16.6
This type of volt-ohm meter has the digital display in one over-sized probe.

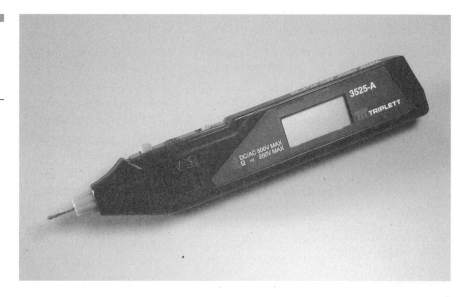

of voltmeter, you do not have to find a flat surface for the meter—it is right in your hand.

Beeper and LED continuity testers are also available and may be effectively used to test wires and connections. Continuity testing merely verifies that an electrical contact has been made, and does not really measure how good that contact is. However, it is rare in these systems to get an "ohmic" contact (one that has a significant resistivity). Since IDC connectors are used almost exclusively these days in LAN wiring, the metallic wire is scraped clean of contaminants during the termination process and the actual point of contact remains virtually gas-free, eliminating oxidation of the contacting surfaces that would lead to increased resistance. Dusty and corrosive environments are an exception to this rule.

Cable Wire Map Testers

A very good device to test an installed cable link is the wire map tester. These little testers give an indication of cable continuity and proper pin-out. They are generally used for installed cables or assembled cords. One type has a main tester and a remote unit. The two units communicate across the cable to perform a wire map test and, of course, a basic continuity test. Most of these testers can easily find all of the standard connection errors and problems, such as opens, shorts, reversed wires, and crossed connections. (These cable faults are all explained in the next section.) Simple miswirings are probably the most common cable installation problem, and the easiest to fix.

The simple wire map tester performs a simple DC continuity test and checks for undesired connections. It does not check for proper pairing, as with a split-pair fault. A *split pair* is a condition where one of the two wires in one pair is accidentally exchanged for one of the wires in another pair. For example, the white-green and the white-brown might be exchanged. This actually means that two pairs are mixed and the balance and self-shielding properties of the paired wires is lost. In this type of miswire, DC connectivity tests fine and it would pass a simple wire map test, but the cable would fail if tested for AC signal balance by a cable scanner. In many instances, a cable with a split-pair fault will cause a LAN link to fail, which is why it is important to test for this with the more sophisticated scanners.

A specialized version of the wire map tester is used to test assembled cables. This cable continuity tester has two modular jacks where each end of the cable is plugged in. When a button on the tester is pressed, LEDs light to indicate continuity of each wire in the cable, one by one.

A Pass is indicated by observing that the LEDs each light up in order through the number of pins in the cable.

This type of tester may be used for 6-pin modular connectors in addition to the 8-pin connectors used for LAN wiring. Frequently the same test jack is used for both sizes of connectors. This creates a potential false-failure problem for pins 1 and 8 of the 8-pin plug. Although the 6-wire plug will fit into the jack of the tester, the plastic sides of the plug may permanently bend the jack's first and eighth pins, making them no longer contact Pins 1 and 8 of an 8-position plug and giving a false indication of failure. The solution is to never use 6-pin plugs in a tester you intend to use for 8-pin plugs! If you believe a false failure has occurred, you can usually form a small stiff wire (such as a paper clip) to bend the jack wires back into position. Then test the cable again.

Cable Tracers

How do you find a single cable in a bundle of cables? How can you quickly verify that an outlet is properly marked? The answer is to simply use a high-impedance tone generator and an inductive cable tracer, shown in Fig. 16.7. These are two of the handiest tools around, and no serious cable installer should be without them.

The tone generator was originally developed by the telephone industry to assist its installers in tracing wires into and out of the distribution frame. Any unmarked jack could quickly be traced by placing a "tone" on the jack and using the telephone test set (commonly called a *butt-in set* or just *butt set*) to find the tone. The tone's distinct warbling sound could be easily distinguished from other sounds on the telecommunications wiring. However, the butt-in set had to be physically connected to one of the cable's wires to hear the tone. Something else was needed.

That something is called an inductive pickup. This device uses a coil and a probe point to magnetically couple to a cable so that you can pick up the tone without having to make electrical contact with any of the wires. There are two types of inductive cable tracers: the passive pickup and the amplified pickup. The passive pickup contains only a pickup coil. Often called a "banana probe" because of its shape, this tracer requires the use of a butt-in set to hear the tone. The amplified probe contains a battery-powered amplifier in addition to the pickup coil. A momentary contact switch conserves battery power when the pickup is not in use. The amplified probe can be quite sensitive, as long as the area is electrically noise-free. However, certain cable areas may generate too much electrical noise for the amplified probe to function effectively. This is

Figure 16.7
A cable tracing tone generator and inductive amplifier. (*Courtesy of Progressive Electronics, Inc.*)

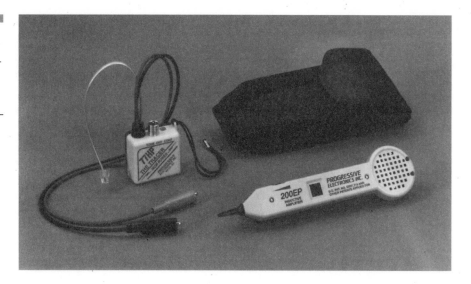

particularly true of "live" telephone circuits, where the residual tone will be faint at best. Many installers carry both types of probes to the job.

Cable Scanners

The advent of structured wiring standards has increased the need for very sophisticated portable field test sets that thoroughly test a cable run. These test sets are often called *scanners* because they use time-domain reflectometry (TDR) techniques to scan along the cable's length for anomalies. Reflectometry is somewhat akin to a radar for cable. A pulse is sent along a cable pair and then the cable pair is monitored to see what comes back (and precisely when it does). The type of reflection that returns is related to whether there is a short, an open, a termination, or a deformity along the cable. Of course, a cable may have several conditions that reflect part of the pulse. The length of time that it takes for a pulse reflection to return is directly related to the distance from the test set. A simple formula defines the distance to the reflection:

$$d = tkc/2$$

where d = distance
k = ratio of the medium's nominal velocity of propagation (NVP) to the speed of light, expressed as a decimal fraction.
c = speed of light (units/second)
t = time (seconds)

The units of the distance will be the same as the distance units of c. In other words, if c is in meters per second, distance will be in meters. The NVP is the decrease in actual propagation of the pulse from the ideal free-space speed of light, when traveling through the cable. The NVP is alternately quoted as a decimal fraction or as a percentage. It ranges from about $0.6c$ to $0.9c$ for electronic cable. Dividing by 2 is required because the time interval is a round trip, but we are only interested in the one-way distance.

Fortunately, all of these portable scanners are computerized, so they make the calculations for us. Typical cable scanners are shown in Figs. 16.8 to 16.10. Scanners can automatically check the wire map for wiring faults, measure NEXT and attenuation, and check the cable impedance and return loss. Cable scanners also check for wire length, to ensure it is below the appropriate limits.

The TIA/EIA-568-B, TSB-67, and TSB-95 standards specify the types of tests that must be made, the way in which the reporting must be done, and the accuracy levels of these field test sets. Accuracy levels, reporting requirements, and testing methods are explained earlier in this chapter.

Many cable scanners have an additional LAN traffic function that is very useful to the LAN manager. This function allows the passive monitoring of a LAN connection to measure traffic and gather some basic LAN statistics. Most of these testers, for example, can determine LAN utilization (expressed as a percentage). Some also measure other LAN traffic parameters and show potential error conditions such as collisions. Even though this goes beyond the cable testing and certifications for a generic cable system, it can be a very useful feature to monitor the basic network functions.

Analog versus Digital Field Testers

A difference exists between two measurement technologies for high-performance cable scanners. The two techniques are generally called "analog" and "digital." The analog testers use a conventional continuous waveform to test such link parameters as attenuation and near-end crosstalk (NEXT). The frequency of the waveform is changed (stepped or swept) and the measurement repeated at each of hundreds of discrete frequency steps that are required for full bandwidth testing. The digital testers measure these parameters by generating a series of pulses, then using digital signal processing techniques to derive the same parameters at all step frequencies more or less simultaneously.

The testing requirements of TIA/EIA-568-B have been drawn from TIA/EIA TSB-67, *Transmission Performance Specifications for Field Testing of Unshielded Twisted-Pair Cabling Systems*. They push the performance limits of field tester technology. The TSB defines strict accuracy levels for the testers and defines two test configurations for testing a horizontal cabling link, the Basic Link and the Channel. In the bulletin, swept/stepped techniques are used to illustrate the measurements required of field testing for cable links.

Figure 16.8
A cable scanner for Category 5 testing. (*Courtesy of Datacom Technologies, Inc.*)

Figure 16.9
A cable scanner for Category 5, 5e, and 6 testing. (*Courtesy of MICROTEST.*)

However, the TSB allows other methods "using frequency domain or time domain [digital/pulse/DSP] measurements" that are equivalent to the swept frequency [analog] methods.

In addition, TSB-67 follows the interesting practice of excluding the actual connection (for example, the plug and jack) that attaches the tester to the link under test. In the case of the Basic Link, this connector is not specified, as it is at the tester end of the 2-m (maximum) test cable. Some tester manufacturers use a high-performance, low-NEXT connector at the tester so as to not introduce adverse levels of NEXT into the link under test. However, at the Channel interface, the actual user cable must connect to the tester at each end. The user cable must be terminated in the standard 8-pin modular plug, which unfortunately has relatively poor NEXT performance.

The most intriguing claim of the digital tester advocates is that the pulse/DSP technique allows them to totally exclude the contribution to NEXT and other parameters that result from the modular connector (mating plug and jack) at the tester itself. The modular connector is the required test interface for a Channel measurement, so an alternative low-NEXT connector cannot be used, as it could for the Basic Link. Remember that a very significant portion of the detrimental crosstalk occurs at this modular connector interface. Digital signal processing can actually exclude this connection measurement, unlike conventional analog methods.

Which type of field tester should you choose? Sorry, but we are going to leave this decision to the reader! The technical positions of each tester manufacturer are strong and emotions run high. There has basically been an agreement to disagree. You may wish to look at several different models and make your own judgment. Frankly, the many testers on the market actually differ in features, options, reporting format, price, and measurement speed in addition to the measurement technique. Either style of tester is acceptable, both to the TIA working committees and to qualified independent evaluators. The bottom line is: Choose the specific tester that will do the best job for you.

TDRs/OTDRs

Although cable scanners use a form of time-domain reflectometry to perform the important certification measurements, a traditional TDR may provide additional insights into cable operation. Expensive laboratory TDRs may not be needed, as some of the field testers mentioned in the previous section now incorporate TDR graphics in their instruments. The actual graph of a cable's reflections can reveal many things about the handling and installation of the LAN wiring that simple Pass/Fail reports and raw measurements obscure. A typical TDR screen is shown in Fig. 16.11.

Optical fiber has a more sophisticated testing requirement than does metallic cabling. The same low-loss characteristics that make fiber optics important in transmitting longer distances make loss measurements

Figure 16.10
Another cable scan-
ner for Category
5, 5e, and 6 testing.
(*Courtesy of Fluke
Networking.*)

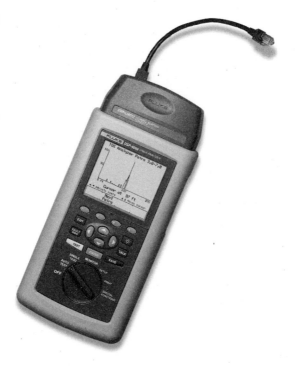

critical to ensuring fiber cable performance. Optical cable is also subject
to impairments from bending and stress that cause additional signal loss.
It is often much more expensive to replace an entire run of fiber than
simply to find the location of the fault and repair it by splicing the
cable, particularly if outside plant cable is involved. The problem is how
to find the fault.

An optical version of the TDR, called an *optical time-domain reflectometer*
(OTDR) is the solution. This device operates in the same manner as the
conventional TDR but uses a pulse of light, rather than an electrical
pulse. Modern OTDRs display a graphic that shows the signal return of
the fiber along its length. By moving a cursor, you can pinpoint the distance from the source to any anomaly. Then the area of the fault can be
located by direct measurement. If the fault is obvious from the outside
of the cable, the bad fiber may simply be cut and spliced, via fusion or
coupling. Otherwise, it may be necessary to replace a section of cable, similarly spliced.

OTDRs may provide built-in connections for both single mode and
multimode fiber. Since the TIA/EIA-568-B standard recognizes both
fiber types, both will need to be tested in large systems. The standard

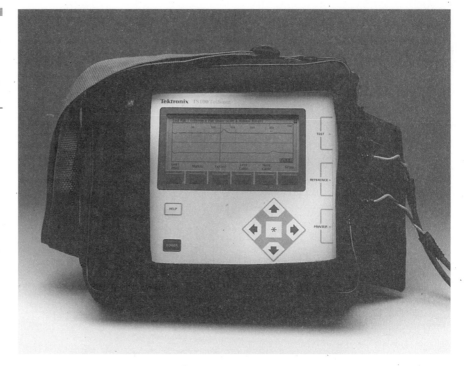

allows several types of fiber optic connectors, including the new SFF
types as well as traditional types such as ST and the 568SC, for use
with both fiber types. Single mode fiber requires special laser-type
sources to couple light effectively into the fiber. Since part of the measurement involves optical loss, it is essential that the coupling be to the
proper type of source. Adaptation to different styles of fiber optic
connectors may be required.

Basic Fiber Optic Testers

Optical fiber may also be effectively tested with simpler equipment.
Very basic fiber test equipment couples light into one end of a fiber
link and detects it at the other end. Some of these testers use visible
light, so you can theoretically see the "light at the end of the tunnel."
However, some higher-power links may have enough infrared light
output that it could be dangerous to the eye. Since the infrared light
is invisible, you should never look directly into a fiber from any close

distance. A good practice with one of these testers is to use a detector or to observe the light, if visible, on another surface, such as a sheet of paper.

Optical loss measurement may also be used to test optical fiber. This equipment measures simple optical loss over a fiber link by placing a calibrated light source at one end of the cable and a sensitive optical power meter at the other end.

17

Monitoring and Administering LAN Wiring

Chapter 17 Highlights

- Monitoring methods
- Physical cable monitoring
- SNMP monitoring and control
- Remote control
- LAN system documentation
- EIA/TIA 606

A LAN wiring system is a dynamic entity in many ways, changing and growing much like a living thing. If the LAN operating system is like the life blood of your network, the signal wiring certainly is like the nervous system, carrying important messages from point to point. LAN wiring must be monitored and tracked, so that you can constantly know the status of your interconnections.

LAN wiring systems are not truly passive systems. They are active in the sense that they can change characteristics over time and can disrupt the LAN if they fail. When a failure occurs, you will need to make quick repairs. How well you are able to respond to a cable problem depends upon the methods available to you to isolate the problem and find the offending cable system component. Proper documentation of your wiring will assist you in your installation and troubleshooting efforts. This is the essence of monitoring and administration.

Monitoring Methods

Monitoring of a cable system can be accomplished either directly or indirectly. For most installed LAN systems, the networking equipment (servers and workstations) can do a part of the job indirectly. We call this "indirect" monitoring, since it is a part of the normal operations of the connected systems, which sense the status of the network links indirectly. If a workstation cannot connect to a server, it may mean that there is a connectivity problem, such as a bad cable or other wiring component. Unfortunately, indirect monitoring of the channel may indicate a communications problem even when the cable connectivity is fine. For example, a bad workstation network interface or a software problem at either the workstation or server would appear as a network communication problem, even though the cabling was good.

"Direct" monitoring uses the capabilities of the hubs, patches, or network interface cards (NICs) to directly monitor the cable link. This type of monitoring senses the condition of a cable and allows direct control and/or testing of the channel. Direct monitoring may be either local or remote. Local monitoring simply uses the built-in status indicators on hubs and NICs. To monitor link status, one simply looks at the LED indicators on the devices. Of course, this assumes that the devices you are using have indicator lights.

Remote monitoring uses intelligence built into the hubs or other network devices to report status to a monitoring station. Remote monitoring can use standard network monitoring software, such as SNMP, described later in this chapter. Direct monitoring is rather new and offers an exciting new view of the LAN wiring as a separable entity. It is, unfortunately, still somewhat too expensive to implement on the low-end networking devices. However, as network speeds increase, and as the costs of networking equipment decrease, it will become more sensible to use monitoring devices to determine the health of the wiring system.

Physical Cable Monitoring

Physical monitoring of installed LAN wiring presents some very special problems. Most of the simple ways that can be used require that the monitoring be performed by the connected LAN devices (such as Token-Ring or 10/100 BaseT hubs). The hub monitors the condition of the link and reports it by indicator lights on the front panel of the hub, or by sending messages to a remote monitoring station. These devices and their capabilities are specific to the type of LAN you are using.

A Token-Ring MSAU, for example, uses DC connectivity on the link to set or reset a sensitive relay in the MSAU. Even a simple MSAU can easily provide an LED indicator on the front panel to indicate link condition. Intelligent MSAUs are able to report the status of any lobe to a remote monitor. Of course, the indication shows more than just the integrity of the cable, since the workstation operation is also a factor in lobe status. Nevertheless, this is an effective means of physical monitoring.

Likewise, the 10BaseT and the 100BaseTX Ethernet twisted pair systems provide a positive indication of link operation. Standard 10/100BaseT actually only evaluates the integrity of the received signal from the other end of the link, but enhanced systems can indicate integrity of the complete link and report to a remote monitor.

Some monitored patching systems can provide connectivity monitoring with virtually any type of LAN. These systems typically use pairs on the cable that are unused by the LAN. For example, 10BaseT, 100BaseTX, and Token-Ring use only two of the pairs on the cable. The other two are available for other purposes, in this case connectivity monitoring. While this does not truly indicate that the network pairs are connected, it certainly will indicate when a patch cable is disconnected. Some systems can even detect where a patch cord has been moved to and report an alarm to a monitoring station. The monitored patch system in Fig. 17.1 is an example of this capability.

Figure 17.1
The monitored patch
system. (*Courtesy of
RIT Technologies Ltd.*)

SNMP Monitoring and Control

Effective remote access to network devices may be provided by using the
Simple Network Management Protocol (SNMP). This protocol commu-
nicates over the network to a management workstation via an extension
of the popular TCP/IP suite. Monitoring of all sorts of hubs, bridges,
routers, and other devices can be accomplished by using SNMP. SNMP
management can be expanded to almost any device with a network connection. Some devices that have no native LAN ports, such as T1 communication equipment, are now being equipped with a LAN port just
for an SNMP interface.

SNMP-based monitoring and control techniques may also be
applied to cable management. Particularly with hub-based cable
management, SNMP is a welcome tool for both management and
control. Some of the intelligent patch panel equipment can communicate via SNMP for both monitoring and control. As networking
hardware becomes more complex, global network management solutions will become more important. Hopefully, these management
tools will offer some degree of cable management along with their
other functions.

Remote Control of Physical Connections

In some circumstances, you may have need for remote control of your physical network connections. For example, you might want to disconnect or reconnect a workstation from your network monitoring location. This can be effective in preventing a single workstation from causing a global network problem.

Many intelligent hubs and similar devices can provide this remote control capability. The most common method is to use in-band signaling via SNMP or a proprietary protocol. However, this method does have one disadvantage. If your network is being jammed by an offending workstation, it may be impossible to communicate over the LAN to command the hub.

Another remote control option is an out-of-band control system using a control port. This control port remains free of any network disruptions and can keep you in the game. A control port can also be used to make a modem connection to a remote system, regardless of the network connectivity.

Some remote-controlled equipment can perform other monitoring functions such as environmental monitoring and access control. A remote temperature alarm could alert you to a condition that might eventually cause the devices in an equipment room to fail. Or you could simply have a remote system start the coffee before you arrive.

LAN System Documentation

Good documentation is the key to controlling your LAN wiring environment. Whether you are troubleshooting a LAN outage or simply adding a cable drop, having well-organized records will save you a great deal of time and trouble.

The systematic marking and identification of all of the components of your LAN wiring system is an essential part of good documentation. Each item of the LAN wiring should be clearly marked with an identifying number. This applies to each cable, each termination, and all of the points in between. As a matter of fact, many of the mounting structures and pathways that are used in support of LAN wiring are best identified as well.

In an attempt to standardize the administration of telecommunications cabling, the EIA/TIA has published EIA/TIA-606, *The Administration Standard for the Telecommunications Infrastructure of Commercial Buildings.* This standard covers the identification, nomenclature, and documentation of wiring, pathways, and grounding/bonding in commercial buildings. It also suggests a system of drawings, work orders, and computerized record keeping to help manage the wiring system after initial installation. Figure 17.2 shows some of the wiring system elements that are expected to be identified and recorded. EIA/TIA 606 is quite detailed and specific, going far beyond the norm in some areas. However, it provides a very useful documentation and administration plan from which to model your record keeping.

The guiding rule of a proper LAN documentation system is "mark everything." That is, systematically identify every outlet, cable, termination block, patch panel, cable pathway, and telecommunications space. Then, record those identifications on drawings and in a database for quick reference.

If you do not carefully manage your LAN wiring installation, the constant pace of changes to the system will soon disorganize things so much that you will be forever lost in the cabling maze. The extent to which you need to implement the full complexity of the EIA/TIA 606 system depends upon the size of your cable plant. Large systems need much more careful documentation than small systems (under 50 stations). As a minimum, the wiring outlets and cable terminations (patch panel jacks or punchdown blocks) need to be given identifying numbers. It is also a good practice to permanently label the ends of each cable run.

Telecommunications outlets should be clearly numbered. The standard method is to place a label on the faceplate above each jack. Many manufacturers have responded to this requirement with special faceplates that have a recessed label area below a removable protective plastic cover. An example of these faceplates is shown in Fig. 17.3. Patch panel jacks should also be labeled on the front of the panel. Labeling on the back is optional, but is a very nice feature for troubleshooting.

Patch cords that run from cable terminations at the patch panel to the hub equipment ports are also numbered to make it easy to find which station cable corresponds to which hub port. Patch cords should be numbered with a removable label, if possible, as they are intended to be moved around. This practice makes it easy to identify the proper jack location for each cable end. Perhaps, in a very small installation, you will be able to maintain a one-to-one correspondence between the two patch

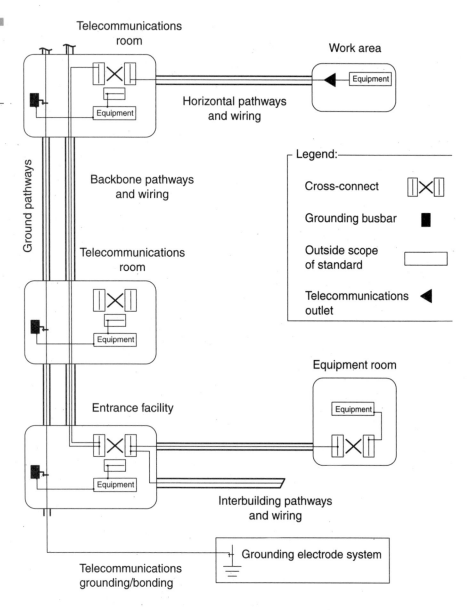

Figure 17.2
Many wiring system elements are expected to be identified and recorded.

jacks, but a larger facility will soon lose this neat arrangement. There is nothing worse than a loose patch cord end hanging down with no idea where to plug it back in.

Large installations will benefit from the labeling of cable pathways, such as ducts, sleeves, cable trays, and conduit. Raised floors, while not directly marked, can also be identified by drawings showing grid numbers for each

floor area. Work area and telecommunications spaces should also be identified, with a marking at the entrance to the area. For many buildings, this requirement is easily handled by existing room numbering.

Additional information is recorded along with the identification number of each element in the wiring system. The exact information required varies with the type of item, but it always includes the identifier and the item type. Additional information may be needed as to the numbers of available positions and whether any are damaged (as in bad pairs in a multipair cable). The administration standard also requires that the information recorded for each wiring element be cross-linked to any of the records that are related to that item. For example, a cable identifier must be linked to records of termination positions, splices, pathways, and grounding. A list of the item types, information to be recorded, and record links is shown in Table 17.1.

Automated database documentation systems can also be very useful. Some cable administration software has a graphical interface that lets you point to items in your cable plant and see the pertinent information that is recorded. A cable administration system can really be operated apart from the network management system (NMS) for your LAN. NMSs tend to concentrate on network node and hub management and control and are not concerned with what cable duct a particular pair might be located in. In addition, the cable administration software can cover telecommunications cable uses that are far beyond the LAN.

TABLE 17.1

Items That Must Be Recorded in the Cable System Documentation

Record	Type	ID?	Type	Other info:	Linkage to records for:
Wiring spaces	Pathway	Y	Y	Fill, loading	Cable, space, pathway, grounding
	Space	Y	Y		Cable, pathway, grounding
Wiring	Cable	Y	Y	Unterminated, damaged, available pairs	Termination position, splice, pathway, grounding
	Termination	Y	Y	Hardware, Damaged position numbers	Termination position, space, grounding
	Position	Y	Y	User code, cable pair numbers	Cable, term. position, term. hardware, space
	Splice	Y	Y		Cable, space
Grounding, bonding	TMGB	Y	Y	Grounding conductor, resistance to earth, date of measurement	Bonding conductor, space
	Bonding cond.	Y	Y	Conductor	Grounding, pathway
	TGB	Y	Y		Bonding conductor, space

Source: EIA/TIA-606

18

Troubleshooting LAN Wiring

Chapter 18 Highlights

- Common failure modes
- Wire faults
- Category and workmanship
- Troubleshooting approaches

Any complex system is subject to periodic failures, and LAN wiring systems are no different. Throughout this book, we have emphasized the installation of a reliable wiring system. We have spent a great deal of time showing the proper components and installation techniques that are required to create a properly functioning, reliable cable plant. The precept for this careful construction is that when anything does go wrong, it should very rarely be the cabling. (Some might say "never" rather than "rarely," but your LAN wiring is in an environment where it cannot totally be protected from those uncaring souls without the proper respect for these thin little wires.)

The fact of the matter is that failures do happen. Fortunately, most of the failures happen during installation, where the link operation is not critical. Still, the failed cable must be repaired, regardless of the circumstances. If the link has been in operation, it is all the more important that we make the repairs quickly.

Cable link failures can be divided into two broad classes with respect to LAN wiring. The classes of failures correspond to the types of test equipment for each. Failures involving basic cable function can be determined by the basic test equipment we talked about in Chap. 16. This equipment can check cable link connectivity and DC wire map. Failures involving performance-based cable function must be determined by the diagnostic and measurement equipment we covered. This equipment tests the performance of the cable link at the level of performance required for your particular LAN application. It tests for split pairs, impedance, NEXT, length, attenuation, and other parameters that are required for proper LAN operation. The basic function can be fine, but the link will still not operate if it doesn't meet your performance needs. For example, you could probably get DC connectivity on a barbed-wire fence, but you wouldn't be able to run data on it.

In this chapter, we will detail many of the types of cable link failures, whether they are from installation errors or cable damage after installation. We will give some approaches to finding the problem and repairing the cable link. Also, we will say a little about a successful troubleshooting approach that you may find useful.

Common Failure Modes

Wire Break

One of the most common wiring failures is a broken wire. Most testers will report the wire as an "open," meaning that the circuit is open. Some testers may actually declare the entire pair bad, although only one wire might be open. A wire that was never connected will also test open, even though the failure is not a result of wire damage. If only one wire of a multiwire cable is open, you should probably look for a bad connection at one of the termination points.

Connections in insulation-displacement connectors, such as 66M or 110 blocks, sometimes are not driven home, resulting in a bad or intermittent connection. What happens is that the insulation is not quite stripped away, and an apparently good connection will actually have a thin layer of insulating plastic remaining. You should never reuse old 66M blocks. The contact jaws can deform outward just enough to cause a problem when new wires are punched. This can also happen if a larger-gauge wire was previously punched down. These blocks may also have the same problem with jumper wires that are frequently moved. You may be able to "rehabilitate" a contact by squeezing the empty contact jaws together with sharp-nosed pliers, or even by rocking the punchdown tool back and forth on the contact. The 110-type blocks are much less prone to this problem, as they use a cutting-displacement method of termination, where the connector contact actually slices through the wire's insulation.

If inspection of the cable termination ends does not reveal the open wire, use a cable scanner to measure the length to the open. Midcable open wires may be very hard to find visually, unless there is obvious mechanical damage to the cable. You never want to find out that an installed cable has an open due to a manufacturing defect, since it is almost impossible to prove that the break did not occur during installation. It is a good idea to do a continuity check on each reel of cable prior to installation. During installation, be careful not to pull the cables over sharp edges, as the cable might be cut. See the next section for advice on how to repair a twisted pair cable open or short.

Fiber optic cables may also be open or fractured, indicating that the cable has been severely bent at some point, or that a connector is bad. The only way to find the break, if it is not obvious by inspection, is to use an OTDR (optical time-domain reflectometer). Fiber optic cables may be spliced by fusion of the two fiber ends at the break or by connectorizing both ends and using a coupler. Fusion of the fiber generally results in less loss, which is an important factor in an optical cable splice.

Wire Short

A wire short in some ways is easier to deal with than an open. A short is rarely caused by a bad connection at a termination, so you can initially skip that inspection. A cable scanner may be able to locate the short within a few feet. If you do not have access to a scanner, use a sensitive ohmmeter and the specification for cable resistance per linear foot (26 ohms per 1000 ft for AWG 26 wire) to find the approximate location of the short. If more than two wires are shorted, this test may be useless, as the linear resistance will be much less than expected.

Shorts are often caused by other contractors trapping the cable between sharp objects during construction. For example, the cable might be pinched between metal studs as a wall is assembled. The ubiquitous plasterboard screw also does a very nice job of piercing right through a cable, shorting the wires in the process. Sometimes (very infrequently, we hope) cable installers will pull the cable past a sharp edge, cutting through the insulation and shorting wires in the cable.

The method used for repair of a short or an open depends upon the type of cable and the performance requirements. Coax cable may be cut in half, the two ends connectorized, and a barrel connector used to join the ends. Fiber may be spliced, as mentioned in the previous section.

Twisted pair cable repair methods depend upon the category of performance desired. Technically, the TIA/EIA-568-B standard allows no splices in a horizontal cable run. However, practical considerations may dictate some variance with Category 3 cable runs. Category 3 has a more gentle twist than Category 5/5e, and it is possible to carefully perform a splice using a shrink-wrap solder splice available from several sources. This hollow, shrinkable-plastic splice is about one inch long and contains a narrow ring of low-temperature solder in the middle. The insulation is removed on the wires to be joined. They are placed in the splice from opposite ends and the splice heated with a proper heat-gun tool to melt the solder and shrink the plastic. (A pocket lighter is not a proper tool, but it works well just the same if used with care.) The next conductor from the same pair is wrapped around the first wire in the same direction as the natural twist and the wire ends are joined in the same manner.

An alternative method will form a temporary splice, even for Category 5e. TIA/EIA-568-B allows one transition point from round to flat cable in a horizontal cable run. This means that an allowance has been made for the inevitable untwist and NEXT that would result from terminating each cable end at the point of transition. Logically, you should be able to insert a plug and jack coupling to repair a cable cut or short, but that is not really allowed as it amounts to a splice. However, if a quick solution will get the network back running, this method can be used temporarily until new cable can be pulled. You should only use a properly rated plug and jack for the cable's category of operation. It may seem easier to use two jacks and an 8-pin coupler for this, but that is actually two plug/jack combinations, not one. Also, it is very unlikely that the 8-pin coupler will meet any standard higher than Category 3. In addition, many of these couplers also reverse the order of the pins, causing yet a different type of failure.

Finally, mark the point of the splice on your cable map. If the splice is temporary, you will want to go back and pull a new cable as soon as you can, as the need to replace the cable will be inversely proportional to how long you wait. If you want to have a Category 5e facility, but are using only Category 3 applications now, the splice may work now, but may fail when you upgrade to higher-speed applications.

Kinks, Bends, and Breaks

The effect of kinks and bends in cable also varies with the type of cable. Coax cable is probably the most resistant to this type of damage, because of the size and durability of the cable. However, severe kinks and very sharp, sustained bends will affect the transmission loss on the cable and may even cause an impedance reflection that could cause a failure.

Fiber optic cable, on the other hand, may be permanently damaged simply by overflexing the cable. Fiber cable will typically show no outward sign of damage, but the optical fiber may have small cracks, called microfractures, that adversely affect link loss. Severe bending and sharp impact may even fracture the fiber completely. Breaks may also be caused by excessive pulling tension during installation. Damaged fiber cable will show poor loss performance even with the simplest fiber optic testers, but you will need an OTDR to find the location of the fracture. Fractures must be repaired using fiber splice techniques we covered in previous sections. Use particular care when pulling fiber not to kink, bend, or overpull the cable.

Twisted pair cable may also be subject to transmission impairment due to kinks or sharp bends. As unbelievable as it may sound, even tight tie-wraps may distort the cable enough to cause this problem. The real problem with kinks or bends in twisted pair cable is that the defect actually distorts the geometry of the cable. This can easily be observed on the screen of a TDR (time-domain reflectometer). Distorting the cable shows up as a reflection hump in the returned signal. Severe bending can actually cause a permanent signal impairment. How serious this impairment is will depend upon the category of operation and just how near the operating margins of the link you are operating.

A major source of kinks in cables is the use of boxed (rather than reeled) cable. These cable boxes are designed to allow the cable to feed in a spiral fashion, without the need for conventional cable spools and reel holders. Unfortunately, the cable can sometimes feed more than one loop of cable and the loops get mixed together trying to get out of the feed hole in the box. Other times, a single loop twists so that it does not feed in a spiral fashion; as the loop collapses, a kink results. Of course, kinks can occur easily in loose coils of cable that have been unboxed or unreeled. Try to avoid kinks, particularly in Category 5e/6 operation. If a kink occurs, attempt to smooth out the kink as much as possible, but be aware that you may have to replace the run if it later fails in testing.

The TIA/EIA-568-B standard recommends a minimum bend radius of four times the cable diameter. For most cables, that radius is about $^3/_4$ to 1 in. Try to avoid pulling cable tightly around corners and use cable management retainers to provide a smooth transition to the point of termination.

Connector Opens

Modular connectors may fail because of open connections. The most common conditions are bad plug crimps, dislodged wire, bent jack pins, and bad connector seating. The popularity of modular plugs in telephone wiring has resulted in the availability of low-cost, 6-pin crimp tools. These tools may be made from plastic or lightweight steel. Unfortunately, the flimsy design has been extended to 8-position crimpers. The 8-pin modular plug is much more difficult to properly crimp than plugs with 6 pins. The low-cost tools often cannot properly seat the connector contacts in the center of the plug. This may result in one or more of the stranded wires not making contact or, worse, making intermittent contact. You can easily identify plugs crimped with one of the inferior tools. When viewed from the front, the middle contacts are visibly higher than the contacts near the edge of the connector plug. Recrimping rarely makes any difference, unless it is done with a proper crimp tool.

You should spend the extra money to purchase a high-quality crimp tool. These tools are heavily constructed and will be quite a bit more expensive than the low-cost tools. Some may have interchangeable dies for different sizes of modular connectors. A high-quality crimp tool for 8-pin modular plugs should cost $100 to $200, depending upon additional features.

Bad crimps may also allow a wire to be pulled out of the plug contact area. The same type of problem can happen in a cable termination at a workstation outlet jack or at the patch or punchdown in the telecommunications room. These connections are not made for very large pulling forces, and may fail when overstressed. User cords in the work area are particularly subject to this kind of damage, since they often are run along the floor in the path of chair legs, furniture, and feet. Always inspect both cable ends of a cord that is suspected of having a bad connection. Replace the cord if there appears to be damage.

A source of intermittent connections in modular plugs is the use of solid wire. These plugs were really designed for stranded wire, and solid wire does not work well. Special versions of modular plugs are allegedly designed to work with solid wire, and they do work better, but you will be much better off if you use only stranded wire with these plugs.

In rare cases, modular connector jacks may not seat properly in the connector faceplate. Some designs include the plug channel in the molded plate, with a separate jack assembly that snaps into the plate. If the rear module is not seated properly, the jack wires may be too far back to properly connect with the plug contacts, even though the plug clip seats with a sharp click.

In rare cases, modular jack pins may get crossed or bent. This will usually test as a shorted wire from the far end and as an open connection from the near end. The design of these jacks places the pins (which are really bent wire pins) in narrow plastic slots so they can move as the plug is inserted. At the top of the tracks, most jacks have an open area where the pins could slide from side to side if the wire pin is pushed upward too far. If the pin is allowed to slip into another pin's track (either during assembly or use), the pins will short together and one or both of them may be prevented from making any contact at all with their corresponding plug contact. If you suspect this problem, visually inspect the plate and replace the jack if it is damaged. This defect tends to permanently bend the wire pin and you may have continuing problems. It is a good idea to glance at the pins in each jack insert before you terminate the station cables. This only takes a second, and will save you hours of frustration later.

Modular jacks should be mounted so that the pins are on the top side of the jack opening. The theory is that this keeps dust and dirt from contaminating the pins. Jacks are also available with dust covers, which should be left in place until the jack is used and replaced when a plug is removed.

A punchdown block or patch panel punch block is a special type of connector. Some older-style 66M punchdown blocks require the use of a metal clip to connect between contacts in different columns. The standard 66M block has four columns of contacts across each punchdown position. Generally the contacts are internally wired as pairs, so that Column 1 and Column 2 contacts are connected in common, as are Column 3 and 4. However, some connectorized designs make all four columns independent, so that a clip must be placed across the contacts in adjacent columns to complete the circuit. A missing clip will cause a problem with the circuit.

An additional problem with punch blocks of any type concerns the use of stranded wire. These side-displacement or insulation-cutting terminals are designed for solid wire only. Do not use stranded wire with these blocks. The wire may not make good contact, since the contacts press in from the side, instead of clamping the wire or piercing into the strands. Even if an initial contact is made, the wire may pull free more easily, since much of the strain resistance of the contact depends upon trapping a solid wire in the contact.

Occasionally, a connector block open can occur when the punchdown tool cuts the cable end of the wire being terminated, rather than the scrap end. A tool may sometimes cut an adjacent wire to the wire being terminated. It is a good idea to check the terminated wires to be sure none seems loose. Cut or loose wires should be reterminated.

Coax cables are subject to having their connections pulled open by connector stress. Crimped or soldered center conductor pins are fairly sturdy, but the pin may sometimes be pulled back into the connector body and not make contact with the mating jack. A visual inspection will show a pin that seems too short. The center pin should be approximately even with the edge of the shield sleeve (not the bayonet sleeve). Cables that are pulled on while mated may also break the shield wires. If a coax shield is improperly stripped, many of the tiny wires that form the braid will be severed, with only a few remaining to make the connection. When pulled, these last few wires will break easily, leaving an open. The proper solution is to cut off the end, prepare the wire properly, and replace the connector.

Any coax LAN wiring that uses BNC-T's or terminators may be subject to opens due to failure of the T or terminator. As a matter of fact, many technicians consider the T component to be the primary cause of coax network problems. The connector may fail by literally falling apart, or it may exhibit intermittent connections that just add to the frustration. Low-cost connectors are the worst offenders. You would be wise to use only high-quality connectors, T's, and terminators on these systems.

Wrong Pin-out

All of the wiring problems that we discussed in the test equipment in Chap. 16 are common failures of cable links. Among these are reversing the order of some of the wires, crossing two pairs, flipping the wiring order, splitting pairs, and total miswires. Reversing the order occurs in pairs. A common mistake is to reverse the primary and tracer colors of a single pair. Another common mistake is to mix up the green and brown pairs, or split these pairs, since the color difference may be subtle in some cable. If you try to count pin numbers in a modular jack, be sure to position the jack correctly (see Chap. 7) Flipping the wiring order of a user cord is a common error with flat (silver satin) telephone-style wire. Inspect the wiring order by holding both cord ends side by side. The colors (to the extent you can see them) should be in the same order straight across the plug. Better yet, use a cable checker.

The standard color code for 4 pair cables should be followed in wiring all jacks and telecommunications room terminations. This code is different for punchdown blocks and modular jacks. Modular jacks actually have two color code standards, T568A and T568B. You must use the same standard throughout a horizontal cable run. This means that you must use the T568A wiring pattern at the workstation outlet if T568A was used at the patch panel termination. Keeping this straight is much more difficult if you use preassembled octopus (fan-out) cables for your equipment connection. Be certain which wiring pattern standard is used with the octopus cable and use the same at the workstation outlet jack. Using different wiring patterns results in a reversed-pairs indication on a tester. See Part 2 of this book for more details on these color codes and wiring patterns.

There are nearly as many ways to connect station wires to modular jacks as there are modular jack manufacturers. The most confusing jack wiring problems stem from the fact that the pairs are not connected to the pins of the modular jack in the same order as they are punched down in the telecommunications room. For example, in the telecommunications room, the cable is terminated on the 66M or 110 block in the pair order 1, 2, 3, and 4. That means that the blue-white Pair 1 gets punched down first on the first two pins of the cable position, the orange-white Pair 2 on the next two positions, and so on. At the modular jack, however, Pair 1 is terminated such that it connects to Pins 4 and 5. The orange-white or the green-white pair is terminated on Pins 1 and 2, depending upon whether you are using the T568A or T568B wiring pattern. If the terminations on the jack module are numbered the same as the pin order, you must be sure to place the wires into the correct slots before you terminate them. This order will be very different from that on the punchdown block. Fortunately, many jack modules have color-coded wire slots, so you can ignore the confusing position numbers.

If you choose to split your station cables into two jacks, with two pairs per jack, you will have extra fun figuring out what color code to use at each jack. If you must do this, one approach is to call each pair in a 2 pair set Pair A or Pair B. Plan which pins on the modular jack should respond to each pair, and draw a connection diagram, skipping the unused wire positions on the jack. This avoids the obvious difficulty you would have placing your Pair 1 in the Pair 2/3 position at the jack. Of course, all of the color codes on the jack will be completely wrong, but you wanted to do it this way, didn't you?

A far better approach, and one that is allowed by TIA/EIA-568-B, is to wire up the jack with the normal T568A or B pattern and use an adapter cable external to the outlet jack. Simply make (or have made) an adapter that connects two 2 pair cables, terminated in two plugs at one end, to a single modular plug, with each wire placed appropriately. That will split out the station cable to the two applications and avoid making your jack plate totally nonstandard.

Wrong Category

Another common failure of high-performance applications results from the use of the wrong category of cable and/or connectors. Either mistake will cause most performance-critical applications to fail. Remember that the category of performance is limited by the lowest category or component or cable used in a link. In addition, of course, you must maintain the distance below the limit, route the wire away from potential interference sources, and maintain the minimum twist of the pairs.

Many older cabling systems can support Category 3 operation. Cable that is 4 pairs and has a minimum of 2 to 3 twists per foot should operate, although not necessarily at the distance limit. Multipair cables of 25 or more pairs should be limited to 6 circuits (12 pairs) to minimize the effects of crosstalk that occurs on such cables. Analog telephone circuits can usually be accommodated on pairs in the same cable sheath, as their frequency of operation is much lower than LAN frequencies. However, you must use caution when placing digital telephone signals, including

ISDN, in the same cable jacket as LAN data, as the two may very well interfere with each other. Many modern phone systems are digital systems, and fall into this area of caution. The best rule is to use only one application per cable, even with analog systems. A later upgrade of your phone equipment might inadvertently shut down your LAN.

Excessive Untwist

The amount of untwisted cable that is allowable when operating at Category 3 is so great that it is easy to view the Category 5/5e/6 requirements with skepticism. After all, how could you go from a generous allowance of two or more inches of loosely twisted wire to a ridiculous requirement of less than $1/2$ in (13 mm)? Well, the answer is easily found when you consider that the maximum frequency of "interest" of the two categories jumps from 10 to 100 MHz. Obviously, the Category 5/5e/6 requirements are much more stringent because they have to be to assure cable link performance at this level.

Good cabling practice is to limit the amount of untwist to only that required to terminate the wires. This is good practice whether you are terminating Category 6 or Category 3. Another good practice is to strip back only enough of the outer jacket as is needed to terminate the pairs. This maintains the cable twist and the positioning of the individual pairs in relation to each other. The better connectors and cable terminations have additional cable management devices to secure the wire in place and to minimize stress and bending of the wire.

Wrong Impedance

Throughout this book we have talked about UTP cabling with 100-ohm impedance and STP cabling with 150-ohm impedance. Any mixing of these two types of wire in the same cable link will result in an impedance mismatch that may cause the link to fail. The different impedances cause a signal reflection at the point of intersection that will affect even lower-speed network links. At the higher speeds, the impedance match is critical to overall circuit performance. The standards generally allow for the expected variations in impedance from one lot of cable to the next. A variation of ± 15% is allowed and more may be acceptable on short cable runs, although it would technically not meet the standard.

The real problem is that network equipment is designed to expect a cable impedance of 100 (or 150) ohms. Using a cable with a higher or lower impedance will cause less power to be coupled to the cable pair and result in a lower received signal arriving at the other end. If that end also sees an impedance mismatch, even more power will be lost in the transfer. Receiver performance depends upon a certain minimum signal strength to override the crosstalk interference from the near-end transmission. If a lower-than-expected signal strength is received, link performance will suffer and may even cause the link to fail.

One final note of caution: Other twisted pair cables exist with different impedances than are in the standard. For example, in some parts of the world, 120-ohm twisted pair cable is widely available. This cable should not be mixed with 100-ohm cable in LAN wiring applications. Be sure your cable is the proper impedance for standard LAN installations.

Troubleshooting Approaches

Troubleshooting involves active problem solving in an operating environment. To those of us who do troubleshooting on a daily basis, the process may seem obvious. However, to one learning about complex system operation, or to one who has to supervise such operations, it may be useful to cover some of the important troubleshooting concepts.

If you had to reduce troubleshooting methodology to just four phrases, they would surely be as follows: Observe the problem, logically divide the component parts, test each portion of the whole, and make no assumptions. The first three items are logical, sequential steps. The last item ("make no assumptions"), really should be included at every step. Let's examine each step one at a time.

First, you should observe the problem. Many times the problem you will be asked to troubleshoot will involve more than just cable and wiring. To distinguish the problem from normal operation, you should be familiar with the normal operation. If not, you should at least get a description of normal operation that is as detailed as possible. In many cases, the best source of this description is an equipment manual or an expert source. The user may be of little help. Here is an example. You are told by a user that the link to the server is down, but the user's computer is working fine in all other respects. You walk to the equipment room, find the link's hub connection (from your excellent documentation), and observe that the hub card status lights are all off. Familiar with the hub operation, you know that at least one light should always be on, even with the patch cable removed. You wouldn't really need to check the cabling, because you have found a hub card failure.

The next step is to logically divide the problem. Let us say that we found the hub card lights normal, except for the one indicating a bad link to our problem workstation. Now we divide the problem logically. A workstation computer connects to the wall outlet, horizontal cable runs to the telecommunications room, the terminated cable is cross-connected to a patch panel, and a patch cord runs to the hub equipment. The hub and computer both look fine, so let's check the Channel. Disconnect the user cord (at the workstation!) and the patch cord (at the hub!), then test the cable for shorts or opens. The cable link was working before the failure, so we don't have to be too concerned about other miswires. Of course, if the application is a Category 5e 100 Mbps one, we would want to use a cable scanner set for Category 5e tests, since cable operation at the higher frequencies could be impaired without affecting DC continuity. If the cable checks good, the physical connection is no longer suspect. Next, you could look at the computer setup, interface card, server setup, and all the other network items that could cause a logical connection to fail.

Another way to logically divide the problem is through substitution of known-good components. Alternatively, you could use substitution to verify the proper operation of suspected components including the computer and hub. For example, you could move a known-good computer to the suspected-bad outlet, or move the suspected computer to a known-good outlet. Substitution causes you to guess which component might be bad and then test your guess. It can be done with very little test equipment, but it requires you to proceed very methodically and with no preconceived assumptions.

The third step, testing each portion of the whole, you might notice has followed immediately after we logically divided the problem. You must be very certain what you test, or your tests may be invalid. For example, what if we had tested the Basic Link instead of the Channel in our prior example? Does that test all of the cable? No, it does not! If the problem had been a failed user or patch cord, we would have missed it.

So, be methodical, be sure you understand the problem, carefully structure your testing so that it eliminates as many variables as possible, and make no absolute assumptions.

19

Training and Certification

Chapter 19 Highlights

- Training versus certification
- Job descriptions and skill levels
- Training resources
- Certification programs
- Sources of information

LAN wiring technology has become increasingly complex as network speeds have increased. The proper installation of any type of network cabling requires a substantial amount of knowledge and experience. This chapter will outline the skill sets needed in the LAN wiring business and will describe the levels of education, as well as the certification process.

Computer Network History

The complexity of modern LAN wiring is most obvious to those of us who have been in the computer-networking field since its infancy and have observed its evolution, depicted in Fig. 19.1. One could well assert that the first computer networking began with the computer-terminal systems of the '60s and '70s. Initially, these computer terminals connected to the mainframe computer using home runs of coaxial cable. Runs of up to 2000 ft (615 m) were common, and network speed actually seemed lightning fast, since the line rate of several megabits per second was used merely to paint plain text to the screen (usually no more than 2000 characters per screen).

This direct connection of cable could produce a rather massive concentration of cable as the cable runs neared the computer room, if the terminal controllers were centralized. You can imagine the bulk and the sheer weight of 500 to 1000 coax cables if run together into the computer room. Such a bundle was several feet in diameter and had to be supported structurally. In those days, computer managers were as often worried about the weight of their network as its speed!

In some installations, the controllers were placed on each floor to minimize the cable runs. This arrangement was essentially the precursor of modern telecommunications rooms, as the controllers were often placed in the same utility room or closet as telephone cross-connects and electrical distribution panels. A driving motivation was the sheer cost of hundreds of lengthy coax runs (at plenum-rated cable prices), plus the very high costs of moves, adds, and changes.

In many ways, the apparent speed and extreme cable lengths of these primitive computer connections created the set of expectations for modern cabling. To this day, centralized fiber optic cabling (at speeds below gigabit) is the only networking method that approaches old coax-terminal connection distances. Fortunately, the practice of placing remote terminal servers (controllers) in each workspace coincidentally limited the run length from controller to terminal. Even in a large building, a controller

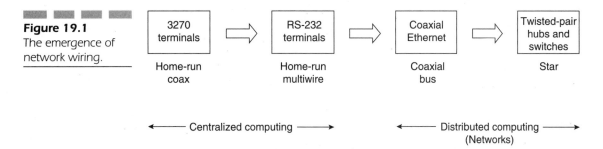

Figure 19.1
The emergence of
network wiring.

was rarely over 200 cable-feet from its terminals. This has made the subsequent 90-m TIA limit much more palatable to the mainframe manager, when forced into the role of overseeing a distributed computer network.

In addition to the coax terminals, some computer installations used a new type of computer-terminal serial data interconnection often referred to by the applicable standard, RS-232. These connections required multiconductor cable, rather than coax, and more severely limited the distance to the terminals. Eventually, we found that you could easily exceed the recommended 50-ft operating limit of RS-232 by utilizing "special" cable with twisted pair stranded wires. In fact, the similarities to telephone wire were inescapable. Network training at this point included knowledge of which pins of the 25-pin interface to use, how to connect them in the proper pattern (including the crossover or "null modem"), and how to arrange the pairing of the unbalanced interface to maximize distance. Using the right techniques, you could easily get 200 to 300 ft (about 60 to 90 m) reliably.

At about the same time, distributed computer networks arose, with the advent of coaxial Ethernet. Actually, early Ethernet was a real bother to install. The coaxial-bus topology required a thick and bulky coaxial cable to be run from each workstation to another, and then required an equally thick AUI cable from the coax tap to the workstation. Not only was it difficult to deal with, but an inexperienced installer could create not-so-subtle network problems, from flaky connections to total network outages.

Ethernet made the shift to thin coax in the mid-1980s at about the same time we were beginning to do RS-232 over twisted pair. Naturally, transmission engineers soon learned how to convert Ethernet's unbalanced thin coax signal to a balanced twisted pair and an industry was born. Thin Ethernet was also unreliable, whether it was true coax or was converted to twisted pair by balun transformers. However, the twisted-pair concept was a natural path to structured wire concentration in telecommunications rooms. Several manufacturers began to insert active hubs into these networks, located in the telecommunications room.

Finally, the first modern hub-and-spoke, or star, networks were a reality, and network wiring technology began to get serious. Standards were developed for structured cabling, originally by the Electronic Industries Association (EIA), and later by a subgroup, formed from an amalgam of industry groups and named the Telecommunications Industries Association (TIA). Network technology was no longer a trivial endeavor, but had become the basic currency upon which our network economy is founded. Now, education and training were necessities.

Wiring Complexity

To understand the nature of wiring complexity, one needs to appreciate the science and engineering involved in modern networking. To some, LAN wiring may seem to be a simple matter of connecting tiny wires in the right order. This approach is similar to simple telephone wiring, which almost anyone can do. Perhaps the simplicity is seemingly apparent to anyone who has connected a basic telephone jack in a home, or who knows how to plug the modular cord into an inexpensive phone, but the simplicity wanes when you approach the first cross-connect. A quick glance at the mass of cables and wires that exist in the average telephone room will convince anyone that even telephone wiring is not a simple matter. That is exactly why we have spent so much of this book describing the twisted pair wiring devices and their corresponding wiring patterns and installation practices.

In fact, high-speed network wiring adds one or two layers of complexity to the already complex field of telephone wiring. At the speed of networks today, the simple twisted pair has become a radio-frequency transmission line. The design of the components that create this high-speed network is a marvel, as well as a radical change for the cabling component manufacturers. They routinely use RF-engineering tools and practices in the creation of everything from low-crosstalk connectors to geometry-stabilized cables. In fact, most of the new terms, such as attenuation and return loss, come from that RF domain. Just ask any microwave engineer. Microwave? Well, the definition of microwave frequencies is any frequency at or above 1000 MHz, which certainly includes the base frequency of Gigabit Ethernet. Now, in reality, we do not place frequencies quite that high on our copper cables—yet—but we are close enough for many of the same concepts to apply.

In addition to proper design of these near-microwave network components, their proper installation is critical if we are able to reach the very high performance levels needed for our modern networks. We no longer have the luxury of simplicity. We must lay out and install our networks to very exacting standards. Although every installation is different, we have a system of rule-based design standards and installation practices that help us through the fray. However, rules are only as good as the players, and we need the certification referee to ensure the players are properly trained.

Training versus Certification

Network installations have required a certain amount of training since the days of twist-on coax connectors. Twisted pair connections require knowledge and skill to reach near-microwave performance levels. Fiber optic connections also require understanding and skill. We need to know what types of components are appropriate for each network installation, and how to properly install those components to achieve the expected network performance.

The process of gaining that knowledge and those skills is called *training*. Training takes many forms. You are clearly gaining the knowledge phase through reading this book. Some of the skills can be gained by following the installation and termination techniques here, particularly in conjunction with instructions that many manufacturers include with their products. In addition, many courses and resources are available to help supply cabling knowledge and skill. Of course, experience is the best teacher, particularly when supplemented by a skilled coworker or instructor. Many of us have used trial and error to successfully compensate for the lack of formal training. As long as the errors are quickly remedied, there is really nothing wrong with this approach. As a matter of fact, most approaches to wiring techniques (including the very design of successful connecting hardware) have been the result of educated trial and error, or at least trial and testing.

Now that the best of these techniques and practices are known, it is paramount that we quickly teach the best methods to the new installers and teach new methods to the experienced ones. This process is done through the dissemination of training manuals, training classes, and videos to the widest audience possible. Many organizations can provide cable installation training, as we will see later. However, as a network technician, how can you prove your level of expertise to a potential employer? Likewise, how can you as a contractor or business manager recognize the credentials of prospective new employees or contractors?

The credentials can best be supplied by a training certificate or a testing certificate. Formal training courses normally issue a "certificate of completion" to the attendee. In most cases, this certificate indicates that the student has successfully completed the course of instruction, which may or may not include an end-of-course exam. The testing certification is provided by an organization that may or may not have actually provided the training. In some cases, testing may be done by a government agency, as a requisite part of formal licensing. This is the same process that many skilled workers go through, from truck drivers to aircraft pilots, from automotive mechanics to airframe technicians, and indeed from programmers to network engineers. In an end-user organization, the particular level of training and certification that is necessary is up to the employer and the employee. Additional training and certification can be a source of job security and advancement, as well as potential job mobility, to the employee.

In outside contracting, some form of certification is almost mandatory. At the least, the lead technicians should be certified as installers for the type of installation being done, and the contractor should have a staff member with the appropriate cabling design certification. The certification is the proof to the customer that the contractor can perform to the level of services required for a successful network cabling installation. It also ensures that the installation will meet the entire scope of the specification, not just the modest network performance for the current networking need. (For example, a Category 5e installation should meet that performance level, even though the customer is using a lower network speed now.) With a proper installation, the network manager will be able to confidently use newer and faster technology, up to the limits to the performance spec, without concern about the wiring integrity.

Job Descriptions and Skill Levels

A very good way to determine the knowledge and training levels that are required for a career in any aspect of computer network cabling is to delineate the various jobs that must be done. Many of the training courses define their course offerings by using sample job titles. The job responsibilities and skill set are easily divided into the same categories, although in any situation, an individual may perform at more than one level. In fact, most cabling and certification programs presume that any particular level of knowledge includes the levels below it. However, in reality, the "designer" job tasks need only a basic understanding of detailed installation techniques and skills, as others perform those tasks. Conversely, the installer levels require a high degree of installation skill, but not much knowledge of the finer points of network and cabling design.

For this discussion, we have divided the job descriptions into four levels: cable system designer, network designer, cabling technician, and cabling installer (Table 19.1). These levels basically describe the different areas of technical knowledge required of someone who works in the field of LAN wiring. The knowledge levels are more easily understood if we characterize each level with a job title and description. We will describe each of these jobs briefly, but keep in mind that there are areas of specialization within each level. For example, a person may be an experienced and knowledgeable copper cabling installer, with no experience in fiber optic cabling. Thus, within each level, appropriate skills must be included for the specific network-wiring task.

Cable System Designer

The cable system designer is the person responsible for the overall design and specification of a LAN wiring system. This person is also responsible for understanding the entire cable plant, both inside and outside. The cable system designer uses the general workplace and network requirements to determine the number and types of cables, locations of workstation outlets and telecommunications rooms, and number and location of wire management devices, including racks, cable trays, ducts, hangers, and consolidation points (or multiuser outlets).

The designer works with architects, consulting engineers, builders, wiring contractors, and computer system managers in designing and implementing a cabling system for an enterprise. The designer may be

	Job description	Experience level	Typical roles
TABLE 19.1 Cable Technologist Levels	Cable system designer	5 years and up	Supervisor, consultant, estimator, sales, management
	Network designer	3–5 years	Cable System Designer, Network Engineer
	Cabling technician	5 years and up	Supervisor, technician, lead installer, team leader
	Cabling installer	1–3 years	Installer, helper/trainee

employed by, contracting for, or consulting with the data center or project manager. Frequently, a person with designer skills is employed by wiring contractors to provide expertise, interpret requirements, and respond to proposal and bid requests from clients.

The cable system designer must have a thorough knowledge of the standards, technology, and practices of LAN wiring. The designer will frequently have additional training and education in technical fields such as electrical engineering and the electrical and telecommunications trades. In many cases, the designer will also have experience in all aspects of network system installation, including specific experience in network and telephone wiring. The training and testing organizations generally recognize the cable system designer, often by a similar name, as the highest level of cable system expertise. We cover some of these specific certifications in a later section.

Network Designer

The job of network designer is to understand the technical aspects of the networking equipment. This designation is often used to indicate an addition to the cable system designer's skills and knowledge. For our purposes here, this job description and knowledge set is primarily intended to complement the wiring knowledge of the cabling system.[1]

The network designer must thoroughly understand the operating parameters of the network technology and topology that the enterprise will use. This includes knowledge of the standards-compliant cable technology, the allowable cabling distances, basic network hardware setup, and network troubleshooting. The actual networking standards, such as those for Fast Ethernet and Gigabit Ethernet, are the focus of this job. Network wiring must be chosen that is appropriate for the network technology being used.

This person may be employed either by the user or by the cabling contractor, but the skill set is particularly appropriate for the contractor. It is important that the wiring contractor ensures that an installation will operate properly for the network topology planned, not just that it was installed according to a particular set of bid specifications.

---

[1]Many other jobs and knowledge levels exist for the fields of network engineering and computer operation, but training and certification for these network-operation jobs are handled by product-specific training programs from the major vendors of applications software and operating systems.

The network designer is the wiring person who knows the difference, for example, between 1000BaseSX and 1000BaseLX and understands the distance limitations and the cabling requirements for each. It is every contractor's nightmare to install a cabling system that is according to spec, but inoperable. At the design stage, it is the network designer's job to cross-check the run lengths, the layouts, and the number and placement of telecommunications outlets, and to question the assumptions that others may make.

A network software technician may possess the skills and knowledge of the network designer, particularly if that person is responsible for the hardware connectivity of the network. However, one should not necessarily assume that the certification testing for software skills prepares one for networking hardware layout or design. Many of the cable-skills certification programs now recognize the importance of this networking job category in the overall success of a computer network installation. In fact, the membership organizations that offer certification may offer the only manufacturer-independent certification of generic networking skills.

Cabling Technician

The cabling technician's knowledge and experience is at the very heart of LAN wiring implementation. This is the person who follows the best workmanship practices, has vast experience with a wide variety of cabling components, and has a working-level knowledge of the appropriate installation standards. The cabling technician should know how to install all of the cabling components and supporting structures used in LAN wiring. This includes general experience with most types of cabling devices, from plugs and jacks to patch panels and cross-connects.

A cabling technician should know the proper workmanship procedures, including those practices that are crucial to proper operation of the higher wiring categories. This person should be able to design and install simple cabling systems that will meet all applicable performance specifications. In addition to installation tasks, the cabling technician should be an expert with the appropriate cable scanning gear, and should be able to quickly diagnose and remedy link failures. This is the person who will generally supervise and informally train the cabling installer, described in the next section.

A cabling technician typically has considerable experience and training in cable installation, and will often be a current or former lead installer. This person will often order the cabling components, select the appropriate components for each stage of the job, supervise installers, inspect each stage of the job, direct or perform link testing, and act as the main point of contact with the customer during the job. In many ways, there is a good degree of overlap between the experienced cabling installer and the cabling technician. The level of knowledge, experience, organization skills, and people skills are probably what most differentiate the two job categories.

Training and certification testing for the cabling technician are more extensive than for the installer. This job level is relied on for expertise, rather than just competency. Consequently, the training programs emphasize this top-end knowledge in all aspects of cabling technology. This is the person who makes the decisions about components and techniques to use for plenums, fire stops, grounding and bonding, cable management, and worker safety on a daily basis. For the most part, the training programs leave the people skills to the individual or to management. But they do an excellent job of transferring technical knowledge.

Cabling Installer

The entry level to cabling technology is the cabling installer. Actually, this level may encompass several steps in knowledge and experience. Many of the training programs recognize these steps by dividing this job category into two or more levels.

The distinguishing characteristics of the cabling installer, as compared to the cabling technician, are the level of knowledge and experience and the need for supervision in the overall job. The beginning installer may essentially be merely a helper, working directly with experienced installers as part of a two- or three-person team. At this stage, the helper learns some of the basic terminology and practices of cable installation. Most training at this level is on-the-job-training. Any formal training is introductory in nature. The pay scale for this entry level is frequently well below that of an installer with a reasonable amount of experience. However, with additional training and an aptitude for cabling, the beginner may advance rapidly.

Additional formal training for the new installer is generally manufacturer-specific. Most cabling contractors install a limited number of component brands, and can easily arrange for training in one or two component styles. Once an installer has had exposure to the basic workmanship concepts of copper cabling, that knowledge easily transfers to other similar component designs. For example, many manufacturers use 110-type wire termination in their outlet jacks, as well as in patches and cross-connects. So, once an installer understands the proper cable dressing, untwist limits, and punchdown techniques on one brand of components, that knowledge can be easily applied to other similar brands, albeit with minor refinements.

As the cabling installer progresses in knowledge and experience, it may become increasingly important to document that competence through formal training and certification testing. Many jobs require that someone on the team be certified to a particular level, and that all installers have appropriate manufacturer-specific training. The cable contractor realizes that these classes and tests are necessary and should be considered mandatory for the employee to progress in responsibility. The classes are a lot of work, and the installer may need additional motivation to complete the training. Some contractors include a pay-rate incentive for certification in particular areas. Certainly, a well-trained installer is more valuable, both to the employer and to the customer, and should be rewarded for taking additional responsibility.

A cabling installer with a high degree of experience and formal training may become a team leader or lead installer. As such, this person becomes a working coach for the installation team, supervising, inspecting, motivating, and training the others on the team. This concept allows the contractor to place multiple teams of installers into a job site with a distribution of management responsibilities.

Training Goals

There are two general approaches to training programs for LAN wiring: general-purpose and product-specific. The two types of training have slightly different purposes and goals. General-purpose training is essentially more conceptual in approach. It covers both the details of installation standards and practices, and the broad, general philosophy of cable design and installation. Don't assume that general-purpose training is simple or easy. It can be quite rigorous. Some training organizations enumerate and delineate wiring rules in a particular fashion, and you must memorize and reproduce their methodology in order to pass the certification tests.

The goal of a general-purpose training program is to give the installer or designer the knowledge to create standards-compliant cabling systems using all the proper layout, fire-protection, electrical, grounding, and workmanship rules. Such a program avoids specific product references, other than for examples of hardware type or for generic installation training. The idea is to give the cabling installer/designer the general rules and techniques for LAN wiring, without focusing on any particular manufacturer's products or methodology. It is assumed that the manufacturers or their designated training organizations will provide specific product training.

Manufacturer-specific training is rather self-defining. This type of training is provided or sponsored by individual manufacturers, particularly those with complete wiring systems. The advantage of this type of training is that you learn the techniques and methods for installing a particular brand of wiring components. Most manufacturers use similar wire-termination methods for copper cable. However, very specific assembly techniques may be required to ensure performance to Category 5e, Category 6, and beyond. In addition, the proper assembly of fiber optic connectors is quite manufacturer-specific, as designs and assembly tools vary widely.

Manufacturer-specific training may be provided by component suppliers or by third-party training organizations, in addition to direct manufacturer-provided training. In many instances, the training is offered at no cost, or for a nominal fee. It is in the best interest of the manufacturer that prospective installers have the proper training and experience in using their wiring components. Part of the product-specific training methodology occurs because installers will develop a clear preference for products on which they have been trained. The training gives manufacturers an opportunity to stress the benefits and presumed superiority of their products, and allows them to show the shortcomings of their competitors' products.

Training Resources

The availability of training for LAN wiring is a constantly changing subject. Training sessions are scheduled throughout the year in a variety of locations. You may want to choose from several appropriate training organizations, depending on the locations, the dates, and the type of training offered. As with all our discussion on training, both general and specific classes are offered.

General-Purpose Training

General-purpose training is provided by educational institutions and by industry organizations. Many colleges and universities have begun to respond to the need for advanced education in wiring technology. Some notable institutions that offer such programs include Mississippi State University, Texas A&M University, and Washington State University. Many others offer such education, but are too numerous to mention here. Also, the programs are constantly changing and expanding. You may consult the LAN-Wiring web site, **www.lan-wiring.com**, for links to these and other training resources.

The granddaddy of general-purpose training is BICSI (pronounced "bik'-see"). BICSI is an industry association that was originally called Building Industry Consulting Services, International. It was formed in the 1970s as an association with ties to the corresponding Bell System units that provided commercial building telecommunications system designs. In the 1990s, the association realigned its official name to BICSI: A Telecommunications Association to emphasize the tie to the telecommunications industry.

BICSI offers extensive training and certification testing for the telecommunications cabling industry. In addition to direct training, BICSI also licenses organizations to give BICSI-approved training to individuals. The BICSI Telecommunications Distribution Methods Manual (TDMM) is one of the most comprehensive standards and practices manuals available. In many areas, the methods go far beyond the general-purpose international standards to illustrate and describe virtually all installation techniques that BICSI endorses.

BICSI offers training at several levels. The RCDD (Registered Communications Distribution Designer) track includes courses in introductory voice/data cabling, telecommunications design, grounding and protection, LAN and internetworking, fiberoptic design, and project management. The installation program has three levels, Apprentice, Installer, and Technician. The installation track includes extensive cable pulling, termination, testing, and cabling technology. Attendees receive a certificate of completion for each course.

BICSI also hosts a number of seminars and conferences that include basic and supplementary training as well as manufacturers' displays. One of BICSI's major goals is to educate its members, and the meetings are organized to forward this aim.

The Association of Cabling Professionals (ACP) in conjunction with the Cabling Business Institute offers another general-purpose training program. Installer training includes information on all aspects of copper and fiber cabling, from basic cable termination to detailed installation practices. The training is layered into phases, with intensive classroom training interspersed with self-study sessions. After completion of the final session, an end-of-course exam is administered. Additional courses are offered in commercial and residential structured cabling design, fiber optic cabling, local area networks, outside plant design, estimating, sales, and project management.

The ACP offers several seminars, conferences, and trade shows throughout the year. Some of these are in association with industry publications and manufacturers. The Cabling Business Institute is associated with the publisher of *Cabling Business Magazine*. The magazine holds an annual conference that includes comprehensive training sessions that are requisites to the institute's certification program. Training videos, seminars, and trade shows are also offered by *Cabling Installation & Maintenance*, a trade magazine.

In addition to these training programs from industry associations, short courses are available from a variety of private training organizations. Several excellent technology training organizations offer appropriate training for cabling installers and designers. Consult the LAN-Wiring web site for current information on companies and offerings. In some instances, these outside training resources may include the requisite information for formal certification. Such training may be offered more frequently or in more convenient locations than the training from the industry associations.

Manufacturer-Specific Training

Product-specific training is offered through a number of sources, but this training is always sponsored or sanctioned in some way by a particular manufacturer or consortium. The best way to obtain this type of training is through a business relationship with a cable-component supplier or manufacturer. Many of the component distributors have their own training programs, which are often underwritten by the various manufacturers. In a few of these, training in several products may be offered, particularly when those products require similar installation techniques. For example, training on 110-type outlet connectors is essentially the same from one manufacturer to the next. Likewise, many fiber optic connectors are assembled in the same manner, and the basic techniques for preparing the fiber for termination are quite similar.

Links for a number of manufacturer/supplier training resources are included on the LAN-Wiring web site.

Training may also be offered through multimedia courses. Video tapes and CD-ROM training is available through several sources. Videos offer a great way to introduce new installer-trainees to the concepts and practices of cable installation and are a good source of general information.

Certification Programs

Formal certification plans match up with the types of training. There are several levels of certification within each plan, and you may or may not need to progress linearly from one to another. In several cases, the installer certification levels have been developed more recently, and no installer precertification may be needed for the designer level certificate. The formal certification of cabling technology knowledge is rather like a diploma from one of the training programs. In many cases, the certification also implies a certain amount of on-the-job experience as well. Certification is important as an employment qualification, as well as a testament to prospective clients that you and your installers have the proper knowledge and training to install the cabling components properly, thus ensuring the proper level of performance. In some areas, such training and certification may be required for local licensing and permits.

General-Purpose Certification

General-purpose training usually offers a test-based certification of competence. These training program certifications are often required by contracts for wiring jobs. In addition, employers may require a specific certification as a job prerequisite, or may offer additional pay and status to employees with these industry certifications. In many ways, the certifications offer the cabling technologist status similar to that of the certified network engineer in LAN technology.

BICSI offers certification of Apprentice, Installer, Technician, Registered Communication Distribution Designer (RCDD), and RCDD/LAN Specialization. ACP offers similar completion certificates for courses from Installer and Journeyman to Designer and LAN Specialist, although the only formal program at present is the Installer designation.

Other general-purpose programs offer a completion certificate as evidence of completion. The certificate may in some cases be obtained only after an end-of-course exam is successfully passed. Some college classes offer college credit for the courses, while others merely offer a completion certificate.

Some of the certification programs, such as those offered by BICSI, require continuing education to maintain the designation. This is similar to the continuing recertification requirement of software training. Continuing education requirements recognize the changing face of wiring technology and the need to stay current in the field.

Manufacturer-Specific Certification

Manufacturer-specific product training usually offers a certificate of completion as evidence of certification. Naturally, all that is certified is that the person has attended the class, and presumably has received all the relevant information and training. In a few instances, the trainee will be given an end-of-course test, and may have to demonstrate assembly competence on the product being covered.

Some manufacturers require that the organization performing an installation be trained on their products in order for the end user to receive a long-term warranty. These warranties are valuable to the end user because they guarantee cable link performance to the prescribed level for up to 15 years. The value to the manufacturer is that an additional incentive is provided the end user in return for specifying that manufacturer's wiring products. In some cases, two or more manufacturers may coordinate a warranty program, as long as properly trained and certified personnel install each manufacturer's components. This type of program would allow a connector and a cable manufacturer to cooperate, and to compete against manufacturers who offer both components with an extended performance warranty.

Many contractors will offer manufacturer-certified installation technicians as an incentive to the customer to use their services. Conversely, many end users may require such certification for installation work that they contract out. Specific product installation training is an important supplement to general-purpose training at any level.

 **Sources of Information**

Training and certification resources are constantly changing. Fortunately, we have a compilation of training information available on the LAN-Wiring web site at **http://www.lan-wiring.com.** This information is updated regularly from a variety of industry sources, including industry associations, manufacturers, public institutions, and independent training organizations. You may also consult many of these organizations through their contact information in the Appendixes.

Appendixes

Appendix A   Online Resources

- World Wide Web—Internet resources
- Newsgroups
- Online forums

Appendix B   Standards Organizations

- U.S. standards organizations
- Other national standards organizations
- International standards organizations

Appendix C   LAN Wiring Standards

- List of standards for LAN wiring
- Where to get copies of standards

Appendix D   Membership and Training Organizations

- Membership organizations
- Training and certification programs

APPENDIX A

ONLINE RESOURCES

World Wide Web—Internet Resources

For updated information on LAN wiring and links to standards groups, manufacturers, and other resources, refer to

www.lan-wiring.com

Additional Web Resources

American National Standards Institute (ANSI)	web.ansi.org
ATM Forum	www.atmforum.com
Canadian Standards Association (CSA)	www.csa-international.org
Electronic Industries Association (EIA)	www.eia.org
Fiber Optic LAN Section (FOLS) of TIA	www.fols.org
Gigabit Ethernet Alliance	www.gigabit-ethernet.org
Institute of Electrical and Electronics Engineers (IEEE)	www.ieee.org. standards.ieee.org
International Electrotechnical Commission (IEC)	www.iec.ch
International Organization for Standards (ISO)	www.iso.ch
ISO/IEC Joint Technical Committee 1 (JTC1)	www.jtc1.org
Internet Engineering Task Force (IETF)	www.ietf.org
Telecommunications Industry Association (TIA)	www.tiaonline.org
Underwriters Laboratories (UL)	www.ul.com
Underwriters Laboratories Canada (ULC)	www.ulc.ca

Where to Order Copies of Standards on the Web

Global Engineering Documents   global.ihs.com

Document Center Home Page   www.document-center.com

News Groups

Data Communications—Cabling   comp.dcom.cabling

Data Communications—LANs—Ethernet   comp.dcom.lans.ethernet

Data Communications—LANs—Token-Ring   comp.dcom.lans.token-ring

Online Forums

CompuServe Forums

Novell Hardware Forum

Windows Networking-A Forum

And various manufacturer-specific forums

America OnLine (AOL)

Computing area, Telecom & Networking Forum

Telecom message board categories:

    Networks/LAN & Networks/HW: Cabling Questions

    Tips & Tricks/For Comm Systems Designers

APPENDIX B

STANDARDS ORGANIZATIONS

U.S. Standards Organizations

ANSI (American National Standards Institute)
11 W. 42nd St., 13th Floor, New York, NY 10036
212-642-4900
(Originates, recognizes, and promulgates standards. ANSI is the U.S. member to the ISO.)

ASTM (American Society for Testing and Materials)
100 Barr Harbor Drive, West Conshohocken, PA 19428-2959
610-832-9500
(Develops standards and testing methods for materials.)

ATM Forum
303 Vintage Park Drive, Foster City, CA 94404
Info line, 415-578-6860; fax on demand, 415-688-4318; info@atmforum.com
(Industry consortium for the development of ATM standards.)

EIA (Electronic Industries Association)
2500 Wilson Blvd., Arlington, VA 22201
703-907-7500
(An industry organization that provides standards for electronic components and equipment.)

IEEE (Institute of Electrical and Electronics Engineers)
345 East 47th St., New York, NY 10017
212-705-7900
Order publications from IEEE Operations Center; see "Where to Get Copies of Standards" in App. C.
(Professional engineering association that develops standards for electronic technology.)

NFPA (National Fire Protection Association
1 Batterymarch Park, P.O. Box 9101, Quincy, MA 02269-9101
617-770-3000
To order publications, see "Where to Get Copies of Standards" in App. C.
(Provides electrical code that is the basis of most local building codes. Generally

recognized as the authority on permitted uses and proper practice in electrical wiring, including LAN cabling.)

TIA (Telecommunication Industries Association)
2500 Wilson Blvd., Arlington, VA 22201
703-907-7700
(An industry organization that provides standards for telecommunications components and equipment.)

UL (Underwriters Laboratories, Inc. (UL)
333 Pfingsten Rd., Northbrook, IL 60022-2096
847-272-8800
northbrook@ul.com
Other UL offices worldwide.

(Provides standards and testing services, including flammability standards and performance certifications for cable and connecting hardware.)

See Internet **www.lan-wiring.com** for links to other standards organizations.

Other National Standards Organizations

Standards Council of Canada (SCC)
178 Rexdale Blvd., Rexdale (Toronto), Ontario M9W 1R3, Canada
416-747-4000; www.scc.ca

Canadian Standards Association (CSA)
178 Rexdale Blvd., Etobicoke (Toronto), Ontario, M9W 1R3, Canada
416-747-4000; fax 416-747-4149; www.csa-international.org; certinfo2@csa-international.org; sales@csa-international.org
In United States and Canada, call toll free 1-800-463-6727

Deutsches Institut für Normung (DIN)
Burggrafenstrasse 6, DE-10787 Berlin, Germany
+49 30 26 01-0; fax +49 30 26 01 12 31; www.din.de; postmaster@din.de

Underwriters' Laboratories of Canada (ULC)
7 Crouse Rd., Toronto, Ontario M1R 3A9, Canada
416-757-3611;

ulcinfo@ulc.ca; www.ulc.ca

See Internet **www.lan-wiring.com** for links to other standards organizations.

International Standards Organizations

International Organization for Standardization (ISO)
1, rue de Varembé, Case postale 56, CH-1211 Genève 20, Switzerland
+41 22 749 01 11; fax +41 22 733 34 30; central@isocs.iso.ch)

(A university-sponsored Internet site is available at http://www.iso.ch/infoe/order.html.)

International Electrotechnical Commission (IEC)
3, rue de Varembé, Case Postale 131, CH-1211 Genève 20, Switzerland

+41 22 919 02 11

See Internet **www.lan-wiring.com** for links to other standards organizations.

APPENDIX C

STANDARDS FOR LAN WIRING[1]

List of LAN Standards

ANSI (American National Standards Institute) Standards

ANSI Z136.2   American Standard for the Safe Operation of Optical Fiber Communications Systems Utilizing Laser Diode and LED Sources

ANSI also approves standards of other organizations, such as EIA and TIA.

ASTM (American Society for Testing and Materials) Standards

ASTM D 4565-90   Physical and Environmental Properties of Insulation and Jackets for Telecommunications Wire and Cable

ASTM D 4566-90   Electrical Performance Properties of Insulations and Jackets for Telecommunications Wire and Cable

CSA (Canadian Standards Association) Standards

C 22.2 0.3 -1992   Canadian Electrical Code, Part 2

CSA-PCC-FT4 and FT6   Premise Communications Cord, Physical Wire Tests

---

[1]*Note:* Selected standards that apply to LAN wiring are shown. The standards organizations marked with ˙ do not supply copies of standards directly. Standards for these organizations may be obtained from the sources listed in "Where to Get Copies of Standards," later in this appendix.

CSA-T527	Bonding and Grounding for Telecommunications in Commercial Buildings (harmonized with EIA/TIA 607)
CSA-T528	Telecommunications Administration Standards for Commercial Buildings (harmonized with EIA/TIA 606)
CSA-T529	Design Guidelines for Telecommunications Wiring System in Commercial Buildings (harmonized with EIA/TIA 568)
CSA-T530	Building Facilities, Design Guidelines for Telecommunications (harmonized with EIA/TIA 569)

EIA/TIA (Electronic Industries Association and Telecommunication Industries Association) Standards

The EIA and TIA are two separate organizations that provide standards for the electronic and telecommunications industries. The TIA was formed in 1988 from the EIA telecommunications group and the U.S. Telephone Suppliers Association. The standards for telecommunications are jointly issued by EIA and TIA. TIA now has responsibility for continued development of the former EIA/TIA standards. It is these standards that are important to LAN wiring. As these standards are revised, the organization names are reversed; thus EIA/TIA-568 has become TIA/EIA-568-A and -B. Many of the EIA/TIA standards are coordinated or harmonized with other recognized national and international organizations, such as ISO, IEC, and CSA. The telecommunications standards are also ANSI-recognized standards.

EIA/TIA-568[2] (1991)	Commercial Building Telecommunications Wiring Standard *(superseded by TIA/EIA-568-A)*
TIA/EIA-568-A[2] (1995)	Commercial Building Telecommunications Wiring Standard *(superseded by TIA/EIA-568-B)*

---

[2]This standard has been replaced or superseded by a later standard, as indicated. It is included here to help clarify references to the standard number that may appear in specifications and references for cabling.

TIA/EIA-568-B (2000)   **Commercial Building Telecommunications Wiring Standard**

The new TIA/EIA-568-B standard comprises three independent parts, designated B.1, B.2, and B.3. The draft revision process is handled separately for each part. The standard replaces ANSI/TIA/EIA-568-A and incorporates and refines the technical content of TSB-67, TSB-72, TSB-75, and ANSI/TIA/EIA-568-A Addenda A-1, A-2, A-3, A-4, and A-5. The titles of each part are as follows:

TIA/EIA-568-B.1	Part 1: General Requirements (SP 4426)
TIA/EIA-568-B.2	Part 2: 100 Ohm Balanced Twisted-Pair Cabling Components Standard (PN 4425)
TIA/EIA-568-B.3	Part 3: Optical Fiber Cabling Components Standard (SP 3894)
TSB-36[2] (1991)	Additional Cable Specifications for Unshielded Twisted-Pair Cable (incorporated into TIA/EIA-568-A)
TSB-40[2] (1994)	Additional Transmission Specifications for Unshielded Twisted-Pair Connecting Hardware (incorporated into TIA/EIA-568-A)
TSB-53[3]	Additional Cable Specifications for Shielded Twisted-Pair Cable
TSB-67[2] (1995)	Transmission Performance Specifications for Field Testing of Unshielded Twisted-Pair Cabling Systems (incorporated into TIA/EIA-568-B)
TSB-72[2] (1995)	Centralized Optical Fiber Cabling Guidelines
TSB-75[2] (1996)	Additional Horizontal Cabling Practices for Open Offices (August 1996)
TSB-95 (1999)	Additional Transmission Performance Guidelines for 100 ohm 4-pair Category 5 Cabling (November 1999)
EIA/TIA-569-A (1998)	Commercial Building Standard for Telecommunications Pathways and Spaces

---

[3]This TSB was not formally released, but was incorporated into TIA/EIA-568-A.

EIA/TIA-570-A (1999)	Residential and Light Commercial Telecommunications Wiring Standard
EIA/TIA-606 (1993)	The Administration Standard for the Telecommunications Infrastructure of Commercial Buildings
EIA/TIA-607 (1994)	Commercial Building Grounding and Bonding Requirements for Telecommunications

Note: *The term TSB was formerly "Technical Systems Bulletin" when these standards were managed by the EIA. The TIA term was changed to "Telecommunications Systems Bulletin" for TIA standard supplements, beginning with TBSs issued after the 568-A revision to the primary commercial cabling standard.*

All of the above EIA/TIA standards shown are approved by the American National Standards Institute (ANSI).

IEEE (Institute of Electrical and Electronic Engineers) Standards

IEEE/ISO 8802-3 (IEEE 802.3)	Information Technology—Local and Metropolitan Area Networks—Part 3: Carrier Sense Multiple Access with Collision Detection (CSMA/CD) Access Method and Physical Layer Specifications (and Supplements J, K, L, and P-Q)
IEEE/ISO 8802-5 (IEEE 802.5)	Information Technology—Local and Metropolitan Area Networks—Part 5: Token Ring Access Method and Physical Layer Specifications (and Supplements B, C, and J)
IEEE 802.12	Standard for Demand Priority Access Method. Physical Layer and Repeater Specifications for 100 Megabits per second Operation.

Note: *IEEE Standards 802.2, 802.3, 802.4, 802.5, and 802.6 are now ISO-approved and have been renumbered to match the ISO numbering system.*

ISO/IEC (International Standards Organization/International Electrotechnical Commission) Standards

IEC 603-7	Connectors for frequencies below 3 MHz for use with printed boards
IEC 807-8	Rectangular Connectors for Frequencies below 3 MHz, Part 8: Detail Specification for Connectors, Four-Signal Contacts, and Earthing Contacts for Cable, First Edition
ISO/IEC 11801 (1995)	Information technology—Generic cabling for customer premises

Note: *ISO/IEC 11801 corresponds to TIA/EIA 568-B. Additional ISO-approved standards are listed in the IEEE section. ISO standards may be ordered from the ISO member in each member country (ANSI in the United States) or from the document distributors listed in "Where to Get Copies of Standards" later in this appendix.*

NFPA (National Fire Protection Association) Standards

National Electrical Code, 1999	ISBN 0-87765-432-8
National Electrical Code Handbook, 1999	ISBN 0-87765-437-9

Order from NFPA at 1-800-344-3555 or by ISBN from any bookstore.

UL (Underwriters Laboratories) Standards

UL 910	(Flamespread tests)
UL 1863	Standard for Communication Circuit Accessories

Where to Get Copies of Standards

Direct sales from standards organizations:

ANSI and ISO Standards

American National Standards Institute
11 W. 42nd St., 13th Floor, New York, NY 10036
212-642-4900

ATM Forum Standards

The ATM Forum
303 Vintage Park Dr.,
Foster City, CA 94404
Info line, 415-578-6860; fax on demand, 415-688-4318; info@atmforum.com

IEEE Standards

IEEE Operations Center
445 Hoes Lane, P.O. Box 1331, Piscataway, NJ 08855-1331
1-800-678-4333 or 908-562-0060

National Fire Protection Association Standards

NFPA Headquarters
1 Batterymarch Park, P.O. Box 9101, Quincy, MA 02269-9101
1-800-344-3555 or 617-770-3000

UL Standards

Underwriters Laboratories, Inc.

333 Pfingsten Rd., Northbrook, IL 60022-2096

1-800-676-9473 or 1-800-786-9473 or 708-272-8800

CSA Standards

Canadian Standards Association

178 Rexdale Blvd., Rexdale (Toronto), Ontario M9W 1R3, Canada

Orders, 416-747-4044; main, 416-747-4000

Document Distributors

Most of the standards from the above organizations and standards from EIA, TIA, ANSI, ISO, ASTM, CSA, and other standards organizations are available from the following document distributors:

Global Engineering Documents

15 Inverness Way East, Englewood, CO 80112-5766

1-800-854-7179 (United States and Canada); 303-792-2181; global.ihs.com

Document Center

1504 Industrial Way, Unit 9, Belmont, CA 94002

415-591-7600; www.document-center.com

For additional information on availability of standards documents, see the standards sections at **http://www.lan.wiring.com.**

APPENDIX D

MEMBERSHIP AND TRAINING ORGANIZATIONS

Membership Organizations

The following organizations are of general interest to those involved in LANs and LAN wiring.

BICSI

BICSI: A Telecommunications Association

(formerly Building Industry Consulting Service International)

10500 University Center Dr., Suite 100, Tampa, FL 33612-6415

1-800-242-7405; 813-979-1991

(A membership organization for individuals and companies. Provides designer and installer certifications, building guidelines handbook.)

EIA

Electronic Industries Association

2500 Wilson Blvd., Arlington, VA 22201

703-907-7500

(An industry organization that develops standards for a wide range of devices, from semiconductors to complex electronic equipment. The original parent of TIA.)

IEEE

Institute of Electrical and Electronics Engineers

345 East 47th St., New York, NY 10017

212-705-7900

(A professional membership organization for individuals with extensive educational and standards activities. Develops standards for LAN networking of several types, including Ethernet, Token-Ring, and AnyLAN.)

NFPA

National Fire Protection Association

1 Batterymarch Park, P.O. Box 9101, Quincy, MA 02269-9101

617-770-3000

(A membership organization for individuals and companies that develops fire safety and electrical standards. Publishes the National Electrical Code that is used as a basis for building codes worldwide.)

TIA

Telecommunication Industries Association

2500 Wilson Blvd., Arlington, VA 22201

703-907-7700

(An industry organization for the telecommunications industry. Develops for telecommunications wiring, including the TIA-568-A standard. Originally part of EIA.)

Training and Certification Organizations

The following organizations offer workshops, seminars, trade shows, training classes, videos, and formal certification programs for the LAN wiring industry.

BICSI: A Telecommunications Association
(formerly Building Industry Consulting Service International)
10500 University Center Dr., Suite 100, Tampa, FL 33612-6415
1-800-242-7405; 813-979-1991

Association of Cabling Professionals (ACP)
4160 Southside Blvd., Suite 3, Jacksonville, FL 32216-5470
904-645-6018; fax 904-645-3181

Cabling Business Magazine
12035 Shiloh Rd., Suite 350, Dallas, Texas 75228
214-328-1717; fax 214-319-6077

Cabling Business Institute
12035 Shiloh Rd., Suite 350, Dallas, TX 75228
214-328-1717; www.cablingbusiness.com

Cabling Installation and Maintenance
98 Spit Brook, Nashua, NH 03062-5737

603-891-0123; fax 603-891-0587; www.cable-install.com

GLOSSARY

10Base2   The IEEE/ISO 8802-3 (IEEE 802.3) standard for thin ethernet coax networks—10 Mbps transmission, baseband signaling, 185 m per coax segment. Also known as thinnet or cheapernet.

10Base5   The IEEE/ISO 8802-3 (IEEE 802.3) standard for Ethernet backbone (Trunk) Cable networks—10 Mbps transmission, baseband signaling, 500 m per coax segment. Also known as thicknet.

10BaseFL, FB, FP   The IEEE/ISO 8802-3 (IEEE 802.3) standards for fiber optic Ethernet network connections for interrepeater links (see FOIRL), synchronous links, and passive links—10 Mbps transmission.

10BaseT   The IEEE/ISO 8802-3 (IEEE 802.3) standard for unshielded twisted pair Ethernet networks—10 Mbps transmission, baseband signaling, unshielded twisted pair cable. Maximum allowable cable length is 100 m.

100BaseFX   The IEEE/ISO 8802-3 (IEEE 802.3) standard for fiber optic Ethernet network connections—100 Mbps transmission.

100BaseSX   The IEEE/ISO 8802-3 (IEEE 802.3) standard for fiber optic Ethernet network connections over multimode fiber using short wavelengths (850 nm)—100 Mbps transmission.

100BaseT   The IEEE/ISO 8802-3 (IEEE 802.3) standard for Ethernet network connections over Category 5 and above unshielded/screened twisted pair cable utilizing only two cable pairs.

100BaseT4   The IEEE/ISO 8802-3 (IEEE 802.3) standard for Ethernet network connections over Category 3 and above unshielded/screened twisted pair cable utilizing all four cable pairs—100 Mbps transmission.

100VG-AnyLAN   The IEEE 802.12 standard for a token-control, arbitrated network connections—100 Mbps transmission. Media may be UTP or fiber optic.

1000BaseELX/ZX   Proposed standards for fiber optic Ethernet network connections over single mode fiber using long wavelengths (1550 nm)—1000 Mbps transmission.

1000BaseLX   The IEEE/ISO 8802-3 (IEEE 802.3) standard for fiber optic Ethernet network connections over single and multimode fiber using long wavelengths (1300 and 1550 nm)—1000 Mbps transmission.

1000BaseSLX   A proposed standard for fiber optic Ethernet network connection over single mode fiber using long wavelengths (1550 nm)—1000 Mbps transmission.

1000BaseSX   The IEEE/ISO 8802-3 (IEEE 802.3) standard for fiber optic Ethernet network connections over multimode fiber using short wavelengths (850 nm)—1000 Mbps transmission.

1000BaseT   The IEEE/ISO 8802-3 (IEEE 802.3) standard for Ethernet network connections over Category 5e and above unshielded/screened twisted pair cable utilizing bidirectional transmission on four pairs—1000 Mbps transmission.

1000BaseTX   The proposed standard for Ethernet network connections over Category 6 and above unshielded/screen twisted pair cable utilizing unidirectional transmission on four pairs, two transmit and two receive—1000 Mbps transmission.

AC   Abbreviation for alternating current.

Alternating current   An electric current that cyclically reverses the direction of electron flow. The frequency is the rate at which a full cycle occurs in 1 second. (See current and direct current.)

ACR   Attenuation-crosstalk ratio. The value of attenuation, less the crosstalk value, both expressed in dB, at a particular frequency. A quality factor for cabling that expresses the relation of two important measured values. (See also signal-to-noise ratio).

Adapter, fiber optic   A passive coupling device that connects two male fiber optic connectors, aligning the light path.

Aerial cable   Telecommunications cable intended for installation between supporting points, such as poles or building entrance supports. Aerial cable normally uses a metallic strength member, which may be internal, bonded externally to the cable jacket, wrapped along the cable, or otherwise attached to the cable to be supported.

Ampere   A unit of measure of electrical current. The amount of current that flows through a 1-ohm resistance when a 1-V electromotive force is applied. An ampere of current is produced by 1 coulomb of electrical charge passing a point in 1 second.

Amplitude   The maximum value of a signal.

Analog signal   An electrical signal that varies continuously, without discrete steps as with a digital signal—for example, the conversion of voice to a continuously varying voltage by a microphone produces an analog signal. (See digital signal.)

Anomaly   An impedance discontinuity on a transmission cable. The discontinuity causes an undesired signal reflection that may distort signals and disrupt a network connection. Anomalies usually are often caused by distortions in cable geometry that do not affect the dc continuity of the cable.

ANSI American National Standards Institute.

Appletalk A network communications protocol developed by Apple Computer Corporation. Appletalk can operate over a LocalTalk cabling system at 230.4 kbps or over Ethernet at 10 Mbps.

Approved ground A building ground or bonded connection to building ground that meets the requirements specified in the NEC.

ARCNET Attached Resource Computer Network. A token bus local area network standard developed by Datapoint Corporation. An RF signal is used to pass a token between stations connected in a bus or star. Star arrangements may use passive or active hubs for signal distribution. ARCNET can use a variety of cabling types, including coax, twisted pair, and fiber optic cable. Standard ARCNET operates at 2.5 Mbps.

ARCNETPLUS A version of ARCNET that operates at 20 Mbps. (See ARCNET.)

ATM Asynchronous Transfer Mode. A transport protocol based on fast switching of 53-byte cells. With its high-speed operation, fast switching, and guaranteed delay times, ATM can transmit voice, data, and video efficiently along the same transmission path. Conventional data access methods, such as Ethernet and Token-Ring, are supported by LAN Emulation (LANE) in order to compensate for their longer packet lengths. ATM is implemented on a variety of transmission media and at a variety of speeds. It may be used for both backbone and primary workstation connections.

Attenuation The decrease in magnitude of a signal as it travels through any transmitting medium, such as a cable or circuitry. Attenuation is measured as a ratio or as the logarithm of a ratio (decibel).

Audio frequency The range of frequencies within the range of human hearing. This range is generally given as 20 to 20,000 Hz, although few humans can detect sounds below 30 Hz or above 16 kHz.

AWG American Wire Gage. A wire diameter specification. The smaller the AWG number, the larger the wire diameter.

Backboard A wooden board attached to a wall for the purpose of mounting cable terminations and telecommunications hardware. Typically $3/4$-in plywood from 2×4 ft to 4×8 ft size. The backboard may be painted to match room decor or to indicate the function of wiring components that are mounted.

Backbone Cable   Cable that interconnects wire distribution centers in a building or a group of buildings.

Balanced line   A circuit consisting of cable with two conductors that carry voltages of opposite polarity and equal amplitudes with respect to ground. The conductors must be twisted together to maintain balance over a distance. (See also unbalanced line, twisted pair.)

Balun   A transforming device that matches an unbalanced circuit to a balanced circuit. The balun may also transform impedance to maximize power transfer and minimize reflections.

Bandwidth   The range of frequencies required for proper transmission of a signal, expressed as a difference in hertz. A continuous range from zero is said to be a baseband signal, while a range which starts substantially above zero is said to be broadband (or narrowband in the case of RF signals).

Basic Link   The portion of a structured wiring cable connection between the cable termination at the horizontal cross-connect and the work area outlet connector, including the test cables and excluding the test connector. (See Channel, Permanent Link.)

Bel   A unit that represents the logarithm of the ratio of two power levels. (See dB.)

Bend loss   Increased attenuation in an optical fiber caused by excessively small bend radius. Bend loss may be temporary or permanent if the microfractures caused by the bend attenuate the light signal through the affected area.

Bend radius   The radius of curvature of a bend in a cable. Cables may have transmission performance degradation from a bend radius that is below a certain amount.

BICSI   BICSI, A Telecommunications Association, formerly known as the Building Industry Consulting Service International. A membership organization that offers training, certification by test, and standard practice reference manuals.

Bifurcated   Split in two.

Binder   A tape or thread used for holding groups of pairs together within a cable. Multipair cables often bind the pairs in groups of 25 and 100 pairs.

Bit   One binary digit. A bit may assume a binary value of 0 or 1.

BNC   A coaxial cable connector that uses a "bayonet" style turn-and-lock mechanical mating method. Intended for RG-58 and smaller diameter cable.

Bonding   The method used to produce good electrical contact between metallic parts. In the wiring context, bonding describes electrical connection of grounding bars and straps within a building to the central approved ground.

Braid   Fine wires woven into a cylindrical cable layer that shields enclosed conductors. Braid also may be formed into a flattened conductor to be used as a grounding strap.

Buffer   A protective coating over a strand of optical fiber.

Butt-in set   A telephone testing device resembling an oversized handset that is used for temporary connection and operation of a telephone line.

Byte   A group of 8 bits.

c   The symbol for the speed of light in a vacuum.

Cable   A group of insulated conductors or optical fibers enclosed within a common jacket.

Cabling   Installed cables. Also, the process of connecting cable between various points in a building or outside plant.

Capacitance   The ability to store electrical charge between two conductors separated by a dielectric material. In one type of capacitor, foil conductor sheets are separated by a thin insulating plastic layer and the assembly is tightly rolled into a cylinder, which is then encapsulated in a small metal can. In wiring, the capacitance is said to be "distributed," as the effect exists between wire conductors insulated with plastic that acts as the dielectric. Capacitance is specified in fractional farads (such as microfarads or picofarads).

CDDI   Copper Distributed Data Interface. A version of the FDDI access method that uses copper wire media, rather than fiber optics.

Channel   The entire structured wiring connection between equipment in the telecommunications room and the work area equipment, including the actual user cords, patch cords, and jumpers, excluding the test connector. (See Basic Link, Permanent Link.)

Characteristic impedance   A value of impedance (resistance and reactance) of a transmission line measured over a frequency range that would exist if the line were infinite in length. A transmission line of finite length will have perfect power transmission, allowing for absorptive transmission losses, if driven and terminated by an exact conjugate matching load impedance. An inexact match will cause reflections that increase transmission loss.

Cheapernet   Another name for thinnet.

Chromatic dispersion   (See dispersion.)

Circuit   A system of conductors that passes a signal or voltage between two points.

Coax   (See coaxial cable.)

Coaxial cable   A type of cable consisting of a conductor centered inside a tubular conducting shield, separated by dielectric material and covered with an insulating jacket. The shield may be a foil, wire braid, or solid metal, and is typically at ground potential.

Collision   A simultaneous transmission by two stations on a shared medium. In practical terms, a condition where a second device begins to transmit before another device's transmission has completed.

Conductivity   The ability of a material to allow the flow of electrical current. Conductivity is the reciprocal of resistivity. The unit of measurement is the mho (ohm backward).

Conductor   A material that offers low resistance to the flow of electrical current.

Consolidation point   A location between a telecommunications room and user work areas at which multiple station cables are brought together for cross-connection or interconnection and consolidation. Typically used with modular offices as a permanent point of termination between the TR and the movable office modules.

Cord   A very flexible insulated cable. User and equipment cords are typically made with stranded conductors and are terminated with the proper connector plug.

Core   The light-transmitting central portion of an optical fiber. The core is surrounded with a cladding that has a higher refractive index and thus helps channel the light along the core, as a waveguide.

Coupling   The transfer of energy between two components of a circuit or cable. The unwanted transfer of signals between pairs of a cable or between a cable and outside objects.

CPE   Customer premises equipment. Equipment that is usually provided by the user or subscriber of a telecommunications service. (See customer premises.)

Crossed pair   A wiring error in twisted pair cabling where both conductors of one pair are incorrectly exchanged for conductors of another pair at one end of a cable.

Crosstalk   The coupling of unwanted signals from one pair within a cable to another pair. (See NEXT.)

CSA   Canadian Standard Association.

CSMA/CD   Carrier Sense Multiple Access with Collision Detection. The access method used by Ethernet.

Current   The flow of charge in a conductor. Current flow is usually considered the flow of positive charge, whereas electron flow is the flow of negative charge. (See alternating current and direct current.)

Customer premises   Buildings, offices, and other structures under the control of the telecommunications customer.

dB   Decibel abbreviation. A logarithmic unit of measure expressing the ratio between two power levels. The value is 10 times the value in bels. (See bel.)

DC   Direct current. (See direct current.)

DEC   Digital Equipment Corporation. The former DEC was split up and its products and services acquired by other companies.

Demarc   Demarcation point. The point of telecommunications circuit termination by a telecommunications carrier within a building or residence. The user (subscriber or customer) has the responsibility for circuit operation beyond the demarc.

Dielectric   An insulating material between two conductors.

Dielectric constant   The ratio of the capacitance of a capacitor whose plates are separated by a vacuum dielectric. Permittivity.

Digital signal   An electrical signal that uses two or more discrete physical levels or signal phases to transmit information. For example, the conversion of a voice waveform to a digitally encoded representation of the waveform produces a data stream in which instantaneous amplitude samples of the voice are represented by a binary value. The data stream may then be transmitted as discrete elements and decoded at the far end to reproduce the original analog signal.

Direct current   An electric current that flows in one direction and does not reverse the direction of charge flow. (See current and alternating current.)

Dispersion   The phenomenon in an optical fiber whereby light photons arrive at a distant point in different phase than they entered the fiber. Mode dispersion occurs when elements of the optical signal take slightly different paths along the fiber and the differing path lengths cause the received signal to lose definition. Chromatic dispersion is caused by differing transmission times of different wavelengths of light, which are refracted differently, according to each wavelength. Dispersion causes signal distortion that ultimately limits the bandwidth and usable length of the fiber link.

DIX   Abbreviation for DEC/Intel/Xerox, the original consortium that developed the standard for classic Ethernet 1.0.

Drain wire   An uninsulated wire in contact with a shield braid of foil along its length. At terminating points of the cable, the drain wire may be used to connect to the shield.

Duplex   Consisting of two parts; bidirectional transmission. Duplex fiber optic cable consists of two fiber strands that are separately jacketed and joined together. (See full duplex, half duplex.)

E   Electronic symbol for voltage, the electromotive force.

E1   A specially framed 2.048 MBps wide area transport system that uses two twisted pairs for transport between a telephone central office and a customer's premises. Similar to T1, but with 30 channels at 64 kBps. The primary deployment in European telecommunications systems. The system was originally designed for digitized voice transmission, but is often used to link digital data systems such as LANs.

Earth   A term for zero-reference ground.

EC   European Community.

EIA   Electronic Industry Association. A membership association of manufacturers and users that establishes standards and publishes test methods. (See TIA.)

Equipment cord   See patch cord.

Electromagnetic field   The combined electric and magnetic fields caused by electron motion in conductors. EM fields may exist at great distances from the conductors; however, near-field effects are of more concern in wiring.

Electrostatic   The electrical charge that exists when the charge is at rest.

Electromagnetic interference   An interfering electromagnetic signal. Network wiring and equipment can be susceptible to EMI and emit EMI as well. (See radio frequency interference.)

ELFEXT   Equal-level FEXT. (See FEXT.)

Equipment room ER. An area that contains telecommunications or LAN equipment. An ER may be collocated within a telecommunications (wiring) room.

EMI (See electromagnetic interference.)

Ethernet A LAN topology and access method that uses the CSMA/CD transmission method. Ethernet may be used with two types of coax cable, twisted pair, or fiber optic cable. Two varieties of Ethernet exist, see text.

f Frequency.

FCC Federal Communications Commission.

FDDI Fiber Distributed Data Interface. A LAN topology and access method that passes tokens over dual counterrotating fiber optic rings at 100 Mbps.

FEP Fluorinated ethylene propylene. A thermoplastic with excellent dielectric properties which is often used for conductor insulation in fire-rated cables.

FEXT Far-end crosstalk. Crosstalk between two twisted pairs, measured at the opposite end of the cable from the disturbing signal source.

Fiber (See optical fiber.)

Fiber optics The transmission of light through optical fibers for communications purposes.

Firestop A material that prevents the passage of flame or smoke through an opening in a wall or floor.

FOIRL Fiber optic interrepeater link. An Ethernet fiber optic connection method intended for connection of repeaters. FOIRL is standardized as 10BaseFL and is defined in IEEE/ISO 8802-3 (IEEE 802.3).

Frequency The number of times a periodic action occurs in a unit of time. The unit of measure is the hertz, abbreviated Hz. One hertz equals 1 cycle per second.

Frequency response A range of frequencies within which a device operates within expectations. The limits are usually given as the frequencies for which the normal signal transfer of a device falls 3 dB below the nominal level. For example, an audio amplifier may have a frequency response of 20–20,000 Hz, while a telecommunications circuit may have a frequency response of 300–3000 Hz.

Full duplex   A communications method where both transmitted and received signals are simultaneously present (e.g., the common telephone instrument and full-duplex Ethernet.) (See half duplex.)

Gbps   Gigabits per second.

Giga   Prefix meaning 1 billion (1000 million) (10 raised to the 9th power).

Gigabit   One billion bits of data. Abbreviated Gb.

Gigahertz (GHz)   A unit of frequency equal to 1 billion hertz.

GIPOF   Graded-index plastic optical fiber.

Ground   An electrical connection to a point of zero potential, relative to earth ground. Earth ground is generally obtained through one or more grounding rods driven into the earth. Often called safety ground. Signal ground is a common return to a point of zero signal potential and may differ from the safety ground at the metal chassis in equipment.

Ground loop   A condition whereby an unintended connection to ground is made through an interfering electrical conductor.

Half duplex   A communications method where both transmitted and received signals are *not* simultaneously present, but alternate in presence (e.g., commercial two-way push-to-talk radio and standard, half-duplex Ethernet.) (See full duplex.)

Hertz   The unit of frequency, 1 cycle per second. (See frequency.)

Home run   A cable run that connects a user outlet directly with the telecommunications room, with no intermediate splices, bridges, taps, or connections.

Horizontal wiring   Telecommunications cabling from the user station to the first termination point in the telecommunications room.

HVAC   Heating, ventilation, and air conditioning system.

Hz   Abbreviation for hertz.

ICEA   Insulated Cable Engineers Association.

IDC   Insulation-displacement connector.

Impedance   The total resistance and reactance offered by a component. The unit of impedance is the ohm and is specified with resistive and reactive values as a complex number. The common symbol for impedance is Z.

Impedance match   A condition where the impedance of a device is a conjugate (equal resistive component and opposite reactive component) match to the device to which it is connected.

Innerduct   A flexible plastic raceway for protection of optical fiber cables. Innerduct is usually round and corrugated.

Insulator   A device or material that has an extremely high resistance to current flow.

Insulation   A material that is nonconductive to the flow of electrical current. The coating (usually thermoplastic) of a conductor that insulates it from other conductors.

Interference   Undesirable signals which interfere with the normal operation of electronic equipment or an electronic transmission.

ISDN   Integrated Services Digital Network. A transmission system for digital circuit distribution. The Basic Rate Interface (BRI) uses one pair for subscriber delivery of two 64 kbps "bearer" (user data) channels and one 16 kbps "data" (used for signaling and sometimes user data) channel. The Primary Rate Interface (PRI) delivers twenty-three 64 kbps bearer channels and one 64 kbps data channel.

ISO   Abbreviation for International Organization for Standardization, the actual name in English, although the ISO abbreviation is often translated as International Standards Organization.

Jacket   The outer protective covering of a cable.

Jumpers   Unjacketed cables of one or more twisted pairs, used to cross-connect circuits terminated at punchdown blocks; also connectorized lengths of 25-pair jacketed cable.

Kilo   Prefix meaning thousand (10 raised to the 3rd power).

LAN   Local area network. A computer network that typically exists within a single building or office or among a group of nearby buildings.

Laser   A coherent source of light with a narrow beam. The word comes from "light amplification by stimulated emission of radiation." Laser sources are normally used for single-mode fiber optic transmission. The laser source for most LAN connections is a semiconductor laser diode.

Lay   The distance required for one conductor or conductor strand to complete one revolution about the axis around which it is twisted. Sometimes referred to as the "lay length" or "twist pitch."

Leakage   An undesirable passage of current over the surface of or through an insulator.

LEC   Local exchange carrier (the local telephone company).

LED   Light-emitting diode.

Line voltage   The value of the electrical potential power line, generally given in volts RMS.

Link pulse   A single-bit pulse which is transmitted every 2 to 150 ms during idle periods on a 10BaseT link segment to verify link integrity.

Load   A device that consumes power from a source.

Lobe   An arm of a Token-Ring that extends from an MSAU to a workstation adapter.

Loss   The portion of power from a source that is dissipated before being consumed by the load.

Manchester coding   A method of LAN signal coding that maintains a null DC level so as to allow the detection of collisions; the coding method of Ethernet and certain other networks.

MAU   Media Attachment Unit. The transceiver in an Ethernet network; also, a common name for the MSAU in a Token-Ring network.

Mbps   Megabits per second. A bit is one unit of a digital signal.

Media filter   An impedance matching arrangement used in Token-Ring networks to transform between the 100-ohm impedance of UTP and the 150-ohm impedance of media interface connections.

Mega   Prefix meaning million (10 raised to the 6th power).

Megahertz (MHz)   Unit of frequency equal to 1 million hertz (1 million cycles per second).

Micro   Prefix meaning one-millionth (10 raised to the −6th power).

Microfared   One-millionth of a farad (commonly abbreviated μF, and less commonly μfd, mf, and mfd).

Micron (μm)   Millionth of a meter (10^{-6} meter).

Mode   A single electromagnetic wave traveling in a waveguide, such as an optical fiber.

Mode dispersion   (See dispersion.)

Modular furniture   A type of office furniture that consists of work desks, shelves, file cabinet modules, and drawer modules that are connected together and to wall dividers to form work spaces in open area offices. The wall dividers may be of varying heights to provide privacy and noise reduction. Divider walls are generally joined to form an array of modular office cubicles. (See modular office, open office.)

Modular office   A system of office work space arrangement using modular office furniture and divider walls to form individual work spaces (often called office cubicles), rather than the traditional arrangement of desks in an open area. Often used to increase worker density and provide a measure of privacy, without the use of permanently constructed individual offices. Also called open offices. (See modular furniture, open office.)

MSAU Multistation Attachment Unit. The device for interconnecting station lobe cables to form a ring in a Token-Ring network.

Multimode fiber A fiber waveguide that supports the propagation of multiple modes. Multimode fiber may have a typical core diameter of 50 to 100 μm with a refractive index that is graded or stepped. It allows the use of inexpensive LED light sources, and connector alignment and coupling is less critical than single mode fiber. Distances of transmission and transmission bandwidth are less than single mode fiber because of dispersion of the light signal. (See also Single Mode Fiber.)

Multiuser telecommunications outlet adapter A multiple-jack arrangement that terminates home-run cables from the TR at a consolidation point and allows the interconnection of user cords from work area outlets in modular furniture.

MUTOA Abbreviation for multiuser telecommunications outlet adapter.

N connector (See Type N connector.)

Nano One-thousandth of one-millionth (10 raised to the −9th power).

Nanometer (nm) One-billionth of a meter.

Nanosecond One-billionth of a second.

NEC National Electrical Code.

NEMA National Electrical Manufacturers Association

Neoprene A synthetic rubber that has good resistance to chemicals and flame. The chemical name is polychloroprene.

NEXT Near-end crosstalk. Crosstalk between two twisted pairs measured at the same end of the cable as the disturbing signal source.

NFPA National Fire Protection Association.

Noise An extraneous, random signal that interferes with the desired signal. In most measurement situations, a nonrandom undesired signal is counted as noise.

NVP (See nominal velocity of propagation.)

Nominal velocity of propagation The speed of signal propagation through a cable, expressed as a decimal fraction of the speed of light in a vacuum. Considered to be "nominal" because variations in geometry along a cable may produce absolute variations in the value. The NVP is used in reflectometry to closely estimate the length of a section of cable by the formula length = $\frac{1}{2}t \times \text{NVP} \times c$, where t is the transit time for the reflected pulse and c is the speed of light in a vacuum. NVP is sometimes given as a percentage, but must be converted to a decimal fraction to be used in the formula.

Ohm The electrical unit of resistance. The value of resistance through which a potential difference of 1 volt will cause a current flow of 1 ampere.

Open A break in the continuity of a circuit.

Open office A work area for multiple workers, usually in a large room. An open office may contain an array of desks or of modular furniture. Open offices present special problems for structured cabling because of the lack of permanent walls near user workstations in which to mount the work area outlet and terminate the horizontal (station) cable. Detailed standards have been developed to address the issues and standardize wiring practices. (See modular furniture, modular office.)

Optical fiber A thin filament of glass used for the transmission of information-bearing light signals.

Outside plant Cabling, equipment, or structures that are out of doors.

OTDR A version of TDR intended for use with fiber optic cables. (See TDR.)

Packet A group of bits grouped serially in a defined format, containing a command or data message sent over a network.

Patch panel A panel containing jacks or connectors that is used to connect or provide access to telecommunications or data circuits.

Patch cord A flexible piece of cable terminated at both ends with plugs. Also referred to as "user cord," equipment cord," or simply "cord" in conjunction with LAN use.

PCC Premises commmunication cable, CSA (Canadian Standards Association) cable designation.

Permanent Link The portion of a structured wiring cable connection between the cable termination at the horizontal cross-connect and the work area outlet connector, excluding the test cables entirely. (See Basic Link, Channel.)

Pico Prefix meaning one-millionth of one-millionth (10 raise to the −12th power).

Plenum   Any space, whether closed or open, that is used for air circulation. For example: air ducts, return ducts, air shafts, above-ceiling air return spaces. Work spaces are generally not considered plenums.

Plenum cable   Cable that has been certified by the manufacturer as meeting accepted standards for installation in plenum spaces without an enclosing metallic conduit. Plenum cable is generally fire resistant and has low emission of smoke and toxic fumes in contact with a flame.

Polymer   A substance made of repeating chemical units or molecules. The term may refer to general types of plastic or rubber.

Polypropylene   A thermoplastic material similar to polyethylene but somewhat stiffer and with a higher softening point.

Polyethylene   A thermoplastic material with excellent electrical properties.

Polyurethane   Broad class of thermoplastic polymers with good abrasion and solvent resistance. Can be solid or cellular (foam).

Polyvinyl chloride (PVC)   A general-purpose thermoplastic used for wire and cable insulation and jackets. PVC is known for high flexibility. Often used in nonplenum wire insulation and cable jackets. A modified version of the material may be found in jacketing of some plenum-rated cables.

POTS   Plain old telephone service.

Prewiring   Wiring installed before walls and ceilings are enclosed.

Protector   A device that limits damaging voltages on metallic conductors.

PS-   Power sum. A term that indicates that a measured parameter includes the sum of the contributions from all pairs in a cable, excluding the pair under test, with the other pairs excited simultaneously. This is in contrast to the normal method of measuring only the pair-to-pair parameter. Examples are PS-NEXT and PS-FEXT.

PVC   polyvinyl chloride

R Symbol for resistor. The Greek symbol omega (Ω) is often used to indicate a resistance value in ohms.

RF   Abbreviation for radio frequency. (See radio frequency.)

Raceway   A channel used to enclose telecommunications and electrical wiring. Raceways may be metallic or nonmetallic and may totally or partially enclose the wiring.

Radio frequency   The frequencies in the electomagnetic spectrum that are used for radio communications. Generally considered to be frequencies above 300 kHz.

Radio frequency interference   Electromagnetic interference at radio frequencies. (See electromagnetic interference.)

RCDD   Registered Communication Distribution Designer. A certification of BICSI, an industry organization, for individuals qualified to consult and design in telecommunications distribution systems. (See BICSI.)

Receiver   A device whose purpose is to capture transmitted signal energy and convert that energy for useful functions.

Reflection   A return of electromagnetic energy that occurs at an impedance mismatch in a transmission line, such as a LAN cable. Reflections in an operating circuit are undesirable and may result from a physical problem, such as crushing or untwist, or from an electrical connectivity problem, such as a short or an open.

Refractive index   The ratio of the speed of light in a vacuum to its velocity in a transmitting medium, such as an optical fiber core.

Repeater   A device that connects two segments in an Ethernet network, or two portions of a Token-Ring network ring so as to extend and regenerate the LAN signal.

Resistance   The opposition a conductor offers to current flow in a DC circuit. Resistance is measured in ohms, where 1 ohm is the resistance that allows 1 ampere to flow when a potential of 1 volt is applied. Resistance is the real component of impedance, which includes opposition to current flow resulting from capacitive and inductive reactance.

Reversed pair   A wiring error in twisted pair cabling where the conductors of a pair are reversed between connector pins at each end of the cable.

RFI   Abbreviation for radio frequency interference. (See radio frequency interference.)

RG/U   RG is the common military designation for coaxial cable and U stands for universal.

Ring   A polarity designation of one wire of a pair indicating that the wire is that of the secondary color of a 5-pair group (e.g., the blue-white wire of the blue pair); a wiring contact to which the ring wire is connected; the negative wiring polarity. (See also tip.)

Riser cable   A type of cable used in vertical building shafts, such as telecommunication and utility shafts, or between utility rooms via cable ports. Riser cable typically has more mechanical strength than general-use cable and has an intermediate fire protection rating.

RJ-45   A USOC code specifying a particular analog line connection that uses an 8-pin modular telecommunications connector jack. The term RJ-45 is often used simply to identify the 8-pin, 8-wire/contact modular plug or jack that is used with twisted pair cable.

Rubber (wire insulation)   A general term used to describe wire insulations made of thermosetting elastomers, such as natural or synthetic rubbers, neoprene, or butyl rubber.

Runt packet   An Ethernet data packet that is shorter than the valid minimum packet length, usually as a result of a collision.

Scanner   A cable testing device which uses TDR methods to detect cable transmission anomalies and error conditions.

Screened cable   (See shield.)

Segment   A portion of an Ethernet network containing workstations, but only at the endpoints; a network cable terminated at both ends.

Sheath   (See jacket.)

Shield   A metallic foil or multiwire screen mesh that is used to prevent electromagnetic fields from penetrating or exiting a transmission cable. Also referred to as a "screen."

Shield coverage   The physical area of a cable that is actually covered by shielding material, often expressed as a percentage.

Short   A near-zero-resistance connection between two wires of a circuit. A short is usually unintended and considered a failure; however, conductors may be intentionally connected by a "shorting bar," as in a data connector plug.

Signal   The information conveyed through a communication system.

Signal-to-noise ratio   The ratio of received signal level to received noise level, expressed in dB. Abbreviated S/N. A higher S/N ratio indicates better channel performance. (See also ACR.)

Single mode fiber   A fiber waveguide in which only one mode will propagate. Single mode fiber has a very small core diameter, in some fibers approximately 8 μm. It allows signal transmission for long fiber distances with relatively high bandwidths and is generally driven with a laser diode. (See also multimode fiber.)

Single-ended   An unbalanced circuit or transmission line, such as a coaxial cable transmission line. (See also balanced line.)

Sinusoidal   A signal that varies over time in proportion to the sine of an angle. The description is often given to contrast with a nonsinusoidal waveform, which may contain significant harmonics above the fundamental frequency of the waveform. Ordinary alternating current is sinusoidal.

Skin effect   The tendency of alternating current to travel on the surface of a conductor as the frequency increases.

S/N   Abbreviation for signal-to-noise ratio. (See also ACR.)

SNMP   Simple Network Management Protocol. A remote management protocol that is part of the TCP/IP suite of protocols. Devices that support SNMP can be monitored and controlled remotely over the network.

SNR   Abbreviation for signal-to-noise ratio; less common than the abbreviation *S/N*. (See also signal-to-noise ratio, ACR.)

SONET   Synchronous optical network. Sonet is a fiber optic technology that is used to transport telecommunications at 155 MBps and greater. It is the physical layer for several ATM implementations.

Source   In fiber optics, the device that converts the electrical information-carrying signal to an optical signal for transmission over an optical fiber. A fiber optic source may be a light-emitting diode (LED) or a laser diode.

Speed of light   In a vacuum, 2.998×10^{-8} meters per second. Abbreviated *c*. See also nominal velocity of propagation.

Splice   The interconnection of two segments of cables with or without a termination or cross-connect device. A copper cable may be spliced by connecting the pairs using a cross-connect block, splice and tap clips, solder splices, or even twist-and-tape methods. Splices are prohibited in all copper wiring above Category 2. In fiber cable, a splice refers to a low-loss fusion of the two fibers, and is allowed in most instances.

Split pair   A wiring error in twisted pair cabling where one of a pair's wires is interchanged with one of another pair's wires. Split pair conditions may be determined with a transmission test. Simple DC continuity testing will not reveal the error, because the correct pin-to-pin continuity exists between ends. However, the error may result in impedance mismatch, excessive crosstalk, susceptibility to interference, and signal radiation.

Step-index fiber   An optical fiber in which the core is of a uniform refractive index with a sharp decrease in the index of refraction at the core-cladding interface.

Stitching   The activity of terminating multiconductor cables on a punchdown block.

STP   Shielded twisted pair. (See also UTP.)

Structured wiring   Telecommunications cabling that is organized into a hierarchy of wiring termination and interconnection structures. The concept of structured wiring is used in the common standards from TIA and EIA.

Surge   A temporary increase in the voltage in an electric circuit or cable. Also called transient voltage.

T1   A specially framed 1.544 MBps wide area transport system that uses two twisted pairs for transport between a telephone central office and a customer's premises. The data stream is channelized into 24 channels at 64 kBps. The system was originally designed for digitized voice transmission, but is often used to link digital data systems, such as LANs.

T carrier   The system of telecommunications circuit multiplexing that utilizes T1, T1c, T2, and T3 framing methods. (See T1.)

TCP/IP   Transmission Control Protocol/Internet Protocol. A protocol used in the interconnection of computers. Although TCP/IP is actually a single protocol, the term is generally used to refer to a suite of closely related protocols for networking. The TCP/IP suite is used extensively in the Internet.

TDR   Time domain reflectometry. A technique for measuring cable lengths by timing the period between a test pulse and the reflection of the pulse from an impedance discontinuity on the cable. The returned waveform reveals many undesired cable conditions, including shorts, opens, and transmission anomalies due to excessive bends or crushing. The length to any anomaly, including the unterminated cable end, may be computed from the relative time of the wave return and the nominal velocity of propagation of the pulse through the cable. (See also nominal velocity of propagation, speed of light.)

Teflon   A DuPont Company trademark for fluorocarbon resins. (See FEP and TFE.)

Telecommunications room   A separate room or area within a building that contains the telephone and/or data cable terminations for nearby work areas and for backbone cabling to other telecommunications rooms in the structured wiring system. The room may also contain equipment for the networking of telecommunications devices. Formerly referred to as a wiring closet. (Abbreviated TR.)

Terminator   A resistor/connector combination that is connected to the end of a transmission line, such as a coaxial cable, to match the characteristic impedance of the cable. The terminator absorbs all signal energy that reaches it, thereby eliminating unwanted signal reflections from the end of the cable.

TFE   Tetrafluoroethylene. A thermoplastic material with good electrical insulating properties, chemical resistance, and combustion resistance.

Thicknet   Ethernet coaxial trunk cabling. (See 10Base5.)

Thinnet   Ethernet thin coaxial cabling. (See 10Base2.)

TIA   Telecommunications Industry Association. A membership association of manufacturers and users that establishes standards and publishes test methods. The body that authored the TIA/EIA 568-B "Commercial Building Telecommunications Wiring Standard" in conjunction with EIA. (See also EIA.)

Tip   A polarity designation of one wire of a pair indicating that the wire is that of the primary (common) color of a 5-pair group (e.g., the white-blue wire of the blue pair); a wiring contact to which the tip wire is connected; the positive wiring polarity. (See also ring.)

Token-Ring   A LAN topology and access method controlled by passing a digital token along a wiring ring formed between connected devices. Rings may operate at either 4 or 16 Mbps and device connections to the ring are made by means of a Multistation Attachment Unit (MSAU or MAU). Implemented originally by IBM, Token-Ring is standardized in IEEE/ISO 8802-5 (IEEE 802.5).

Topology   The physical or logical interconnection pattern of a LAN.

TR   Abbreviation for telecommunications room.

Tracer   The contrasting color-coding stripe along an insulated conductor of a wire pair.

Transmission line   An arrangement of two or more conductors that is used for signal transmission.

Transmitter   A device that converts electrical signals for transmission to a distant point.

Twisted pair   A communication cable using one or more pairs of wires that are twisted together. When a pair is driven as a balanced line, the twisting reduces the susceptibility to external interference and the radiation of signal energy.

Type N connector   A threaded-barrel constant-impedance coaxial connector for large-diameter cables.

UL   Underwriters Laboratories, Inc.

Unbalanced line   A transmission line in which voltages on the two conductors are unequal with respect to ground. Generally one of the conductors is connected to a ground point. An example of an unbalanced line is a coaxial cable. (See also balanced line.)

Underground cable   Cable that is intended to be placed beneath the surface of the ground in ducts or conduit. Not necessarily intended for direct burial in the ground.

User cord   (See patch cord.)

USOC   Universal Service Order Code.

UTP   Unshielded twisted pair, a type of transmission cable.

V   Symbol for volt.

Vampire tap   The method of connecting a transceiver to an Ethernet trunk cable by drilling an access hole in the cable, inserting a contact to the center conductor, and clamping the transceiver onto the cable at the tap.

Velocity of propagation   (See nominal velocity of propagation.)

VHF   Very high frequency. The portion of the electromagnetic spectrum extending from 30 to 300 MHz.

Video   A signal that contains visual information, such as a picture in a television system.

Volt   The unit of electrical potential. One volt is the electrical potential that will cause 1 ampere of current to flow through 1 ohm of resistance.

Voltage   Electrical potential or electromotive force expressed in volts.

Voltage drop   The voltage developed across a component by the current through the resistance of the component.

W   Symbol for watt.

WAN   Wide area network. An extension of a local computer network (a LAN) over data communications lines provided by telecommunications common carriers. A typical WAN connects one or more remote office sites to the central corporate site over dedicated analog or digital circuits.

Watt   A unit of electrical power. One watt is equivalent to the power represented by 1 ampere of current through a load with a voltage drop of 1 volt in a DC circuit.

Waveform   The amplitude of a signal over time.

Wavelength   The physical distance between successive peaks of a wave in a transmission medium. Wavelength may be calculated from the frequency of a signal and the nominal velocity of propagation. In a vacuum, the NVP is simply the speed of light.

Wiring closet   A common term for a small room where telecommunications cabling is interconnected. The term has been replaced by telecommunications room. (See telecommunications room.)

Z   Symbol for impedance.

INDEX

Note: **Boldface** numbers indicate illustrations.
* Indicates a table.

519

Index

Index

ABOUT THE AUTHOR

James Trulove has accumulated more than a quarter century of experience in LAN and data communications systems with companies such as Alteon, Lucent, AT&T, Motorola, and Intel. Trulove has an extensive background in designing, installing, and troubleshooting LAN cabling and networks, and currently works for a leading manufacturer of networking systems and equipment.